AF616205

ROYAL BOTANIC GARDENS KEW

Kew Bulletin Additional Series IV

The *EUPHORBIACEAE* *of BORNEO*

H. K. Airy Shaw

LONDON
HER MAJESTY'S STATIONERY OFFICE

First published 1975

ISBN 0 11 241099 5*

Contents

Introduction

This enumeration is planned on similar lines to those adopted for the account of the *Euphorbiaceae* of Siam in Kew Bull. 26: 191–363 (1971). It will make, I hope, an appreciable contribution towards the eventual complete account of the *Euphorbiaceae* of the whole of Malesia.

Part of that extensive area has already been covered by various works during the past half-century. Merrill's Bornean (1921) and Philippines (1923) enumerations, however, lack keys and descriptions, and Ridley's and Gagnepain's descriptive treatments for Malaya (1924) and Indochina (1925–7) respectively are somewhat marred by inaccuracies. Corner's *Wayside Trees of Malaya* (1940), though full of interesting and valuable information of every kind, is of course incomplete as regards species. On the other hand, Backer & Bakhuizen's excellent account of the family in the first volume of the *Flora of Java* (1963) and Whitmore's very recent treatment of the group (for foresters) in the *Tree Flora of Malaya*, vol. 2 (1973)—both works taking the form of elaborate 'descriptive keys'—are complete for their respective areas, and provide reliable means for specific identification. My own account of the *Euphorbiaceae* of Siam (1971, *l.c. supra*) sought to bridge the large gap that still remained between the accounts for Burma (Hooker, *Flora of British India*, 5 (1886–8)), Indochina, and Malaya.

The present enumeration of the *Euphorbiaceae of* Borneo fills in yet a further gap. Floristically and phytogeographically it is an area of prime importance —almost the 'hub' of Western Malesia—lying as it does midway between the Philippines, Indochina, Siam, Malaya and Java. The islands of Sumatra, Celebes, New Guinea, the Moluccas and the Lesser Sunda Islands remain to be tackled, but it may be doubted whether the more limited extent to which they have so far been explored and collected would yet warrant the attempt.

A useful summary of the characteristics, history, relationships and exploration of the Bornean flora was provided by Merrill in the prefatory remarks to his Bibliographic Enumeration (*Journ. Straits Br. Roy. As. Soc.*, Spec. No.: 1–29 (1921)). During the fifty or more years that have elapsed since then very much further exploration and collecting has taken place. From the collecting point of view the most constant and prolific sources of herbarium material have been the various Forestry Departments, particularly those of Sabah (formerly North Borneo), East Indonesian (formerly Dutch) Borneo, and Sarawak. But private collectors and expeditions have also made invaluable contributions, notably Elmer (1921–3), the Clemenses (1929–33), Oxford University Expeditions (1932, 1953), Kostermans (1951),

Jacobs (1958), etc. To very varying degrees the greater part of the island has now been botanically 'sampled', with the exception of what might be termed the great 'empty quarter' of south Indonesian Borneo—the large, mostly low-lying, roughly triangular area lying between the Schwaner and Müller mountains to the north-west and the Meratus range in the south-east. This area, except for some collecting in the regions of Bandjermasin and Sampit, is botanically virtually untouched, probably because, from the physical point of view, it is unpleasant terrain, both monotonous and difficult of access.

I may repeat here some of the explanatory remarks prefaced to the Siamese enumeration, which apply equally to the present account. The keys, both generic* and specific, are based so far as possible on macroscopic characters, or on features likely to be found on incomplete material, e.g. size and pubescence of leaves, colour on drying, etc. The keys are therefore to be regarded as a rough general guide, rather than as a precise scientific determinator. In view of the small size of the flowers and their frequent dioecism, it is hoped that this more general approach may prove helpful. In many cases, however, it has obviously been impossible to construct workable keys without recourse to more abstruse characters.

In order to save space, the most frequently cited literature references have been reduced to author (often abbreviated), page and date. The titles and volume numbers should be supplied in each case as follows.

Backer & Bakh. f.—Fl. Java 1 (1963).
Beille (*Phyllanthus*, *Glochidion*, *Breynia*, *Agyneia*, *Sauropus*)—in Lecomte, Fl. Gén. Indoch. 5 (1925–7).
Boerl.—Handl. Fl. Nederl. Ind. 1(2) (1900).
Boiss.—in DC., Prodr. 15(2) (1862).
Corner—Wayside Trees of Malaya (1940).
Engler—Pflanzenr. IV. 147 (1910–24).
Gagnep.—in Lecomte, Fl. Gén. Indoch. 5 (1925–7).
Hook. f.—in Hook. f., Fl. Brit. Ind. 5 (1886–8).
Jabl(onszky) (*Bridelia* & *Cleistanthus*)—in Engl., Pflanzenr. IV. 147. viii (1915).
Meijer—in Botanical News Bulletin, Herbarium, Forest Department, Sandakan (1967–8).
Merr.—A bibliographic enumeration of Bornean plants, in Journ. Straits Branch Roy. As. Soc., Spec. No. (1921).
,, —Enum. Philipp. Fl. Pl. 2 (1923).
,, —Pl[antae] Elmer[ianae Borneenses], in Univ. Calif. Publ. Bot. 15: 1–316 (1929).
Muell. Arg.—in DC., Prodr. 15(2) (1866).
Pax & Hoffm.—in Engl., Pflanzenr. IV. 147 (1910–24).
Ridley—Fl. Malay Penins. 3 (1924).
J. J. Sm.—in Koord. & Valet., Bijdr. No. 12 Boomsoorten Java, in Meded. Dep. Landb. No. 10 (1910).
Whitmore—Tree Flora of Malaya 2 (1973).

In the accounts of each species the accepted name and synonyms (if present) are followed by a brief indication of distribution within Borneo, as

* The generic key includes all Malesian genera, and also [in square brackets] a few from Australia that might occur in the Malesian area (especially New Guinea).

follows: **SR** (= Sarawak), **B** (= Brunei), **SB** (= Sabah), **K1, 1a,* 2, 3** (= Kalimantan = Indonesian Borneo, East, South-East,* South and West respectively; see outline map, p. 22), followed again by a statement of general distribution. 'SE. Asia' includes Burma, Siam and Indochina; 'Malesia' the whole area from Sumatra, Malaya and the Philippines to eastern New Guinea.

The next paragraph is a summary or digest of the information given in the collectors' field notes, and applies (unless otherwise stated) to Bornean collections only. Details include habit, height, ecology and altitude.

A thumbnail sketch of the salient features of the plant then follows, often including important diagnostic features in relation to its closest relatives. In a final paragraph economic notes or items of miscellaneous information are sometimes added.

An alphabetical arrangement of genera and species has been adopted, not only for ease of reference, but also because no scheme of classification hitherto published seems entirely satisfactory, and some are open to serious criticism. A very tentative scheme, eliminating some previous weak points, but obviously still leaving much to be desired, was published on pp. 193–4 of my Siamese enumeration; an enlarged and modified version, covering the whole of Malesia, together with some Australian genera, is given below. Attention is drawn to the fact that the genera *Buxus*, *Daphniphyllum* and *Bischofia* are excluded from the present account.

It is interesting to note that, whereas in 1921 Merrill listed 52 genera and 165 species of Bornean *Eurphorbiaceae* (*sensu stricto*, i.e. excluding *Antidesma*, *Galearia* and *Microdesmis* as well as the three above-mentioned), the present account includes roughly 80 genera and 340 species, representing increases of over 50% and 100%, respectively, over the past half-century.

I am most grateful to my colleague Mr. A. Radcliffe-Smith for again providing the account of *Euphorbia*, and to Dr. T. C. Whitmore for his kindness in undertaking the difficult genus *Macaranga*.

* This small south-east corner of Borneo has been treated as a separate area in view of the distinctive character (dry scrub, etc.) of much of the vegetation.

Tentative scheme for possible natural grouping of genera*

BORNEODENDREAE
Borneodendron

PETALOSTIGMATEAE
Petalostigma

AUSTROBUXEAE
Austrobuxus

DISSILIARIËAE
Dissiliaria
Choriceras

PHYLLANTHEAE
Phyllanthinae
Phyllanthus
Glochidion
Securineginae
Margaritaria
Securinega
Richeriella
Andrachninae
Actephila
Leptopus
Chorisandrachne
Sauropodinae
Breynia
Sauropus
Synostemon

DRYPETEAE
Drypetes

BRIDELIËAE
Cleistanthus
Bridelia

DICOELIËAE
Dicoelia

BACCAUREËAE
Baccaurea
Aporusa
Ashtonia
?Clonostylis

CROTONEAE
Crotoninae
Croton
Chrozophorinae
Sumbaviopsis
Doryxylon
Chrozophora
Mallotinae
Rockinghamia
Ptychopyxis
Trewia
Neotrewia
Mallotus
Octospermum
Melanolepis
Macaranga
Endospermum
Homonoiinae
Spathiostemon
Lasiococca
Homonoia
Blumeodendrinae
Blumeodendron
Botryophora
Agrostistachydinae
Agrostistachys
Chondrostylis
Alchorneinae
Alchornea
Wetria
Cleidion

* The 'tribal' and 'subtribal' names here used are without nomenclatural significance.

Mercurialinae
Claoxylon
Micrococca
Plukenetiinae
Pachystylidium
Cnesmone
Megistostigma
Pterococcus
Dalechampiinae
Dalechampia
Epiprininae
Adriana
Epiprinus
Koilodepas
Cladogynos
Cephalomappa
Erismanthinae
Erismanthus
Syndyophyllum
Moultonianthus

ACALYPHEAE
Acalypha

RICINEAE
Ricinus

JATROPHEAE
Jatrophinae
Annesijoa
Jatropha
Aleurites
Reutealis
Vernicia
Tapoïdes
Loerzingia
Hylandia
Elateriospermum
Omphalea

Manihotinae
Manihot
Alphandiinae
Alphandia
Codiaeinae
Codiaeum
Blachia
Strophioblachia
Baliospermum
Ostodinae
Fahrenheitia
Ostodes
Dimorphocalyx
Trigonostemon
Fontainea

BALOGHIËAE
Baloghia

CHAETOCARPEAE
Chaetocarpus
Trigonopleura

SUREGADEAE
Suregada

CHEILOSEAE
Cheilosa
Neoscortechinia

EUPHORBIËAE
Pimelodendrinae
Pimelodendron
Hippomaninae
Sapium
Excoecaria
Stillingia
Sebastiania
Homalanthus
Euphorbiinae
Euphorbia

MARGINALLY RELATED FAMILIES APPENDED TO THE ENUMERATION

STILAGINACEAE
Antidesma

PANDACEAE
Galearia
Microdesmis

DOUBTFULLY RELATED OR UNRELATED FAMILIES EXCLUDED FROM THE ENUMERATION

BUXACEAE
Buxus

DAPHNIPHYLLACEAE
Daphniphyllum

BISCHOFIACEAE
Bischofia

Artificial key to the genera of *Euphorbiaceae* of the Flora Malesiana area

1 Twining plants, sometimes with stinging hairs:
  2 Inflorescence very condensed, capitate, bisexual, long-peduncled, with 2 large membranous external bracts forming an involucre, and several pectinate inner bracts (Sumatra, Java) . **Dalechampia** *L.*
  2 Inflorescence slender, ± racemose, bisexual or occasionally unisexual, not enclosed in 2 involucral bracts:
    3 Coarse glabrous plants, with coriaceous leaves and geniculate petioles; stamens connate into a pileiform body (W. & E. Malesia) **Omphalea** *L.*
    3 Slender plants, with membranous or chartaceous leaves and obscurely geniculate petioles; stamens free:
      4 Stamens 8–13; capsule 4-locular, alate or appendiculate (W. Malesia) . . . . **Pterococcus** *Hassk.*
      4 Stamens 2–3; capsule 3-locular, without appendages:
        5 Stamens 2; filaments very short and slender; connective neither thickened nor produced; base of flower not hollowed (Philippines, Java) . . **Pachystylidium** *Pax & Hoffm.*
        5 Stamens 3; filaments slightly longer and thicker; connective much thickened, truncate above; base of flower ± hollowed, with raised rim:
          6 Connective not produced into a deflexed apical appendage; styles thickened and fused into a large subglobose mass (W. Malesia) . . . . **Megistostigma** *Hook. f.* (*Clavistylus* J. J. Sm.)
          6 Connective produced into a slender deflexed apical appendage; styles free except at base, stigmas plumose-papillose (W. Malesia) . . . . **Cnesmone** *Bl.*

1 Plants not twining, but sometimes scrambling; no stinging hairs:
  7 Leaves and ♂ flowers in regular whorls of 3; indumentum stellate (N. Borneo) . . . . **Borneodendron** *Airy Shaw*
  7 Leaves opposite or alternate, sometimes crowded into false whorls (cf. *Blumeodendron*, *Lasiococca*, *Croton*, etc.), but not truly verticillate:
    8 Leaves digitately compound, with free leaflets; petals present, conspicuous (New Guinea) . . . **Annesijoa** *Pax & Hoffm.*
    8 Leaves simple, though sometimes deeply lobed:
      9 Leaves opposite or apparently so:
        10 Stellate or fascicled hairs present (sometimes very small):
          11 Fruit a capsule:

12 Leaves pseudo-verticillate; inflorescence unisexual or bisexual (Malaya, Sumatra) . . . **Epiprinus** *Griff.*

12 Leaves truly opposite; inflorescence unisexual:

12*a* [Styles branched; anther-connective acutely subulate (Australia) . . . . **Adriana** *Juss.*

12*a* Styles unbranched; connective not subulate (widespread genus) . . . . . . **Mallotus** *Lour.*

11 Fruit a drupe, indehiscent (genera closely related to *Mallotus*):

13 Leaves ovate, cordate (rarely broadly rounded) at base; ♀ inflorescence 1–4-flowered; stigmas 2–4, elongate (W. Malesia) . . . . **Trewia** *L.*

13 Leaves ± elliptic, narrowed at base; ♀ inflorescence many-flowered; stigma 1(–2), circinate (Philippines, Celebes) . . . . . **Neotrewia** *Pax & Hoffm.*

10 No stellate hairs present:

14 Each pair of leaves accompanied by two large, cordate-ovate, stipuliform structures at the base; ♀ flowers very long-pedicelled (Sumatra, Borneo) **Moultonianthus** *Merr.*

14 Leaves without such structures:

15 Capsules densely echinate-muricate; leaves in false whorls, not truly opposite; stamens united in phalanges (Malay Peninsula) . . **Lasiococca** *Hook. f.*

15 Capsules not echinate-muricate; stamens not in phalanges:

16 Capsules large, 2–5 cm. diam.:

17 Leaves entire; petioles often elongate; flowers often in very condensed cymes; capsules tardily dehiscent (widespread genus) **Blumeodendron** (*Muell. Arg.*) *Kurz*

17 Leaves crenate; petioles very short; flowers in narrow racemes; capsules dehiscing normally **Excoecaria** *L.* (*p.p.*)

16 Capsules smaller, dehiscing normally; petioles short or very short:

18 Leaves unequally cordate at base; petioles often red; ♂ flowers on long pilose pedicels (Malaya, Borneo, Sumatra) **Erismanthus** *Wall. ex Muell. Arg.*

18 Leaves symmetrical; petioles not red; ♂ flowers otherwise:

19 Inflorescence very elongate, pendulous, bisexual; capsules shortly golden-sericeous (Sumatra, Borneo, New Guinea) **Syndyophyllum** *Lauterb. & Schum.*

19 Inflorescence otherwise; capsules not sericeous:

20 Inflorescence spicate or racemose; milky juice in all parts (W. Malesia) **Excoecaria** *L.* (*p.p.*)

20′ Inflorescence composed of cyathia; milky juice present (widespread genus) (see also 20″) . . . **Euphorbia** *L.*

20″ Inflorescence loosely cymose; no milky juice:

21 Leaves entire; branchlets glabrous; styles in fruit short and inconspicuous (W. Malesia) **Austrobuxus** *Miq.*

21 Leaves finely and shallowly crenulate; branchlets pubescent; styles free, forming 3 conspicuous apical horns on the capsule (New Guinea) **Choriceras** *Baill.*

9 Leaves alternate:

22 Leaves densely or sparsely granular-glandular below (occasionally above also) (cf. also the minutely lepidote leaves of *Homonoia* and *Cephalomappa* spp.):

23 Anthers 3–4-locellate (widespread genus) **Macaranga** *Thou.*

23 Anthers regularly 2-locular:

23*a* Fruit a 2–3-locular capsule (widespread genus) **Mallotus** *Lour.*

23*a* Fruit an 8-locular 8-angled 8-seeded drupe (New Guinea) **Octospermum** *Airy Shaw*

22 Leaves not granular-glandular:

24 Leaves pellucid-punctate; inflorescence leaf-opposed, fascicled (widespread genus) . **Suregada** *Roxb. ex Rottl.*

24 Leaves not pellucid-punctate; inflorescence not leaf-opposed:

25 Stellate hairs or scales present (sometimes sparse or minute):

26 Inflorescence a ± diffuse cyme or thyrse:

27 Inflorescence unisexual; petals absent; fruit a small, indehiscent or partly dehiscent, externally thinly fleshy capsule (widespread genus) **Endospermum** *Benth.*

27 Inflorescence bisexual; petals present in ♂ flower, exceeding the calyx:

28 Leaves rather hard and rigid, shining and reticulate above, glaucescent beneath; petals somewhat thick and fleshy; fruit a large capsule (New Guinea) (see also p. 10) **Alphandia** *Baill.*

28 Leaves not very hard or rigid, dull on both surfaces, not glaucescent; petals not fleshy:

29 Leaves often cuneate at base, sometimes lobed; inflorescence ochreous-tomentellous; stamens 15–20, in 3 whorls; ovary and fruit mostly bilocular (widespread genus; also cult.) **Aleurites** *J. R. & G. Forst.*

29 Leaves cordate, never lobed, thin; inflorescence densely grey-tomentellous; stamens 7–10, in 2 whorls; ovary and fruit trilocular (Philippines)
Reutealis *Airy Shaw*

26 Inflorescence a raceme or panicle, or indeterminate, the ♂ flowers sometimes aggregated in dense globose heads:

30 ♂ flowers aggregated in dense globose heads, which may be borne on long or short axes:

31 Leaves conspicuously white-tomentellous below, coarsely repand-dentate (N. Malaya, Philippines, Java, Lesser Sunda Is.)
Cladogynos *Zipp. ex Span.*

31 Leaves not white-tomentellous below, crenate or entire:

32 ♂ flower-heads terminal on stiff branches of a rather short inflorescence; capsules muricate (W. Malesia)
Cephalomappa *Baill.*

32 ♂ flower-heads sessile, arranged closely or distantly along an elongate axis; capsules smooth:

33 Leaves short-petioled; filaments thickened and connate below; axis slender; glomerules distant (W. Malesia, N. Guinea)
Koilodepas *Hassk.*

33 Leaves (in our species) long-petioled; filaments free; axis stout; glomerules crowded (Sumatra, Malaya)
Epiprinus *Griff.* (*Symphyllia* Baill.)

30 ♂ flowers individually racemose or spicate, or narrowly thyrsiform in arrangement, not aggregated into globose heads:

34 Inflorescence unisexual:

35 Petals present; ♂ sepals imbricate; inner filaments united into a column; inflorescence a long, narrow, pendulous thyrse (W. Malesia)
Fahrenheitia *Reichb. f. & Zoll.*

35 Petals wanting:

36 Leaves conspicuously palmate-lobulate:

37 Inflorescence (♂) a panicle; indumentum often floccose (widespread genus, close to *Mallotus*)
Melanolepis *Reichb. f. & Zoll.*

37 [Inflorescence (♂) a false spike (Australia) . **Adriana** *Juss.*]

36 Leaves not palmate-lobulate; inflorescence not paniculate; indumentum rarely floccose:

38 Leaves willow-like, narrowly elliptic-oblong, densely lepidote (not truly stellate) below, very short-petioled; inflorescence a spike; stamens united in phalanges; ovules 1 per loculus (widespread) (see also p. 15)
Homonoia *Lour.*

38 Leaves not willow-like, mostly ovate, not lepidote, ± long-petioled; inflorescence a raceme or very narrow thyrse; stamens free; ovules 2 per loculus (widespread) . . **Baccaurea** *Lour.*

34 Inflorescence bisexual:

39 Petals absent; disk present in ♀ flower; ovary and fruit mostly bilocular (widespread genus)
Melanolepis *Reichb. f. & Zoll.*

39 Petals present; ovary and fruit mostly trilocular:

40 Filaments inflexed in bud; ♂ sepals ± imbricate (widespread genus)
Croton *L.*

40 Filaments erect in bud:

41 ♂ sepals connate to midway, lobes open in aestivation; outer filaments free, inner ± connate, thickish, abruptly deflexed at apex; inflorescence thyrsiform; tree with smooth minutely lepidote leaves (New Guinea)
Alphandia *Baill.*

41 ♂ sepals valvate; filaments all free; inflorescence a long or short raceme; plants tomentellous:

42 Stamens united below; woody herb or small shrub (Java)
Chrozophora *Neck. ex Juss.*

42 Stamens free: shrub or tree:

43 Styles entire, undivided; petals 5; shrub (Philippines, Lesser Sunda Is.)
Doryxylon *Zoll.*
(*Sumbavia* Baill.)

43 Styles bifid; petals 5 or 10 tree (W. Malesia)
Sumbaviopsis *J. J. Sm.*

25 No stellate hairs or scales present:
44 Leaf-base distinctly asymmetrical (flowers fascicled; ovules 2 per loculus):
45 Fruit a dry or fleshy 1–2-locular drupe; sepals of ♂ flower commonly 4, much imbricate (widespread genus) . . . . **Drypetes** *Vahl*
45 Fruit a 3–25-locular capsule or small 3–8-locular berry or drupe; sepals of ♂ flower 5–6 (cf. also *Trigonopleura*):
46 ♂ flower without disk; styles ± united, erect or depressed, often clavate or much reduced, rarely free; ovary 3–25-locular (widespread genus) . **Glochidion** *J. R. & G. Forst.*
46 ♂ flower with disk; styles free or connate below, often bifid, usually spreading; ovary 3–8-locular (widespread genus) **Phyllanthus** *L.*
44 Leaf-base symmetrical:
47 Petals present in ♂ flower:
48 Inflorescence spicate, racemose, or narrowly thyrsiform:
49 Petals thickish, with two collateral oval excavations on inner face (Sumatra, Banka, Malaya, Borneo) **Dicoelia** *Benth.*
49 Petals thin, without excavations:
50 Petals exceeding calyx, sometimes red (and then drying black); sepals free, imbricate (W. Malesia) **Trigonostemon** *Bl.*
50 Petals shorter than or equalling calyx, sometimes minute, never drying black:
51 Sepals almost free, imbricate; petals minute; bracts not glumaceous; leaves entire; primary nerves clearly anastomosing in loops some distance from margin, minor nerves lax, inconspicuous (Philippines to E. Malesia) **Codiaeum** *Juss.*
51 Calyx at first entire, splitting into 2–3 valvate segments; petals ± equalling calyx-segments; bracts glumaceous; leaves entire or serrate; primary nerves very regular, anastomosing inconspicuously very close to margin, minor nerves closely parallel, conspicuous (W. Malesia) **Agrostistachys** *Dalz.*

48′ Inflorescence fasciculate, or flowers solitary, axillary (see also 48″):

52 Calyx-segments clearly valvate; petals small, obovate or spatulate; ovules 2 per loculus:

53 Fruit a 3-locular capsule; lateral nerves rarely distinctly parallel (widespread genus) **Cleistanthus** *Hook. f. ex Planch.*

53 Fruit a 1–2-locular drupe; lateral nerves often conspicuously parallel (widespread genus) **Bridelia** *Willd.*

52 Calyx-segments imbricate:

54 Petals equalling or exceeding the calyx; ovules 1 per loculus:

55 Leaves distichously arranged, base sometimes asymmetric; capsule globose, unlobed, slightly hexagonal, pericarp grey-tomentellous, slightly wrinkled; columella broadly winged; seeds black, shining, arillate (W. Malesia) **Trigonopleura** *Hook. f.*

55 Leaves spiral or pseudo-verticillate; capsule 3-lobed, not wrinkled; columella not winged; seeds mottled, not arillate (W. Malesia) **Trigonostemon** *Bl.*

54 Petals shorter than calyx; ovules 2 per loculus;

56 Shrubs with larger chartaceous or coriaceous leaves; disk-glands episepalous (widespread genus) **Actephila** *Bl.*

56 Subherbaceous undershrubs, with smaller membranous leaves; disk-glands epipetalous (plants of limestone soils or areas with seasonal climate) **Leptopus** *Decne*

48″ Inflorescence cymose or broadly thyrsiform:

57 ♀ sepals fringed with long capitate glandular cilia, at least at base and apex:

57*a* Stamens numerous, free; ♀ flowers apetalous (Philippines, Celebes) **Strophioblachia** *Boerl.*

57*a* Stamens 3, united in a column; ♀ flowers petaliferous (W. Malesia) **Trigonostemon** *Bl.*

57 ♀ sepals without glandular cilia:
58 Calyx closed in bud, splitting spathaceously into 2–5 segments:
58*a* Leaves cordate-ovate, membranous; inflorescence and petals grey-pubescent; filaments glabrous (Philippines) **Reutealis** *Airy Shaw*
58*a* Leaves elliptic, chartaceous; inflorescence and petals glabrous; filaments pubescent (Borneo) **Tapoïdes** *Airy Shaw*
58 Calyx open in bud, segments free or connate into a cup:
59 Stamens all free:
60 Filaments pilose below; ♀ flowers petaliferous; inflorescences often borne below terminal tuft of leaves (W. Malesia) . . **Ostodes** *Bl.*
60 Filaments glabrous; ♀ flowers apetalous:
61 ♀ calyx somewhat accrescent in fruit (W. Malesia) **Blachia** *Baill.*
61 ♀ calyx not or scarcely enlarged in fruit (Philippines to E. Malesia) **Codiaeum** *Bl.*
59 Stamens (at least the inner) shortly connate or forming a column:
62 Leaves broadly ovate or orbicular in outline, sometimes deeply lobed, palmately nerved (cult.) JATROPHA *L.*
62 Leaves elliptic or oblanceolate, occasionally broadly elliptic-ovate, never lobed, pinnately nerved:
63 Petals densely white-tomentose all over; calyx cupular, shortly lobed or dentate; leaves often undulate on the margin, somewhat recalling *Pittosporum* spp. (New Guinea) **Fontainea** *Heckel*
63 Petals variously pubescent but not white-tomentose:

64 [Ovary and fruit bilocular; ♂ disk-glands 5, distinct, subglobose; inflorescence ± pyramidal-thyrsoid (Queensland)
Hylandia *Airy Shaw*]

64 Ovary and fruit trilocular:

65 Inflorescences very varied, usually bisexual, sometimes unisexual, usually monoecious; leaves often triplinerved at base; petals often coloured (yellow, red or violet); calyx very rarely accrescent in fruit (W. Malesia)
Trigonostemon *Bl.*

65′ Inflorescences short, unisexual, flowers usually dioecious; leaves not or obscurely triplinerved; petals usually white; calyx frequently accrescent (W. Malesia, Lesser Sunda Is., New Guinea)
Dimorphocalyx *Thw.*

65″ Inflorescences (at least the ♂) pyramidal-thyrsoid, unisexual and dioecious; leaves not or obscurely triplinerved; petals white; calyx not accrescent (Sumatra)
Loerzingia *Airy Shaw*

47 No petals in ♂ flower*:

66 Stamens united in phalanges:

67 Leaves palmately lobed; stem often glaucous; flowers

* Here will probably come also **Clonostylis** *S. Moore* (Sumatra), of which the ♂ flowers are still unknown.

large, in terminal bisexual racemes (introd.)
RICINUS *L.*

67 Leaves undivided, penninerved; stem not glaucous; flowers smaller, at least the ♂ in axillary unisexual racemes:

68 Leaves narrowly oblong, willow-like, densely arranged along the branches, densely lepidote below; ♀ flowers short-pedicelled in dense axillary racemes; capsules smooth (W. Malesia)
Homonoia *Lour.*

68 Leaves neither willow-like nor lepidote; ♀ flowers long-pedicelled, solitary or loosely racemose; capsules echinate:

69 Leaves crowded in false whorls, elliptic-oblanceolate, narrowly cordate at base, very shortly petioled; ♀ flowers solitary, with conspicuous sepals (Malay Peninsula)
Lasiococca *Hook. f.*

69 Leaves scattered, broadly elliptic or ovate or obovate, not cordate, petiole 1–4 cm. long; ♀ flowers loosely racemose, with very small sepals (widespread genus) **Spathiostemon** *Bl.*

66 Stamens free, or united in a column:

70 Fruit an echinate capsule:

71 Flowers in dense axillary clusters; ♂ sepals imbricate; filaments connate in a staminal column; capsule covered with very dense, contiguous bristles (W. Malesia) . . . . . **Chaetocarpus** *Thw.*

71 Flowers in racemes or cymes; ♂ sepals valvate; stamens free:

72 Inflorescences unisexual; leaves opposite or alternate, but not crowded into false whorls; stamens not interspersed with small processes (widespread genus) . . . **Mallotus** *Lour.*

72 [Inflorescence bisexual; leaves crowded into false whorls; stamens interspersed with small processes (Queensland)
Rockinghamia *Airy Shaw*]

70 Fruit a smooth capsule or drupe:

73 ♂ flower compressed, with 2 suborbicular sepals:

74 Leaves crenate, coriaceous; stipules minute or obsolete; inflorescences axillary; no glands beneath each ♂ flower; fruit large, massive, drupaceous (widespread genus)
Pimelodendron *Hassk.*

74 Leaves entire, membranous to chartaceous; stipules usually conspicuous, elongate, membranous; inflorescences terminal; 1–2 glands beneath each ♂ flower; fruit small, glaucous, dehiscent or indehiscent (widespread genus)
Homalanthus *Juss.*

73 ♂ flower not as above:

75 ♂ flowers fascicled, or solitary, axillary:

76 Fruit a hard, massive, tardily dehiscent capsule; leaves rather long-petioled; ♂ calyx closed in bud, splitting into 3–4 valvate segments; ovules 1 per loculus (widespread genus) **Blumeodendron** (*Muell. Arg.*) *Kurz*

76 Fruit a drupe or a thin-walled dehiscent or indehiscent capsule; leaves short-petioled; sepals imbricate; ovules 2 per loculus:

77 Stamens numerous (8 or more), free:

78 Stigmas ovate, subcordate, subsessile (New Guinea) **Austrobuxus** *Miq.*

78 Stigmas ± flabellate, stipitate:

79 Fruit ultimately dehiscent; pericarp fleshy, wrinkled, shining; stigmas 3–4, elongate, conspicuous (New Guinea) **Petalostigma** *F. Muell.*

79 Fruit indehiscent; pericarp dry, smooth, dull; stigmas 1–2, small (widespread genus) **Drypetes** *Vahl*

77 Stamens few, often 3 or 6:

80 Stigmas 1–2, shortly flabellate (widespread genus) . . **Drypetes** *Vahl*

80 Stigmas 3–6, slender, sometimes coiled or very short, not flabellate:

81 Flowers dioecious; stamens free:

82 Sepals 5; stamens 3–5; fruit a small globose short-pedicelled drupe, with 6 cocci (widespread genus) **Securinega** *Juss.* (*Flueggea* Willd.)

82 Sepals 4; stamens 4; styles 3, bifid; fruit larger, dry, globose, long-pedicelled, thin-shelled, bursting irregularly; leaves deciduous; branches lenticellate (scattered genus) **Margaritaria** *L. f.*

81 Flowers monoecious:

83 Fruit a normal dry capsule, dehiscing completely into 3–6 segments; top of ovary not excavated, without raised rim; stamens connate or free (widespread genus) **Phyllanthus** *L.*

83 Fruit at first slightly fleshy, finally becoming dry and crustaceous, dehiscence sometimes incomplete and irregular; top of ovary sometimes excavated, with a raised rim; stamens connate:

84 Plant often blackening on drying; ♂ calyx turbinate, obconic or hemispheric (widespread genus) **Breynia** *J. R. & G. Forst.*

84 Plant not blackening on drying; ♂ calyx mostly flattened, ± disciform or rotate, the apical half of the segments sharply inflexed, forming short inward-pointing lobes round the stamens (the apparent outer margin or radiating lobes being radial extensions from the region of flexion) (scarcely distinct from *Breynia*) (widespread genera) **Sauropus** *Bl.* and **Synostemon** *F. Muell.*

75′ ♂ flowers cymose or thyrsoid (see also 75″):

85 ♂ sepals valvate:

86 Stipules usually conspicuous or at least evident; stamens fewer than 20, not interspersed with disk-glands:

87 Anthers 3–4-locellate; leaves never stipellate, but often granular-glandular below (widespread genus) **Macaranga** *Thou.*

87 Anthers 2-locular; leaves sometimes stipellate at base of lamina, but never granular-glandular below (widespread genus) **Alchornea** *Sw.*

86 Stipules minute or obsolete; stamens usually more than 20, interspersed with numerous small disk-glands:

87 Leaves crenate, decurrent into non-pulvinate petiole; inflorescence glabrous (W. Malesia) **Chondrostylis** *Boerl.*

87 Leaves entire, not decurrent, petiole pulvinate:

88 ♂ inflorescence ± glabrous, pendulous; buds red; calyx closed in bud, bursting at anthesis; connective ± peltate (W. Malesia) **Botryophora** *Hook. f.*

88 ♂ inflorescence ± pubescent, erect or spreading; buds not red; sepals valvate, distinct in bud; connective not peltate (scattered genus) **Ptychopyxis** *Miq.*

85 ♂ sepals imbricate:

89 Leaves entire or obscurely crenate (cf. *Baccaurea*):

90 Inflorescences terminal, dichotomously branched, bisexual; ovules 1 per loculus; capsule large, massive, tricoccous (W. Malesia) **Elateriospermum** *Bl.*

90 Inflorescences axillary, not dichotomous, unisexual; ovules 2 per loculus; capsule small or medium, tricoccous, rarely dicoccous:

91 Inflorescences lax, with widespreading branches; flowers very small (W. Malesia) **Richeriella** *Pax & Hoffm.*

91 Inflorescences dense, elongate, with very short cymose branches, often apparently racemose; flowers larger (widespread genus) **Baccaurea** *Lour.*

89 Leaves manifestly lobed, sinuate, crenate or glandular-dentate:

92 Male calyx large, coloured; leaves deeply lobed; capsule winged (cult.) . . . . MANIHOT *Mill.*

92 Male calyx small, green; leaves shallowly lobed or dentate; capsule unwinged:

93 Capsules small, clearly tricoccous, readily dehiscent, not whitish-tomentellous; habit sometimes subherbaceous (Malaya, Java, Sumbawa) **Baliospermum** *Bl.*

93 Capsules larger, not or obscurely lobed, tardily dehiscent, whitish-tomentellous; habit definitely woody:

94 Disk present in ♂ flower; capsule large, depressed-globose, thick-walled (W. Malesia) **Cheilosa** *Bl.*

94 Disk 0; capsule smaller, ellipsoid, with thinner wall (doubtfully distinct from *Cheilosa*) (widespread genus) **Neoscortechinia** *Pax & Hoffm.*

75″ ♂ flowers spicate or racemose:

95 ♂ sepals imbricate; ovules 2 per loculus:

96 ♂ flowers minute or very minute, densely crowded in continuous or interrupted spikes; stamens mostly 2–3; pistillode minute or 0; fruits dehiscent (widespread genus) . . . **Aporusa** *Bl.*

96 ♂ flowers not minute, not densely crowded, in racemes or narrow thyrses; stamens mostly 5–6; pistillode large:

97 Sepals 4–8; disk-glands usually present in ♂ flowers; fruits 2–3-locular, dehiscent or more often indehiscent (widespread genus) **Baccaurea** *Lour.*

97 Sepals 3; disk-glands in ♂ flower apparently 0; fruit a dehiscent 3–4-locular capsule (Malaya, Borneo) **Ashtonia** *Airy Shaw*

95′ ♂ sepals valvate; ovules 1 per loculus (see also 95″):

98 Anthers 4-locellate; capsules 2-locular (if rarely 3-locular, 1 loculus abortive), borne singly on long rigid pedicels (widespread genus) . . **Cleidion** *Bl.*

98 Anthers bilocular; capsules usually 3-locular, not borne on long rigid pedicels:

99 Anthers with distinct, free, ± erect or spreading thecae:

100 Anther-thecae elongate, vermiform, twisted; no juxta-staminal glands; capsules without purple sap (widespread genus) **Acalypha** *L.*

100 Anther-thecae short, erect; filaments usually accompanied by short, often pilose 'juxta-staminal glands' at base; capsules often with purple sap:

101 Anther-thecae relatively firm in texture; filaments usu-

ally longer than thecae; disk-glands of ♀ flower ± transverse, usually broader than long (widespread genus) **Claoxylon** *Juss.*

101 Anther-thecae rather soft in texture; filaments not longer than thecae; disk-glands of ♀ flower narrow, longer than broad (Malaya) **Micrococca** *Benth.*

99 Anthers with adnate or pendulous thecae:

102 Stamens 4–8; leaves sometimes stipellate (widespread genus) **Alchornea** *Sw.*

102 Stamens numerous; leaves never stipellate:

103 Styles bipartite; leaves oblanceolate, very short-petioled, with numerous spreading parallel nerves (W. Malesia, New Guinea) **Wetria** *Baill.*

103 Styles undivided; leaves various, rarely oblanceolate or short-petioled; nerves not so numerous or conspicuously parallel (widespread genus) **Mallotus** *Lour.*

95″ ♂ calyx much reduced, aestivation open or obscure; ovules 1 per loculus*:

104 Annual herb of sandy ground, sometimes perennating from woody rootstock; leaves linear to narrowly elliptic; capsule-lobes dorsally spinose-dentate (W. Malesia) **Sebastiania** *Spreng.*, § **Elachocroton**

104 Shrubs or trees, with broader leaves; capsules unarmed:

105 Leaves glaucous or glaucescent beneath:

106 Capsules borne on straight, rigid, very elongate pedicels; ♂ racemes rather short, slender, fascicled, with very minute flowers (W. Malesia) **Sebastiania** *Spreng.*, § **Sarothrostachys**

* A reliable key at the generic level in this group is practically impossible; I have therefore keyed out to sections and even species.

106 Capsules borne on short pedicels; ♂ racemes more robust, not fascicled, with larger flowers (W. Malesia)
Sapium *L.*, § **Triadica**

105 Leaves not glaucous beneath:

107 Capsule seated on persistent triangular base (coccophore) (Lesser Sunda Is.)
Stillingia *L.*

107 Capsule without persistent triangular base:

108 Capsule large, globose or ellipsoid, woody, tardily dehiscent (widespread group) **Sapium** *L.*, § **Parasapium**

108 Capsule small, tricoccous, crustaceous, normally dehiscent (widespread species)
Excoecaria agallocha *L.*

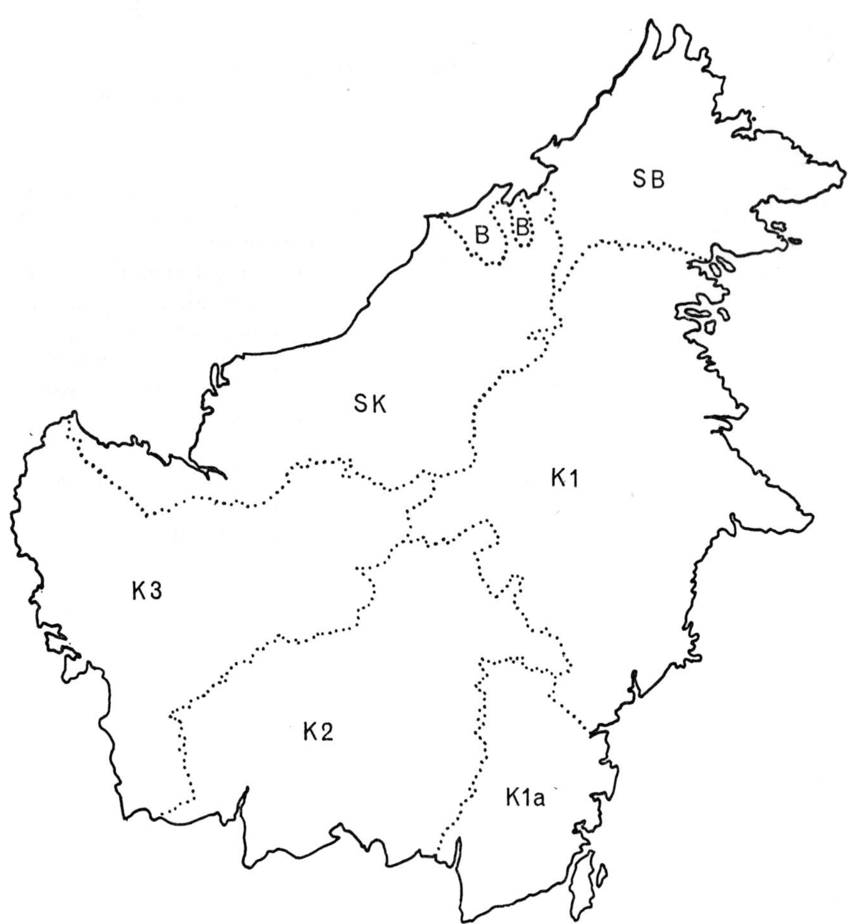

Map 1. Outline map of Borneo, showing regions. SK = Sarawak, B = Brunei, SB = Sabah, K1 = Kalimantan (Indonesian Borneo) East, K1a = Kalimantan South-East (Bandjermasin Region), K2 = Kalimantan South, K3 = Kalimantan West.

Alphabetical enumeration

Acalypha *L.*

1 Annual herb; inflorescences bisexual, with short slender ♂ portion; ♀ bracts small, strongly parallel-veined, with numerous short acute teeth. **A. lanceolata**

1 Shrubs or small trees; inflorescences unisexual:

  2 ♀ bracts entire; ovary and fruit bilocular; ♀ inflorescence slender, lax; flowers dioecious . . . . **A. caturus**

  2 ♀ bracts dentate; ovary trilocular; ♀ inflorescence denser:

    3 ♀ spikes ± equalling leaves; flowers mostly dioecious . **A. amentacea**

    3 ♀ spikes much shorter than leaves; flowers monoecious A. WILKESIANA

Acalypha amentacea *Roxb.*, Fl. Ind., ed. Carey & Wall., 3: 676 (1832); Merr., Interpr. Rumph. Herb. Amboin.: 322 (1917), & Enum.: 343 (1921) & 444 (1923) & in Philipp. Journ. Sci. 29: 384 (1926); Holth. & Lam in Blumea 5: 199 (1942).

A. glandulosa Blanco, Fl. Filipp.: 749 (1837); Muell. Arg.: 888 (1866); *non* Cav. (1800).
A. amboynensis Benth. in Hook., Lond. Journ. Bot. 2: 233 (1843).
A. stipulacea Klotzsch in Nova Acta Acad. Nat. Cur. 19, Suppl. 1: 416 (1843); Muell. Arg.: 807 (1866); Stapf in Trans. Linn. Soc., Bot. 4: 226 (1894); Boerl.: 286 (1900); Hutch. in Journ. Linn. Soc., Bot. 42: 135 (1914); Pax & Hoffm. xvi: 152 (1924).
A. affinis Klotzsch, *l.c.* (1843).
A. grandis var. *amboinensis* [sic!] (Benth.) Muell. Arg. in Linnaea 34: 10 (1865), & in DC.: 806 (1866); Pax & Hoffm. xvi: 150 (1924).
A. fruticosa sec. Muell. Arg.: 822 (1866), Hook. f.: 415 (1887), Pax & Hoffm. xvi: 169 (1924), *quoad synon. A. amentaceam* Roxb., *non* Forssk.

SB (incl. Banggi I.).—Philippines (throughout), Sangi & Talaud Is., Celebes, Moluccas and New Guinea.

Shrub to 6 m. high, locally common in primary and secondary forest, ascending to 1200 m. on Kinabalu.

Leaves ovate to rhombic-ovate, up to 20 × 8 cm.

Acalypha caturus *Bl.*, Bijdr.: 630 (1825); Muell. Arg.: 805 (1866); Scheff. in Ann. Mus. Bot. Lugd.-Bat. 4: 121 (1868–9); Stapf in Trans. Linn. Soc.,

Bot. 4: 225 (1894); Boerl.: 286 (1900); J. J. Sm.: 510 (1910); Hutch. in Journn. Linn. Soc., Bot. 42: 135 (1914); Merr.: 343 (1921) & 445 (1923); Pax & Hoffm. xvi: 141 (1924); S. Moore in Journ. Bot. Brit. & For. 63, Suppl.: 101 (1925); Merr., Pl. Elmer.: 160 (1929); Holth. & Lam in Blumea 5: 199 (1942); Backer & Bakh. f.: 489 (1963); Whitmore: 51 (1973).

A. minahassae Koord. in Meded.'s Lands Plantent. 19: 579, 624 (1898).
A. similis Koord., *l.c.* (1898).
A. cardiophylla Merr. in Philipp. Journ. Sci. 1, Suppl.: 80 (1906).
A. subcinerea Elm., Leafl. Philipp. Bot. 7: 2631 (1915).

SR; SB; K1, 1a.—Philippines, Talaud Is., Sumatra, Java, Celebes, Moluccas, ? Lesser Sunda Is.

Tree to 16 m. high, in primary or secondary forest, frequently on limestone (e.g. Bau series), or on stony ground or reddish soil, up to 1500 m. alt. on Kinabalu.

Leaves broadly ovate or suborbicular, up to 27 × 22 cm.

Timber used for house construction; leaves used medicinally (unspecified); fruit edible.

Acalypha lanceolata *Willd.*, Sp. Pl. 4: 524 (1805); Merr. & Chun in Sunyatsenia 5: 92 (1940); Airy Shaw in Kew Bull. 26: 206 (1971); Whitmore: 51 (1973); Radcliffe-Smith in Kew Bull. 28: 527, 528 (1974).

Urtica pilosa Lour., Fl. Cochinch.: 558 (1790), *non Acalypha pilosa* Cav. (1800).
Acalypha boehmerioides Miq., Fl. Ind. Bat., Suppl.: 459 (1861); Muell. Arg.: 871 (1866); Merr.: 445 (1923); Pax & Hoffm. xvi: 96 (1924); Gagnep.: 337 (1925); Merr. in Trans. Amer. Philos. Soc. II, 24(2): 238 (1935); Holth. & Lam in Blumea 5: 199 (1942); Lam, *ibid.*: 577 (1945); Backer & Bakh. f.: 490 (1963).
A. fallax Muell. Arg. in Linnaea 34: 43 (1865) & in DC.: 872 (1866); Hook. f.: 416 (1887); Ridley: 274 (1924).
A. wightiana Muell. Arg., var. *genuina* Muell. Arg. & var. *lanceolata* (Willd.) Muell. Arg., *ll.cc.*

SR.—India to Polynesia.

Weed (?) in rest-house grounds, Kuching, presumably introduced. Not yet certainly known as a native plant in Borneo, but it should occur there, as it is recorded from almost all the surrounding areas except Sumatra.

Acalypha lanceolata is one of the small number of species that produce allomorphic female flowers: see the paper by Radcliffe-Smith, *l.c. supra.*

ACALYPHA WILKESIANA *Muell. Arg.*: 817 (1866); J. J. Sm.: 20 (1910); Merr.: 446 (1923); Pax & Hoffm. xvi: 153 (1924); Backer & Bakh. f.: 489 (1963); Airy Shaw in Kew Bull. 26: 208 (1971); Whitmore: 51 (1973), *in obs.*

SB (cult.).—Native of Polynesia (? Fiji); widely cultivated as an ornamental.

Shrub to 2 m. high; leaves sometimes variegated.

Actephila *Bl.*

1 Coarse plant; leaves up to 30 cm. long and 12 cm. wide; petioles mostly over 2 cm. long . . . . **A. excelsa** var. **javanica**
1 Slender plant; leaves up to 14 cm. long and 4·5 cm. wide; petioles less than 2 cm. long . . . . **A.** cf. **gitingensis**

Actephila excelsa (*Dalz.*) *Muell. Arg.* in Linnaea 32: 78 (1863) & in DC.: 222 (1866); Hook. f. 283 (1887); Pax & Hoffm. xv: 191 (1922); Ridley: 196 (1924); Gagnep.: 535 (1927); Hend. in Journ. Malayan Br. Roy. As. Soc. 17: 68 (1939); Backer & Bakh. f.: 470 (1963); Meijer, 7: 30 (1967); Airy Shaw in Kew Bull. 26: 209 (1972).

Var. **javanica** (*Miq.*) *Pax & Hoffm.* xv: 192 (1922); Merr.: 390 (1923) & Pl. Elmer.: 139 (1929); Airy Shaw, *l.c. supra*; Whitmore: 52 (1973).

Actephila javanica Miq., Fl. Ind. Bat. 1(2): 359 (1859); Muell. Arg.: 222 (1866); Hook. f.: 283 (1887); J. J. Sm.: 46 (1910); Merr.: 390 (1923); Ridley: 196 (1924); Gage in Journ. As. Soc. Beng. 75 (5): 521 (1936); Airy Shaw in Kew Bull. 25: 499 (1971), *in obs.*
Pimeleodendron dispersum Elm. in Leafl. Philipp. Bot. 1: 308 (1908).
Actephila dispersa (Elm.) Merr. in Philipp. Journ. Sci. 4, Bot.: 276 (1909).

SB; K1.—Peninsular Siam, W. Malesia, New Guinea.
Tree to 11 m. high, scarce in primary or secondary forest on brown sandy soil at low altitudes up to 30 m.

A coarse almost glabrous plant with large (up to 30 × 12 cm.), chartaceous, smooth, obovate or elliptic, shortly caudate leaves; petioles variable, 1–7 cm. long (mostly over 2 cm.). The large size of the leaves and their frequently elongate petioles distinguish var. *javanica* from the continental forms of the species. For distinctions from the closely related *A. lindleyi* (Steud.) Airy Shaw, of Eastern Malesia and Australia, see Kew Bull. 25: 499 (1971).
Only six gatherings from Borneo are present in the Kew Herbarium. There appears to be no evidence of any association of this plant with limestone in Borneo, as there is in Malaya.

Actephila cf. **gitingensis** *Elm.* in Leafl. Philipp. Bot. 3: 903 (1910).

A. excelsa var. *javanica* sec. Pax & Hoffm. xv: 192 (1922); Merr.: 390 (1923); *vix* (Miq.) Pax & Hoffm.

SR.—Philippines.

Two very distinctive gatherings from 'near Kuching', March and April 1893, *Haviland* 3094 and 3238 (K), without ecological information, are tentatively referred here. Both have narrow or very narrow leaves, 6–14 cm. long by 1–4·5 cm. broad, agreeing in this respect with *A. gitingensis*, but whereas in no. 3238 the petioles are 1–2 cm. long and very slender (as in *gitingensis*), in no. 3094 they are only 2–5 mm. long and much stouter. Moreover whereas no. 3238 bears only solitary axillary female flowers and fruits, no. 3094 bears, in addition, numerous greatly abbreviated (less than 1 cm.) but much-branched and densely imbricate-bracteate male inflorescences, mostly arising from the axils of fallen leaves. The bracts are minute, about 0·5 mm. long, arranged in about four vertical ranks, and almost every branch of the inflorescence bears a flower-bud at its apex. These inflorescences appear to be perennial structures of unlimited growth. (The male inflorescences of *A. gitingensis* are described as 'short bracteate tubercles'.) Somewhat comparable inflorescences occur in forms of *Phyllanthus ciccoïdes* Muell. Arg. in New Guinea (*Hoogland* 4424, *Henty* NGF 16998, etc.); also in *Knema tridactyla* Airy Shaw (*Myristicaceae*; cf. note in *Flora Malesiana* I, 4: 353 (1953), under *Gonystylus augescens* Ridl.). It is remarkable that this plant has never been collected again near Kuching (or elsewhere) in the past eighty years. Further material of the population, as also of the Philippine *A. gitingensis*, is needed in order to establish their taxonomic status.

Agrostistachys *Dalz.*

1 Inflorescences (♂ and ♀) very short, with densely imbricate bracts (subgen. *Agrostistachys*); leaves thinner, dentate towards apex . **A. indica**

1 Inflorescences lax, sometimes greatly elongate, bracts not imbricate (subgen. *Sarcoclinium* (Wight) Benth.); leaves more thickly coriaceous, entire or obscurely sinuate:

  2 Stem 4–5 mm. in diameter; insertion of leaves relatively spaced, with longer sections of stem visible between them; lamina decurrent at base, sometimes leaving a clear length of slender petiole, usually drying brown . . . . **A. longifolia**

  2 Stem 6–15 mm. in diameter; insertion of leaves very crowded, with only very short sections of stem visible between; lamina abruptly contracted to a narrowly rounded base, with a very short broad petiole, usually drying green . . . . **A. sessilifolia** var.

Agrostistachys indica *Dalz.* in Hook. Journ. Bot. & Kew Gard. Misc. 2: 41 (1850); Muell. Arg.: 726 (1866); Hook. f.: 406 (1887); Pax & Hoffm. vi: 103 (1912); Gagnep.: 465 (1926); Airy Shaw in Kew Bull. 14: 472 (1960) & 26: 210 (1972) & 29: 312 (1974).

A. indica subsp. *longifolia* (Muell. Arg.) Pax & Hoffm. var. *subintegra* Pax & Hoffm., *l.c.*: 105 (1912).

Heterocalyx laotica Gagnep. in Not. Syst. 14: 33 (1950); *cf.* Airy Shaw, *l.c.* (1960).

SB (Banguey I.).—W. Peninsular India, Ceylon, SE. Asia, Philippines (var. *maesoana* (Vidal) Pax & Hoffm.), New Guinea (Papua).

Shrub; no details of stature, habitat or altitude available for Borneo.

Closely related to *A. gaudichaudii* Muell. Arg., of Lower Siam and Malaya, differing in the leaves being usually sharply dentate, at least towards the apex, and often drying a yellowish-green, and in the shorter and denser inflorescences. The species is so far represented for the Bornean flora by a solitary gathering from Banguey (Banggi) Island, in the Kudat District of Sabah, but it is not listed by Merrill in his accounts of the plants of that island in *Philipp. Journ. Sci.* 24, Bot.: 113 (1924) and 29: 341–427 (1926).

Agrostistachys longifolia (*Wight*) *Benth. ex Hook. f.*: 407 (1887); Pax & Hoffm. vi: 100 (1912); Whitmore: 53 (1973).

Sarcoclinium longifolium Wight, Ic. Pl. Ind. Or. 5: 24, tt. 1887–8 (1852); Muell. Arg.: 727 (1866).
Agrostistachys longifolia var. *latifolia* Hook. f.: 407 (1887).
A. borneënsis Becc., Nelle For. Born.: 331 (1902); Ridley: 268 (1924); Corner: 229 (1940); **synon. nov.**
A. latifolia (Hook. f.) Pax & Hoffm. vi: 100 (1912); Merr.: 337 (1921).
A. leptostachya Pax & Hoffm. vi: 102 (1912); Merr. in Philipp. Journ. Sci. 16: 553 (1920) & Enum.: 337 (1921) & 428 (1923); **synon. nov.**
A. longifolia var. *leptostachya* (Pax & Hoffm.) Whitmore, *l.c.* (1973).

SR; B; K1, 3.—S. India, Ceylon, Malaya.

Shrub or small tree to 6 m. high, in primary lowland Heath (*kerangas*) forest, in Dipterocarp-Fagaceous forest, or Low Dipterocarp forest, or swamp forest, on sandstone, or yellow sandy clay, or clay loam, or on white sand Tertiary raised beach plateau, once on steep limestone slope (Bau, nr. Kuching), ascending to 405 m. on Mt. Santubong.

Between the very broadly obovate leaf of *A. borneënsis* (almost limited to the Kuching district of SW. Sarawak) and the oblanceolate leaf of *A. leptostachya* (described from the Baram district of NE. Sarawak, and more frequent there, but also occurring at high altitudes elsewhere, e.g. summit of Mt. Santubong near Kuching) every gradation seems to occur. There is apparently no taxonomic significance in the slightly more slender inflorescence of *A. leptostachya.*

Agrostistachys sessilifolia (*Kurz*) *Pax & Hoffm.* vi: 102 (1912); Airy Shaw in Kew Bull. 20: 26 (1966), *p.p.*

Sarcoclinium sessilifolium Kurz in Flora 58: 31 (1875).

Var. **graciliflora** *Airy Shaw*, var. nov., foliis exsiccando viridulis, inflorescentiis gracilibus saepe abbreviatis distincta.

SABAH. Tenom Distr.: Mandolom Forest Reserve, primary forest on hillside, sandstone ridge, alt. . . . m., 10 Dec. 1962, *Mikil* SAN 31944:—Tree, height 2 m., girth 12·5 cm.; flower white, stamens yellowish; fruits grow between leaves; native name (Kinabatangan), 'tanamud'. Sandakan Distr.: Leila For. Res., near path to Wireless Stn., alt. . . ., 29 May 1964, *Sam & Madani* SAN 39824:—Small tree; flower [♂] yellowish white. Sandakan Distr.: West of Tampias, scattered in primary forest along ridge top, alt. . . ., 16 Sept. 1971, *Leopold & Dewol* SAN 74342:—Treelet, height 3 m.; flowers [♂] yellowish. Semporna Distr.: Montoritip, primary forest, sandy soil, alt. 15 m., 21 Feb. 1964, *Gibot* SAN 40921:—Small tree, height 6 m., girth 10 cm.; flower on the branch. Tawau Distr.: Tawau River For. Res., alt. 90 m., 6 July 1959, *Meijer* SAN 19457:—Tree, height 3 m.; fruit light green. Tawau Distr.: Mt. Wullersdorf, primary forest on top of mountain, alt. . . ., 20 June 1969, *Talip* SAN 65844 (holotype, K):—Yellowish, 2·5 m. high; fruits greenish.

E. INDONESIAN BORNEO. Nunukan, N. of Tarakan, near S. Binusan, in marshy secondary forest, few m. alt., 5 Dec. 1953, *Meijer* 2416:—Treelet 4 m. high, cauliflorous.

SB; K1.—Endemic.
Shrub or small tree to 6 m. high, on sandstone or sandy soil, in primary or secondary forest, sometimes marshy, up to 90 m. alt.

I am loth to propose another taxon in this difficult group, but this Bornean plant is certainly not identical with typical *A. sessilifolia* of the Malay Peninsula and Sumatra, in which the leaves as a rule turn brown in drying, and the inflorescences are almost invariably robust and elongate (up to 60 cm., according to Corner, *l.c.*).

This is so far the only representative of the subgenus *Sarcoclinium* in Sabah, where the common *A. longifolia* does not seem to occur.

Alchornea *Sw.*

1 Leaves cuneate-obovate to elliptic, penninerved, estipellate, shortly petioled; ♂ inflorescence terminal, compound, paniculate (§ *Cladodes*) **A. rugosa**

1 Leaves ovate or rhombic-ovate, trinerved and stipellate at base, with longer petioles; ♂ inflorescences axillary or even cauliflorous, sometimes almost simple (§ *Stipellaria*) . . . . . **A. borneënsis**

Alchornea borneënsis *Pax & Hoffm.* in Mitt. Inst. Bot. Hamburg 7: 227 (1931).

K3.—Endemic (?).
Tree to 5 m. high, in primary forest, probably low alt.

This may prove to be a form of *A. villosa* (Benth.) Muell. Arg., with which it was compared in the original description. It was based on a single collection from West Indonesian Borneo (**K3**), which I have not seen. The authors state that it differs from *A. villosa* in its longer petioles (up to 9 cm. long; but in a specimen from Sumatra, coll. *Teijsmann*, in Herb. Kew., they are up to 10 cm.!), smaller stipels (1–2 as against 2–5 mm.), linear black-glandular-toothed female sepals (and bracts), and shorter styles (5–7 as against 10–22 mm.).

Alchornea rugosa (*Lour.*) *Muell. Arg.* in Linnaea 34: 170 (1865) & in DC.: 905 (1866); Hook. f.: 422 (1887); J. J. Sm.: 467 (1910); Pax & Hoffm. vii: 243 (1914); Merr.: 538 (1923); Gagnep.: 379 (1926); Merr. in Trans. Amer. Philos. Soc. II, 24(2): 237 (1935); Holth. & Lam in Blumea 5: 199 (1942); Airy Shaw in Kew Bull. 26: 211 (1972); Whitmore: 53 (1973).

Cladodes rugosa Lour., Fl. Cochinch.: 574 (1790).
Conceveibum javanense (!) Bl., Bijdr.: 614 (1825).
Adelia glandulosa Blanco, Fl. Filip.: 814 (1837).
Aparisthmium javense (!) Endl. ex Hassk., Cat. Pl. Hort. Bogor. Alter: 235 (1844); Zoll. in Linnaea 28: 330 (1857).
Tragia innocua Blanco, *op. cit.* ed. 2: 479 (1845).
Alchornea javensis (Endl. ex Hassk.) Muell. Arg. in Linnaea 34: 170 (1865) & in DC.: 905 (1866); Ridley: 278 (1924).

A. hainanensis Pax & Hoffm. vii: 242 (1914).
A. javanensis (Bl.) Backer & Bakh. f.: 485 (1963).

SR; SB.—Nicobar Is., SE. Asia, and throughout Malesia to New Guinea and Queensland.

Shrub or small tree to 6 m. high, apparently almost confined to a few limestone areas, in primary forest or on swampy alluvium or among boulders, under 300 m. alt.

The association of this species with limestone in Borneo is noteworthy, since in other areas (e.g. Java; cf. J. J. Sm.: 469), although it is occasionally noted on limestone, it apparently occurs more frequently on neutral soils. It is not included by Henderson in his list of the limestone-loving plants of Malaya in *Journ. Malayan Br. Roy. As. Soc.* 17: 68 (1939).

Zollinger (*l.c.*: 227–330, sub *Aparisthmium javense*) carried out an extensive investigation of several hundred flowers of this species in Java and found great variation in the number of sepals and stamens, of which he gives details. The plant moreover could be either monoecious or dioecious, and in the former case the inflorescences could be either unisexual or bisexual in varying degrees. I am convinced that there are insufficient grounds for distinguishing the Javanese population specifically from *A. rugosa.*

Aleurites *J. R. & G. Forst.*

Aleurites moluccana (*L.*) *Willd.*, Sp. Pl. 4: 590 (1805); Muell. Arg.: 723 (1866); Scheff. in Ann. Mus. Bot. Lugd.-Bat. 4: 120 (1868); Hook. f.: 384 (1887); Pax in Engler, IV. 147 (Heft 42): 129 (1910); J. J. Sm.: 551 (1910); Merr.: 344 (1921) & 448 (1923); Ridley: 253 (1924); Merr. in Trans. Amer. Philos. Soc. II, 24(2): 239 (1935); Corner: 231 (1940); Backer & Bakh. f.: 478 (1963); Meijer: 26, tab. opp. p. 130 (1967); Airy Shaw in Kew Bull. 20: 393 (1966) & 26: 213 (1971); Whitmore: 54 (1973).

Jatropha moluccana L., Sp. Pl.: 1006 (1753).
Aleurites triloba J. R. & G. Forst., Char. Gen. Pl.: 111, t. 56 (1776); Boerl.: 282 (1900).
Juglans camirium Lour., Fl. Cochinch.: 573 (1790).
Camirium cordifolium Gaertn., Fruct. 2: 195 (1791).
Aleurites commutata Geisel., Croton. Monogr.: 82 (March 1807).
A. ambinux Pers., Synops. 2: 587 (Sept. 1807).
A. lobata Blanco, Fl. Filip.: 756 (1837).
A. lanceolata Blanco, *l.c.*: 757 (1837).
A. cordifolia (Gaertn.) Steud., Nomencl. ed. 2, 1: 49 (1840).
Camirium oleosum Reinw. ex Muell. Arg.: 723 (1866).

SR; B; SB.—India and China to Polynesia and New Zealand.

Tree to 18 m. high, mostly in secondary forest or disturbed situations at low altitudes, occasionally in primary forest, on brown soil, up to 600 m. alt.

The candlenut is cultivated for the oil obtained from the seed, but the fruit is said to be edible, and the seed is mixed with curry in cooking (Sabah).

Aporusa *Bl.*

1 Ovary quite glabrous:
  2 ♀ flowers borne in racemes:
    3 Racemes lax, elongate; fruiting pedicels 1–2 cm. long; ♂ inflorescence unknown . . . . **A. chalarocarpa**
    3 Racemes shorter, more compact; fruiting pedicels less than 1 cm. long; ♂ inflorescence interrupted, with flowers in glomerules **A. symplocoïdes**
      4 Nerves smooth below . . . . var. **symplocoïdes**
      4 Nerves white-granular below under a lens . var. **chondroneura**
  2 ♀ flowers borne in fascicles or singly (rarely in depauperate 2–3-flowered racemes):
    5 Leaves membranous or thinly chartaceous, not shining:
      6 ♂ inflorescences densely fascicled, very slender, 0·5–1·5 mm. thick, up to 11·5 cm. long, the flowers irregularly interrupted **A. stenostachys**
      6 ♂ inflorescences solitary or few together, cylindric, 2–3 mm. thick, up to 2 cm. long, dense-flowered throughout **A. frutescens**
    5 Leaves firmly chartaceous to coriaceous, often ± smooth and shining:
      7 ♂ inflorescence slender, consisting of interrupted glomerules; capsule trilocular . . . . **A. prainiana**
      7 ♂ inflorescence not interrupted, dense-flowered, cylindric, clavate or globose, sometimes stipitate:
        8 ♂ inflorescence borne on a purple stipes or peduncle (0·5–2·5 cm. long), the flowers massed in a ± clavate head; capsule 2·5–3 cm. diam., (4–)5-locular . . . . . . **A. nitida**
        8 ♂ inflorescence cylindric, floriferous to the base; capsule bilocular:
          9 Petiole over 1·5 cm in length; lamina often drying a bright luminous yellow-green; capsule drying light brown **A. aurea**
          9 Petiole less than 1·5 cm. in length; lamina usually drying a dull ochraceous or blackish-brown: capsule drying blackish or fuscous-brown . . . . **A. lucida**
1 Ovary densely or sparsely pubescent, at least at the base (but capsule may be glabrescent):
  10 Capsule ± narrowly bottle-shaped, pale when dry, beaked, bilocular, 1·5–2·5 cm. long, 6–7 mm. thick, ± long-pilose; stipules falcate, 1 cm. long, sometimes ± persistent . . . **A. lagenocarpa**
  10 Capsule ovoid, ellipsoid or globose, or if bottle-shaped then black or fuscous-brown when dry:
    11 ♂ inflorescence broadly ovoid or obovoid, 6–9 mm. thick, very dense-flowered, pale when dry; midrib beneath very shortly and softly tomentellous, ochraceous, the lamina usually green when dry; capsules globose, trilocular, up to 2·5 cm. diam., borne singly . . . . **A. confusa**
    11 ♂ inflorescence not as above, 1–4 mm. thick:
      12 Capsule markedly asperous-granular; ♀ inflorescence shortly spicate, the flowers congested towards the apex; ovary bilocular . . . . **A. granularis**

12 Capsule not or much less strongly granular:
 13 ♀ inflorescence and infructescence very short and congested, the rhachis not exceeding 1·5 cm. in length; capsule bilocular (cf. also *A. penangensis*):
  14 Capsule globose, blackish when dry, dehiscing with a conspicuous whitish suture; leaves up to 30 cm. long, entire, often drying a lurid bluish colour **A. nigricans**
  14 Capsule ellipsoid, brownish when dry, dehiscence not as above; leaves up to 20 cm. long, often distantly and shallowly crenate, never drying bluish **A. dioica**
 13 ♀ inflorescence and infructescence less congested, with a more elongate rhachis:
  15 Ovary and capsule bilocular, relatively small:
   16 ♂ (and ♀) inflorescences lax, slender, ± interrupted:
    17 Leaves up to 30 cm. long; stipules broader; inflorescences up to 7 cm. long; styles short, recurved, densely papillose-laciniate **A. elmeri**
    17 Leaves up to 14 cm. long; stipules narrower; inflorescences up to 2·5 cm. long; styles elongate, arising from a conical boss, abruptly deflexed on to the sides of the ovary, very sparsely laciniate . . . **A. rhacostyla**
   16 ♂ inflorescences thicker, dense-flowered:
    18 Leaves blackish or plumbeous above when dry, almost glabrous; capsule often beaked **A. antennifera**
    18 Leaves brownish above when dry; capsule not beaked (cf. also forms of *A. nervosa*):
     19 Primary nerves 8–9 pairs; plant densely brown-tomentose . . **A. subcaudata**
     19 Primary nerves 11–15 pairs; plant thinly tomentellous or glabrescent **A. acuminatissima**
     Cf. also **A. bracteosa**
  15 Ovary and capsule trilocular, relatively large:
   20 Stipules caducous at an early stage:
    21 Leaves and petioles quite glabrous; petiole up to 3 cm. long . . . . . **A. caloneura**
    21 Leaves and petioles ± tomentellous or pilosulous:
     21*a* Petioles robust, 2–3·5 cm. long **A. illustris**
     21*a* Petioles less robust, 1–2(–2·5) cm. long **A. nervosa**
   20 Stipules ± persistent, often conspicuous:
    22 Leaves up to 15 cm. long:
     23 Leaves with spectacularly bullate nervation **A. bullatissima**
     23 Leaves with not or scarcely bullate nerves **A. falcifera**

22 Leaves up to 30(–37) cm. long:
24 Shortly pubescent; ♂ inflorescence dense-flowered above, interrupted below **A. lunata**
24 Glabrous or almost so:
25 ♂ inflorescence dense-flowered above, ± interrupted or bare below **A. benthamiana**
25 ♂ inflorescence interrupted throughout, bearing distinct glomerules of flowers **A. grandistipula**

Aporusa acuminatissima *Merr.*, Pl. Elmer.: 142 (1929).

SB; KI.—Endemic?
Tree to 17 m. high, in primary or secondary forest on sandstone or sandy loam or blackish soil, or on a tufa plateau, up to 1050 m. alt.

Thinly tomentellous or puberulous or glabrescent; nerves 11–15 pairs; secondaries less conspicuously prominent and scalariform than in *A. subcaudata*; fruit small, ellipsoid, 6–9 mm. long, 4–5 mm. diam.

Aporusa antennifera (*Airy Shaw*) *Airy Shaw*, comb. & sp. nov.

A. nigro-punctata Pax & Hoffm. var. *antennifera* Airy Shaw in Kew Bull. 20: 382 (1966); Meijer, 7: 34 (1967); Whitmore: 61 (1973).

SR; SB; KI.—Malaya, Banka.
Tree to 20 m. high, in primary Mixed Dipterocarp forest on dark brown sandy soil, sandy loam, granodiorite, basalt, tufa, or greyish or brownish soil, up to 1050 m. alt.

Leaves 7–13 × 2–4·5 cm., elliptic to almost oblong, caudate-acuminate, chartaceous, almost glabrous, blackening above when dry, conspicuously transversely nervose below. Male inflorescences fascicled, 1–1·5 cm. long, very dense-flowered, floriferous almost to the base. Infructescence up to 3·5 cm. long, lax, the rhachis ferrugineous-puberulous, pedicels 2–5 mm. long; capsule ovoid to lageniform, bilocular, 5–13 mm. long, acute or beaked at the apex, sparsely puberulous, black when dry; styles elongate, up to 3 mm. long, spreading, flexuous, very sparsely papillose.
Since publishing the description of this plant as *A. nigro-punctata* var. *antennifera* I have seen further material with mature male inflorescences. These are very dense-flowered and floriferous almost to the base, thus differing considerably from the New Guinea *A. nigro-punctata* in which they are sparse-flowered with many gaps. Moreover the ovary is bilocular, not trilocular as in *nigro-punctata*. These facts, coupled with the sparsely papillose styles and disjunct geography, indicate that var. *antennifera* should be raised to specific rank.

Aporusa aurea *Hook. f.*: 351 (1887); Pax & Hoffm. xv: 87 (1922); Ridley: 240 (1924); Merr. in Papers Mich. Acad. Sci. 20: 100 (1935); Airy Shaw in Kew Bull. 26: 214 (1972); Whitmore: 59 (1973).

A. frutescens sec. Merr., Pl. Elmer.: 141 (1929), *non* Bl.

SR; SB.—Lower Siam, Malaya, Sumatra.

Tree to 12 m. high, in primary or secondary lowland forest intermediate between *kerangas* and Dipterocarpaceous types (once noted near a mangrove swamp), up to 90 m. alt.

Leaves firmly coriaceous, up to 25 cm. long, very smooth and shining above, glabrous, usually drying the characteristic lurid yellowish-green of an aluminium accumulator; petiole 1·5–2·7 cm. long. Male inflorescence 1–3 cm. long, about 2 mm. thick, dense-flowered. Ovary bilocular; capsule more or less globose, 1–1·5 cm. diam., glabrous, usually brown when dry.

Aporusa aurea approaches extremely closely to *A. lucida*, and it may be doubted whether the two are specifically distinct. The most constant feature seems to be the elongate petiole of *A. aurea*, and this is usually combined with the marked tendency of the lamina to dry a brilliant luminous yellow-green colour, or at least a pale clear green, with the capsule usually drying a light brown. There are, however, a relatively small number of specimens in which this association of characters does not obtain. There is need for field work on these two taxa. *A. aurea* is also almost indistinguishable from *A. nitida* Merr. except in the male flower and fruit; see note under that species.

Aporusa benthamiana *Hook. f.* in Hook. Ic. Pl. 16: t. 1583 (1887) & Fl. Brit. Ind. 5: 352 (1887); Pax & Hoffm. xv: 84 (1922) (var. *genuina*); Meijer, 7: 33 (1967); Whitmore: 59 (1973).

Antidesma lunatum sec. Benth. in Benth. & Hook. f., Gen. Pl. 3: 282 (1880), *non* Miq.

Aporusa euphlebia Merr. in Philipp. Journ. Sci. 11, Bot.: 62 (1916) & Enum.: 330 (1921), **synon. nov.**

A. stipulosa Merr. in Philipp. Journ. Sci. 16: 547 (1920), **synon. nov.**

A. lunata var. *philippinensis* Pax & Hoffm. xv: 82 (1922), **synon. nov.**

A. lunata var. *stipulosa* (Merr.) Merr.: 410 (1923), *non illegit.*

A. grandifolia Merr., Pl. Elmer.: 143 (1929), **synon. nov.**

SR; SB; K1.—Malaya, Philippines.

Tree to 30 m. high, in primary Lowland or Mixed Dipterocarp forest, on sandstone, alluvial soil, black, red or yellowish-brown soil, or basalt, up to 800 m. alt.

Differs from *A. lunata* chiefly in the perfect glabrescence of its foliage.

Aporusa bracteosa *Pax & Hoffm.* xv: 95 (1922); S. Moore in Journ. Bot. Brit. & For. 63, Suppl.: 98 (1925); Meijer, 7: 34 (1967); Whitmore: 60 (1973).

SR.—Sumatra, Malaya.

No ecological information.

The status of this taxon is very obscure. It was based upon three gatherings: *Beccari* 2516 from Mt. Matang in SW. Sarawak; *Forbes* 3132 from Nepal Litjin, Palembang, in Sumatra; and *Ridley* 6484 from Singapore. (I have not seen the Forbes collection.) The chief point of distinction from *A. subcaudata* emphasized by Pax & Hoffmann was the larger size of the floral bracts of the male inflorescence, 2 mm. diam. instead of 1 mm. I cannot, however, discover any associated distinguishing character. In general features *Beccari* 2516 is more or less intermediate between *A. subcaudata* and *A. acuminatissima*. The relatively numerous lateral nerves (up to 15 pairs) suggest that it may be referable to the latter species, in which larger floral bracts sometimes occur.

Aporusa bullatissima *Airy Shaw* in Kew Bull. 20: 379 (1966).

SR; B; SB; K1.—Endemic.
Tree to 25 m. high, in primary Lowland Dipterocarp forest from low altitudes up to 1200 m.

Close to *A. lunata*, but readily distinguished by its smaller leaves (up to 15 cm. long) with exceptionally bullate nervation.

Aporusa caloneura *Airy Shaw* in Kew Bull. 23: 5 (1969).

SR; SB.—Endemic.
Tree to 30 m. high, in primary hill forest up to 1250 m. alt.

Differs from *A. nervosa* in the completely glabrous condition of the leaves and petioles, the very sparsely and minutely puberulous ovary, and the ellipsoid or ovoid fruit, up to 2 cm. long. The male inflorescences (unknown when the species was described) are fascicled, 1·5–3·5 cm. long, dense-flowered above but interrupted towards the base.
Cf. also *A. penangensis* (*infra*).

Aporusa chalarocarpa *Airy Shaw* in Kew Bull. 20: 380 (1966); Meijer, 7: 34 (1967).

SR; SB.—Endemic.
Tree to 15 m., in primary hill forest up to 150 m. alt.

Near *A. symplocoïdes*, but differing remarkably in the very slender, elongate, lax infructescences, up to 16 cm. long, and in the ovoid-fusiform, obtusely trigonous fruits, up to 2 × 1 cm., borne on slender pedicels 1–2·2 cm. long. Both male and female flowers are still unknown.

Aporusa confusa *Gage* in Rec. Bot. Surv. Ind. 9: 229 (1922); Ridley: 238 (1924); Meijer, 7: 35 (1967); Whitmore: 59 (1973).

A. mollis Merr., Pl. Elmer.: 144 (1929); Meijer, 7: 35 (1967); **synon. nov.**

SR; SB; K1.—Sumatra, Malaya.
Tree to 21 m. high, in primary Mixed Dipterocarp forest on basalt or sandstone or brownish or black sandy soil, up to 1600 m. alt.

The slender, shortly fulvous-tomentellous midrib (beneath), coupled with the usually green colour of the lamina on drying, is almost diagnostic for this species; the leaves are thinly chartaceous or almost membranous. The branchlets and petioles usually have a similar indumentum to the midrib. The male inflorescence is shortly and broadly clavate or obovoid, and almost sessile. The rather large fruits, which are globose, trilocular, 2–2·5 cm. diam., and thinly pubescent, are borne singly (rarely two together) on rather short peduncles (as in *A. prainiana*), not in racemes. They are reported as being edible, with a sourish taste.

Aporusa dioica (*Roxb.*) *Muell. Arg.*: 472 (1866); Pax & Hoffm. xv: 103 (1922); Airy Shaw in Kew Bull. 23: 3 (1969), *in obs.*, & 26: 215 (1972); Whitmore: 60 (1973).

Alnus dioica Roxb., Fl. Ind. 3: 580 (1832).
Scepa stipulacea Lindl., Nat. Syst. Bot. ed. 2: 441 (1836); cf. Airy Shaw, *l.c.* (1969).
S. aurita Tul. in Ann. Sci. Nat. III, 15: 254 (1851).
S. chinensis Champ. ex Benth. in Hook. Journ. Bot. & Kew Gard. Misc. 6: 72 (1854).
Aporosa aurita (Tul.) Miq., Fl. Ind. Bat. 1(1): 431 (1855); Muell. Arg.: 474 (1866); Pax & Hoffm. xv: 100 (1922); Merr.: 409 (1923); Backer & Bakh. f.: 456–7 (1963); Meijer, 7: 34 (1967).
Tetractinostigma microcalyx Hassk. in Flora 40: 533 (1857) & Hort. Bogor. s. Retziae ed. nov.: 55 (1858).
Aporosa roxburghii Baill., Ét. Gén. Euphorb.: 645 (1858); Hook. f.: 347 (1887).
A. microcalyx (Hassk.) Hassk. in Bull. Soc. Bot. France 6: 714 (1859); Muell. Arg.: 471 (1866); Hook. f.: 346 (1887); Hubert Winkler in Engl., Bot. Jahrb. 50, Suppl.: 208 (1914), *in obs.*; Merr.: 330 (1921); Pax & Hoffm. xv: 101 (1922); Ridley: 238 (1924); Gagnep.: 555 (1927).
A. leptostachya Benth., Fl. Hongk.: 317 (1861).
A. frutescens sec. Benth., *l.c.*, *non* Bl.
A. microcalyx var. *chinensis* (Champ. ex Benth.) Muell. Arg.: 472 (1866); Pax & Hoffm. xv: 102 (1922).
A. microcalyx var. *intermedia* Pax & Hoffm. xv: 102 (1922).
A. chinensis (Champ. ex Benth.) Merr. in Lingnan Sci. Journ. 13: 34 (1934).

SB; K1a.—E. Himalaya, Assam, Andamans, SE. Asia, S. China, and throughout W. Malesia.

Tree to 12 m. high, in primary hill forest, or on river banks on black soil or in poor *Imperata* savannah, from sea-level up to 180 m. alt.

Leaves usually very shallowly and distantly sinuate-crenate, sometimes subentire, pale brownish below when dry, sometimes lurid yellowish-green above; capsules ellipsoid, sessile or almost so, up to 13 mm, long.

The scarcity in Borneo of this elsewhere abundant polymorphic complex is remarkable. Three scattered specimens from Sabah and two from the Bandjermasin–Martapura area of SE. Borneo are all that I have seen.

Aporusa elmeri *Merr.*, Pl. Elmer.: 142 (1929); Meijer, 7: 35, tab. (1967).

SB; K1.—Endemic.

Tree to 13 m. high, in primary forest on black or brown sandy soil up to 600 m. alt.

Branchlets shortly fulvous-tomentellous; leaves rather large, obovate or sometimes oblong-elliptic, up to 30 × 12 cm., chartaceous, midrib and nerves fulvous-tomentellous below, glabrous or sparsely pilosulous above; stipules large, obliquely ovate, up to 1 cm. long, caducous; male inflorescences fascicled, slender, up to 7 cm. long, very interrupted, bearing small distant glomerules of flowers; female inflorescences slender, fascicled, up to 6 cm. long in the fruiting stage; fruits ovoid, 10–13 mm. long, puberulous, brown or orange when dry.

The large, broadly ovate, rather thin leaves and large stipules, and the fascicled, elongate, slender and very interrupted male inflorescences, are characteristic for *A. elmeri*, which is almost confined to Sabah. These features, together with the relatively lax female inflorescence, distinguish it from the often rather similar *A. subcaudata*, which is commoner in Sarawak.

Aporusa falcifera *Hook. f.*: 352 (1887); Pax & Hoffm. xv: 83 (1922); Ridley: 236 (1924); S. Moore in Journ. Bot. Brit. & For. 63, Suppl.: 97 (1925); Meijer, 7: 33 (1967); Airy Shaw in Kew Bull. 26: 215 (1972); Whitmore: 60 (1973).

A. hosei Merr. in Philipp. Journ. Sci. 11, C. Bot.: 63 (1916) & Enum.: 330 (1921); Pax & Hoffm. xv: 84 (1922).

SR; SB; K1.—Lower Siam, Malaya, Sumatra.

Tree to 50 m. high, in primary forest (sometimes marshy), on sandstone or sandy loam or loam with lime, or black or greyish soil, ascending to 1200 m. on Kinabalu.

Differs from *A. lunata* and *A. benthamiana* in its much smaller leaves, with not or scarcely bullately impressed nerves. Midrib usually puberulous beneath.

Aporusa frutescens *Bl.*, Bijdr.: 514 (1825); Muell. Arg.: 476 (1866); J. J. Sm.: 229 (1910); Hallier in Meded. Rijks Herb. 1910: 7 (1911); Pax & Hoffm. xv: 91 (1922); Ridley: 241 (1924); Backer & Bakh. f.: 457 (1963); Meijer, 7: 33 (1967); Airy Shaw in Kew Bull. 26: 216 (1972); Whitmore: 59 (1973).

Leiocarpus fruticosus Bl., Bijdr.: 582 (1825).
Aporusa fruticosa (Bl.) Muell. Arg.: 475 (1866).
Baccaurea banahaënsis Elm., Leafl. Philipp. Bot. 4: 1475 (1912); Pax & Hoffm. xv: 70 (1922).
Aporusa similis Merr. in Philipp. Journ. Sci. 9, Bot.: 472 (1914); Pax & Hoffm. xv: 92 (1922).
A. agusanensis Elm., *op. cit.* 7: 2636 (1915).
A. banahaënsis (Elm.) Merr.: 410 (1923).

SR; SB; K1, 1a.—Lower Burma, Lower & SE. Siam, and throughout W. Malesia; Moluccas?

Tree to 18 m. high, in primary or secondary Lowland or Mixed Dipterocarp forest, on yellow clay-rich soil, granodiorite-derived soil, yellowish sandy soil, sandy loam or loam with limestone, or brown, black or moist rocky soil, up to 1620 m. alt.

Very variable; distinguished from *A. prainiana* by the membranous leaves and smaller globose capsules, from *A. aurea* further by the leaves never drying the luminous golden-green colour of that species, and from *A. symplocoïdes* by the dense-flowered male inflorescence and thicker-walled globose fruit.

The fruits (or arillate seeds) are said to be eaten by squirrels.

Aporusa grandistipula *Merr.* in Philipp. Journ. Sci. 21: 521 (1922); Meijer, 7: 33 (1967).

SR; B; SB; K1, 1a, 3.—Celebes.

Tree to 24 m. high, in primary or secondary Lowland Dipterocarp forest, on clay or alluvium, or black or brown sandy or stony soil, or sandy loam or loam with lime or sandstone and limestone, up to 400 m. alt.

Very similar to *A. arborea*, but differing in the large persistent falcate-semiorbicular stipules and very slender glabrous male inflorescences bearing distinct globose glomerules of flowers. The leaves may attain a length of 37 cm. A very puzzling specimen is *Anderson* S. 14711, from Bukit Raya, in the Third Division of Sarawak, which has the large persistent stipules of *A. grandistipula* but the almost continuous-flowered male inflorescence of *A. arborea.*

One collector (Sabah) has noted the plant as poisonous. The fruit is said to have an acid flavour.

Aporusa granularis *Airy Shaw* in Kew Bull. 29: 283 (1974).

A. microstachyae forma? sec. Airy Shaw in Kew Bull. 23: 3 (1969), *in obs., non* (Tul.) Muell. Arg.

SR; K1.—Endemic.

Shrub or small tree to 9 m. high, in primary Mixed Dipterocarp forest on sandstone or yellow sandy clay or tufa up to 210 m. alt.

A slender species, closely related to the Malayan *A. microstachya* (Tul.) Muell. Arg. and *A. isabellina* Airy Shaw, but differing in the ovary being pilose only at the base, otherwise glabrous, and especially in the strongly granular-asperous surface of the almost glabrous fruits. The leaves usually dry greenish, rather than blackish or ochraceous respectively as in those species.

Aporusa illustris *Airy Shaw* in Kew Bull. 29: 281 (1974).

A. nervosa forma, sec. Airy Shaw in Kew Bull. 23: 5 (1969), *in obs., non* Hook. f.

SR.—Endemic.

Tree to 12 m. high, in primary Lowland Dipterocarp or hill forest on brown soil up to 1350 m. alt.

Near *A. arborea* Bl., but differing in the strong development of a tomentellous indumentum on all parts. *A. arborea* is not yet known from Borneo; the inclusion of this island in my geographical data for *arborea* in Kew Bull. 26: 214 (1971) was erroneous.

Aporusa lagenocarpa *Airy Shaw* in Kew Bull. 21: 355 (1968); Meijer, 10: 229 (1968); Airy Shaw in Hook. Ic. Pl. 38: t. 3701 (1974).

SR; SB; K1.—Endemic.

Tree to 45 m. high, in primary Mixed Dipterocarp forest on basalt or sandstone up to 1080 m. alt.

Young parts densely fulvo-tomentellous, branchlets finally almost glabrous; leaves mostly elliptic, sometimes almost oblanceolate, chartaceous, up to 16 × 5 cm.; stipules falcate-lunate, 1 cm. long, sometimes more or less persistent; male inflorescences solitary or 2 together, up to 1 cm. long, bare below, dense-flowered above; female inflorescences very abbreviated, 1–5 mm. long, tomentellous, few-flowered, flowers sessile, crowded, ovary tomentellous, styles spreading, 3–4 mm. long; infructescences 1–3 cm. long, tomentellous; fruits sessile, bottle-shaped, more or less beaked, strongly pubescent.

A most distinct species, not closely related to any other.

Aporusa lucida (*Miq.*) *Airy Shaw*, comb. nov.

Tetractinostigma lucidum Miq., Fl. Ind. Bat., Suppl.: 471 (1861); cf. Pax & Hoffm. xv: 87 (1922), *in adnot.*

Leiocarpus arboreus sec. Miq., *l.c.*: 443 (1861), *non* Bl.

Aporosa miqueliana Muell. Arg.: 474 (1866); Pax & Hoffm. xv: 103 (1922); Ridley: 241 (1924); Meijer, 7: 34 (1967); Whitmore: 61 (1973); **synon. nov.**

A. lanceolata sec. Muell. Arg.: 475 (1866), quoad pl. born.; Merr.: 330 (1921); *non* (Tul.) Thw.

A. microsphaera Hook. f.: 530 (1887); J. J. Sm.: 243 (1910); Merr.: 330 (1921); Pax & Hoffm. xv: 86 (1922); Backer & Bakh. f.: 456–7 (1963); **synon. nov.**

A. borneënsis Pax & Hoffm. xv: 87 (1922), **synon. nov.**

SR; B; SB; K1.—Malaya, Sumatra, Banka, Java.

Tree to 18 m. high, in *kerangas*, primary Lowland Dipterocarp forest, transition from Heath to Mixed Dipterocarp forest (sometimes periodically inundated), on clay or yellow clayey sand or brownish soil, up to 400(–1200) m. alt.

Very close to *A. aurea*, only differing in the shorter petioles, 0·5–1·5 cm. long, in the lamina usually drying a dull ochraceous or blackish-brown colour, and in the capsule usually becoming blackish or fuscous-brown when dry. See note under *A. aurea*.

Aporusa lunata (*Miq.*) *Kurz* in Journ. As. Soc. Bengal 42. ii: 239 (1873); Hook. f.: 352 (1887); J. J. Sm.: 236 (1910); Merr.: 330 (1921); Pax & Hoffm. xv: 82 (1922) (excl. var. *philippinensis*); Ridley: 237 (1924); Backer & Bakh. f.: 456 (1963); Meijer, 7: 33 (1967); Airy Shaw in Kew Bull. 26: 216 (1972); Whitmore: 59 (1973).

Antidesma lunatum Miq., Fl. Ind. Bat., Suppl.: 467 (1861); Muell. Arg.: 251 (1866). [*Non* sec. Benth. in Benth. & Hook. f., Gen. Pl. 3: 282 (1880), quod = *Aporusa benthamiana* Hook. f.]

SR; SB; K1, 1a.—Lower Siam and throughout W. Malesia (exc. Philippines).

Tree to 30 m. high, in primary or secondary Lowland or Mixed Dipterocarp forest, on sandstone or sandy loam or yellow clay-rich soil, or granodiorite-derived soil, or black soil, up to 1290 m. alt.

Distinguished from *A. benthamiana* by its manifest though very short indumentum, and from *A. falcifera* by its much larger leaves and stipules.

Aporusa nervosa *Hook. f.*: 350 (1887); Pax & Hoffm. xv: 97 (1922); Ridley: 238 (1924); Meijer, 7: 34 (1967); Airy Shaw in Kew Bull. 23: 4 (1969); Whitmore: 61 (1973).

A. basilanensis Merr. in Philipp. Journ. Sci. 9, Bot.: 471 (1914); Pax & Hoffm. xv: 94 (1922); Merr.: 410 (1923); **synon. nov.**

SR; SB; K1.—Malaya, Sumatra, Banka, Philippines.
Tree to 21 m. high, in Mixed Dipterocarp forest on basalt or clayey or acid sandy soil or black soil, up to 690 m. alt.

Leaves mostly more or less oblong-elliptic, coriaceous, not brittle, usually more or less brownish-ochraceous when dry, petiole rarely exceeding 2 cm., mostly very shortly tomentellous; male inflorescence (very rarely collected) dense-flowered, up to 2 cm. long, 2–3 mm. thick; female inflorescence racemose; fruit up to 2 cm. diam., ovoid, ellipsoid or globose, 2–3-locular (sometimes in the same infructescence).

Aporusa nervosa shows a great range of variation. An extreme form is represented by *A. basilanensis*, in which the leaves are somewhat oblong, with more numerous lateral nerves, and the capsule is bilocular and ellipsoid, 17–20 × 9–10 mm. But intermediate forms occur which cannot be placed in either taxon, and I am therefore treating them as forms of one species. The apparent rarity of the male plant is remarkable.

Aporusa nigricans *Hook. f.*: 347 (1887); Merr. in Philipp. Journ. Sci. 11, C. Bot.: 65 (1916) & Enum.: 330 (1921); Pax & Hoffm. xv: 97 (1922); Ridley: 239 (1924); S. Moore in Journ. Bot. Brit. & For. 63, Suppl.: 97 (1925); Merr., Pl. Elmer.: 142 (1929); Meijer, 7: 33 (1967); Airy Shaw in Kew Bull. 26: 217 (1972); Whitmore: 61 (1973).

SR; B; SB.—Malaya, Sumatra.
Tree to 11 m. high, in primary Lowland or Mixed Dipterocarp forest or mixed peat-swamp forest or wet *kerangas*, on black moist soil or clay loam, up to 30 m. alt.

Leaves very similar to those of *A. arborea* in shape and texture, but not drying a plumbeous grey above, rather an olivaceous brown, the lower surface also olivaceous but sometimes assuming a livid indigo blue shade; petiole 1–3 cm. long; female inflorescence abbreviated, ovary densely ferrugineous-tomentose; infructescence short and stout; capsule bilocular, pubescent, blackish when dry, splitting with a conspicuous whitish suture.

Aporusa nitida *Merr.*, Pl. Elmer.: 143 (1929); Meijer, 7: 33 (1967).

SR; SB; K1.—Endemic.
Shrub or tree to 18 m. high, in primary Lowland or Mixed Dipterocarp forest (sometimes periodically inundated), on granodiorite-derived soil, reddish stony soil, sandstone, sandy soil, sandy loam, clay loam, loam with lime, tufa, or swampy alluvium, from sea-level up to 600 m. alt.

This very distinct species is vegetatively almost indistinguishable from *A. aurea*, though the leaves often attain a greater breadth (13 cm.) and apparently never turn the same luminous yellow-green colour on drying. The usually ramiflorous male inflorescence, however, is unmistakable. It consists of a bare peduncle (or *stipes*), 0·5–2·5 cm. long, bearing a subglobose, clavate or subcylindrical head, 0·5–1·2 cm. long, densely covered with minute flowers. The peduncle is described as 'hard-brittle and fleshy', dark blue-purple or violet below, passing through reddish and pinkish upward to the yellow head of flowers. The inflorescence suggests at first sight something in *Moraceae* (*Artocarpus*, etc.), and was even thought by one collector to be galled.

The fruit is also characteristic, being large, 2·5–3 cm. in diameter, depressed-globose, glabrous, (4–)5-locular, thick-walled, glabrous, orange-coloured, and borne (ramiflorously) on a rather long violet peduncle. It is described as edible. The seeds are scarabaeiform, 10–12 mm. long, with a sweet juicy aril.

The species was first collected (in Sarawak) by Beccari in 1871 and by Hose in 1895, but was not described until 1929.

Aporusa penangensis (*Ridley*) *Airy Shaw* in Kew Bull. 23: 3 (1969) Whitmore: 61 (1973).

A. maingayi var. *penangensis* Ridley: 242 (1924).

B.—Malaya, Sumatra.

Tree of 19·5 m. in forest at 60 m. alt.

Near *A. nervosa*, but leaves smaller, up to 15 × 5 cm., and petiole slender, up to 2·8 cm. long.

I refer to this species a single Bornean gathering, *Wood, Smythies & Ashton* SAN 17550. from the Andulau Forest Reserve, Kuala Belait District, Brunei. A collection from SUMATRA: Upper Riauw, Pakanbaru, Tenajan R., on sandy loam, alt. 30 m., *Soepadmo* 188, seems also to be referable here.

Aporusa prainiana *King ex Gage* in Rec. Bot. Surv. Ind. 9: 228 (before 20 March 1922); Ridley: 240 (1924); S. Moore in Journ. Bot. Brit. & For. 63, Suppl.: 97 (1925); Meijer, 7: 34 (1967); Airy Shaw in Kew Bull. 25: 476 (1971); Whitmore: 60 (1973).

A. prainiana 'Ridley' ex Pax & Hoffm. (*pro sp. nov.*) xv: 92 (19 Sept. 1922).

SR; B; SB; K1.—Malaya, Sumatra.

Tree to 9 m. high, in primary or secondary Lowland or Mixed Dipterocarp forest, on sandstone, yellow or darkish sandy clay, clay loam, clay-rich shale-derived soil, limestone, sedimentary rocks, or basalt, sometimes on tufa soil, up to 1500 m. alt.

Close to *A. symplocoïdes*, but distinguishable by its female flowers being borne singly or in very few-flowered fascicles, instead of in racemes, by its large stigmas, and by its larger, thicker-walled, trilocular capsules. Male and sterile material of the two species, however, seems indistinguishable.

The fruit is said by one collector to be edible.

Aporusa rhacostyla *Airy Shaw* in Kew Bull. 29: 282 (1974).

SR.—Endemic.

Small tree 3–6 m. high, in hill forest on clay-rich soil up to 200 m. alt.

This is near *A. elmeri*, but differs in its smaller leaves and stipules, weaker indumentum, and shorter male and female inflorescences. Also, whereas the styles of *A. elmeri* are very short, recurved and densely papillose-laciniate, those of *A. rhacostyla* arise from a prominent conical boss on the apex of the ovary and are then abruptly deflexed and more or less adpressed to the sides of the latter; they are also elongate and only very sparsely laciniate.

Aporusa stenostachys *Airy Shaw* in Kew Bull. 29: 285 (1974).

SR (C.).—Endemic.
Small tree in riverain forest, probably at low altitude.

Vegetatively indistinguishable from the common *A. frutescens*, but the male inflorescences are fascicled (up to 23 together), elongate (up to 11·5 cm. long), very slender, and interrupted, instead of being solitary, short, thickish, cylindrical and continuous. The species is so far known only from two gatherings from the Bintulu District of Sarawak. The female plant has not yet been collected.

Aporusa subcaudata *Merr.* in Philipp. Journ. Sci. 11, Bot.: 64 (1916) & Enum.: 330 (1921); Pax & Hoffm. xv: 95 (1922); Merr., Pl. Elmer.: 142 (1929).

SR; B; SB.—Malaya.
Tree to 18 m. high, in primary Lowland or Mixed Dipterocarp or submontane forest on clay loam or yellow sandy clay over Tertiary sandstone, or on basalt, up to 1000 m. alt.

Densely brown-tomentose; primary nerves 8–9 pairs, secondary nerves prominent below, closely scalariform; fruit globose, 10–13 mm. diam.
There appear to be no clear dividing lines between this species, *A. bracteosa* and *A. acuminatissima*.

Aporusa symplocoïdes (*Hook. f.*) *Gage* in Rec. Bot. Surv. Ind. 9: 229 (before 20 March 1922); Pax & Hoffm. xv: 90 (19 Sept. 1922); Ridley: 240 (1924); Meijer, 7: 34 (1967); Airy Shaw in Kew Bull. 26: 218 (1972); Whitmore: 60 (1973).

Baccaurea? symplocoïdes Hook. f.: 376 (1887).

Var. **symplocoïdes.**

SR; SB.—Lower Siam, Malaya, Sumatra, Banka.
Tree to 18 m. high, in primary Lowland Dipterocarp forest on black or brown sandy soil on Tertiary granodiorite, ascending to 1500 m. on Kinabalu.

Leaves glabrous, chartaceous, with a slender petiole; male inflorescences very slender, up to 10 cm. long, glabrous, bearing small distant glomerules of flowers; ovary with small stigmas; infructescence slender, lax, several-fruited, pedicels elongate, fruits ovoid or ellipsoid, subacute, sometimes with 2–3 slender ribs. The racemose infructescence distinguishes this species from the very similar *A. prainiana*, in which the fruits are usually borne singly or in reduced fascicles.

Var. **chondroneura** (*Airy Shaw*) *Airy Shaw*, comb. nov.

Aporusa prainiana var. *chondroneura* Airy Shaw in Kew Bull. 25: 476 (1971).

SR; SB; K1.—Endemic.

Tree to 15 m. high, in Mixed Dipterocarp forest, on basalt, etc., up to 1250 m. alt.

Differs from the typical plant in its larger leaves (up to 30 cm. long), with closer elevate-reticulate venation, the veins themselves conspicuously white-granular beneath under a lens.

In referring this plant to *A. prainiana* in 1971 I overlooked the fact that the female inflorescences and infructescences were racemose. The stigmas, however, are almost as large as those of *A. prainiana*, possibly indicating that *chondroneura* deserves specific rank.

The following specimens may represent aberrant forms of *A. symplocoïdes*, or possibly new species. SARAWAK: Fourth Division, Long Dapoi, *Haji Suib* S. 23427 (leaves stiffly chartaceous, drying a deep lurid brownish-green; young fruits glabrous, trilocular, with small stigmas, borne singly on short densely bracteate and densely ferrugineous peduncles 3–4 mm. long). SABAH: Beaufort Distr., Ulu Buangaya, Mt. Lumaku, *Ahmad & Nicholas* SAN 65318 (leaves oblong or narrowly elliptic, up to 19 × 5·5 cm., chartaceous, petioles 1–1·5 cm. long; male inflorescences fascicled, the rhachis sinuous, up to 2·5 cm. long, the flowers not in compact glomerules but in irregular shortly elongate groups); Sandakan Distr., Sepilok Forest Reserve, *G.H.S. Wood* A 3437 (leaves as the last; female inflorescences solitary or two together, up to 2 cm. long, puberulous, with the flowers crowded towards the apex; young fruits glabrous, with very small stigmas).

Ashtonia *Airy Shaw*

Ashtonia excelsa *Airy Shaw* in Kew Bull. 21: 357 (1968); Meijer, 10: 230 (1968).

Blumeodendron tokbrai sec. Meijer, 10: 229, tab. opp. p. 64 (1968), *non* (Bl.) Kurz.

SR; B; SB; K1, 3.—Endemic.

Tree to 33 m. high, frequently emergent, sometimes dominant or co-dominant, in forest transitional between Heath forest and Mixed Dipterocarp forest, on 'pale leached yellow sandy soils, with accumulations of acid raw humus and frequently a horizon below the humus of pure white sand, overlying the yellow, as yet incompletely leached, sand below', up to 780 m. alt. (Ashton *apud* Airy Shaw, *l.c.*: 359).

The yellow-green colour assumed by the leaves on drying doubtless indicates an aluminium accumulator. The plant superficially resembles a species of *Aporusa*, but differs from that genus in the male flower bearing 5–6 very short stamens and a large peltate pistillode, and in the very large tetramerous capsule (up to 5 cm. diam.). A second species of the genus has recently been described from Malaya.

Austrobuxus *Miq.*

Austrobuxus nitidus *Miq.*, Fl. Ind. Bat., Suppl.: 445 (1861); Airy Shaw in Kew Bull. 25: 506 (1971) & 29: 309 (1974); Whitmore: 63 (1973).

Choriophyllum malayanum Benth. in Hook. Ic. Pl. 13: 62, t. 1280 (1879); Hook. f.: 344 (1887); Boerl.: 219 (1900); Merr.: 333 (1921).
Longetia malayana (Benth.) Pax & Hoffm. xv: 291 (1922); Ridley: 224 (1924).
L. nitida (Miq.) v. Steenis in Blumea 12: 362 (1964) & 15: 155 (1967).

Var. **nitidus.** Capsule up to 1·8 cm. long.

SR; B; SB; K1, 3.—Malaya, Sumatra.
Shrub or tree to 24 m. high, in primary or secondary Mixed Swamp forest or *Agathis* forest on acid waterlogged sand and peat at lower levels, but also in forest on mountain summits, ascending to 1560 m. on Kinabalu.

The male flowers are stated to be fragrant. The timber is used for house construction in Sabah.

Var. **macrocarpus** *Airy Shaw* in Kew Bull. 25: 506 (1971); Whitmore, *l.c.* (1973).

SR.—Malaya.

Capsule up to 2·5 cm. long.—The commonest form in Sarawak, *teste* T. C. Whitmore, but apparently rarely collected.

Baccaurea *Lour.*

1 ♀ inflorescences and infructescences short, rigid, ± erect, borne among the leaves of the current year's shoots (§§ *Everettiodendron* & *Calyptroön*):
  2 Capsule bilocular, almost glabrous, blackish when dry, splitting very readily; leaves smooth, glabrous, ± acute at the base, remaining ± green when dry . . . . **B. sumatrana**
  2 Capsule 3(–4)-locular, puberulous, reddish-brown when dry, splitting very tardily; leaves venose, ± puberulous, mostly rounded at the base, ± brown or ochreous when dry:
    3 Leaves closely fuscous-punctate on lower surface . **B. bracteata**
      4 Leaves less coriaceous, more closely punctate . var. **bracteata**
      4 Leaves more coriaceous, less closely punctate . var. **crassifolia**
    3 Leaves not fuscous-punctate below:
      5 Lower leaf-surface and pubescence of young parts pallid ochreous-grey or occasionally a lurid yellow-green . . **B. reticulata**
      5 Lower leaf-surface and pubescence ± brown or ferrugineous:
        6 ♂ inflorescence lax, many-flowered, up to 14 cm. long, with smaller inconspicuous bracts . . . **B. nanihua** *var.*
        6 ♂ inflorescences less branched, fewer-flowered, up to 6 cm. long, with longer more conspicuous bracts . **B. philippinensis**
1 ♀ inflorescences and infructescences elongate, ± lax, pendulous, mostly borne on the trunk and older branches:
  7 Fruit with distinct keels or wings:

8 Fruit globose or shortly pyriform, 10–13 mm. long, capsular, sharply but shallowly keeled, dehiscing normally; leaves membranaceous to chartaceous, up to 20 cm. long, usually with spreading or adpressed strigose hairs on the midrib below (§ *Calyptroön*) **B. trigonocarpa**

8 Fruit elliptic-ovoid or fusiform, 3–5 cm. long, narrowly 6-winged, red, fleshy, ± indehiscent; leaves coriaceous, glabrous, up to 45 cm. long (§ *Pierardia*?) . . . . **B. angulata**
(Cf. also forms of **B. parviflora**)

7 Fruit without distinct keels or wings (sometimes bluntly angled):

9 ♂ inflorescence short, dense, cylindric, catkin-like, often bare of flowers at the base (§§ *Everettiodendron* & *Calyptroön*):

10 Leaves (at least some) slightly or distinctly cordate at the base:

11 Leaves usually with slight distant marginal indentations; infructescences up to 60 cm. long . . . **B. hookeri**

11 Leaves without marginal indentations; infructescences shorter; secondary veins prominent and closely scalariform below:

12 Leaves very smooth, with immersed venation on the upper surface; indumentum short and inconspicuous **B. maingayi**

12 Leaves more 'veiny', venation not immersed on the upper surface; indumentum densely fulvous-tomentellous **B. kunstleri**

10 Leaves acute to rounded but not cordate at base:

13 Leaves rather small, 5–10(–15) cm. long, usually conspicuously and abruptly caudate; petiole distinctly geniculate at apex **B. minor**

13 Leaves larger, not or less markedly caudate; petiole not or less markedly geniculate:

14 Ultimate venation distinctly minutely subquadrate-tessellate, showing up as a pale reticulum against the dark brown upper surface of the leaves when dry; fruits mostly bilocular, about 3 cm. diam. . . . . **B. edulis**

14 Ultimate venation not as above; fruits mostly trilocular:

15 Leaves usually bearing a series of minute tooth-like tufts of fulvous hair on the margins; fruit often conspicuously pear-shaped when young **B. pyriformis**

15 Leaves without marginal tufts of hair; fruits not pear-shaped:

16 Fruits 2–3·5 cm. diam., leaves ample; plant finely puberulous . . . . . . **B. latifolia**

16 Fruits 1–1·5 cm. diam.:

17 Leaves up to 30 cm. long, smooth, coriaceous, almost glabrous, drying a deep reddish-brown; infructescence up to 30 cm. long **B. macrophylla**

17 Leaves up to 20 cm. long, with prominently scalariform secondary nerves below, less coriaceous, mostly ± puberulous, ± ochra-

ceous when dry; infructescence up to 20 cm. long . . . . **B. costulata**

9 ♂ inflorescence ± elongate, lax, each branchlet rather diffusely 3–many-flowered:

18 Perianth-segments externally glabrous:

19 Leaves large, up to 30 cm. long, together with the stems remaining bright green when dry; ♀ tepals suborbicular, much imbricate; fruit soft, ± baccate, up to 5 cm. diam. (§ *Hedycarpus*) . . . . **B. lanceolata**

19 Leaves 10–20 cm. long, becoming brownish when dry; ♀ tepals narrowly obovate; fruit small, capsular, 1–1·5 cm. long (§ *Pierardia*) . . . . **B. javanica**

18 Perianth-segments externally papillose-puberulous (§ *Pierardia*):

20 Leaves mostly exceeding 20 cm. in length:

21 Leaves coriaceous, smooth, glabrous, with 6–8 pairs of nerves; fruits up to 4 cm. diam. . . . **B. macrocarpa**

21 Leaves chartaceous, with 12–16 pairs of nerves, which are pubescent on the lower surface; fruit up to 2·5 cm. diam.:

22 Fruits very thinly puberulous; pedicels 5–10 mm. long, recurved; leaves broadly cordate; petioles longer . . . . **B. motleyana**

22 Fruits manifestly pubescent; pedicels obsolete or up to 2 mm. long, ascending; leaves narrowly cordate; petioles shorter . . . . **B. brevipes**

20 Leaves mostly less than 20 cm. in length:

23 Branches with a conspicuous tuft of stipules or bud-scales at the apex; leaves with a faint dusty whitish suffusion when dry, never remaining green; average length of ♀ inflorescence greater (20–30 cm.); rhachis often reddish; fruits ± ellipsoid, 1 cm. long . . . . **B. stipulata**

23 Branches with a less conspicuous apical tuft of stipules; leaves often remaining green when dry, or turning brownish, without any whitish suffusion; average length of ♀ inflorescence less (10–25(–30) cm.); rhachis not reddish (the following species are indistinguishable without fruits):

24 Fruits fusiform, 2–2·5 cm. long, 1 cm. diam., sometimes shallowly winged . . . . . . **B. parviflora**

24 Fruits neither fusiform nor winged:

25 Fruit 3-locular, globose or ellipsoid or subtrigonous, 2–2·5 cm. diam., thick-walled, usually bearing numerous white pustules . . . **B. racemosa**

25 Fruit bilocular, transversely dicoccous or subreniform, about 1 cm. long and 1·5 cm. diam., with a shallow longitudinal constriction (occasionally globose and 1-seeded), smooth, thin-walled, with fewer and smaller pustules . . . **B. trunciflora**

Baccaurea angulata *Merr.*, Pl. Elmer.: 148 (1929); Meijer, 7: 35 (1967).

SR (C. & NE.); **SB; K1.**—Endemic. (Java?)

Tree to 21 m. high, in primary Lowland Dipterocarp forest on sandstone or stony yellow soil, or on tufa, up to 780 m. alt.

Almost glabrous except the inflorescences. Leaves oblong or oblanceolate, 15–45 cm. long, 5–14 cm. broad, chartaceous to coriaceous, obtuse to very shortly acuminate, cuneate or occasionally rounded below, drying greenish above and ochraceous to chestnut-brown below, or greenish throughout, surface smooth but rather dull, nervation lax, petiole 5–13 cm. long. Male inflorescences (previously undescribed) very slender, fascicled, cauline, up to 12 cm. long, papillose-puberulous; bracts subulate, 1–2 mm. long, not adnate; flowers in 3-flowered fascicles, pedicels 3–4 mm. long, flower 1·5–2 mm. diam. Female inflorescences robust, cauline (occasionally ramiflorous), in small groups, up to 21 cm. long; rhachis angled, finely puberulous; pedicels spreading or deflexed, 3–5 mm. long; tepals narrowly oblong, 7–8 mm. long, conduplicate, 1 mm. wide (2 mm. when opened flat), densely grey-papillose; ovary 3-locular, lageniform, 3 mm. long, densely sericeous below, puberulous above, styles 3, very short, reflexed, coarsely papillose. Fruit elliptic-ovoid, 3–5 cm. long, conspicuously beaked, red, fleshy, strongly though narrowly 6-winged, wings sinuate.

A very distinct species, probably referable to § *Pierardia*, but perhaps distantly related to *B. lanceolata*, from which it differs in its more oblong, duller leaves, with much laxer and less reticulate venation, its narrow, elongate, externally papillose female perianth-segments, and its 3-locular, 6-winged, beaked fruit.

BACCAUREA BIVALVIS Merr. = **B. sumatrana** (*Miq.*) *Muell. Arg.*

Baccaurea bracteata *Muell. Arg.*: 466 (1866); Hook. f.: 372 (1887); Boerl.: 281 (1900); Merr.: 330 (1921); Pax & Hoffm. xv: 65 (1922); Ridley: 246 (1924); Pax & Hoffm. in Mitt. Inst. Bot. Hamburg 7: 223 (1931); Corner: 239 (1940); Meijer, 7: 36 (1967); Airy Shaw in Kew Bull. 26: 219 (1972); Whitmore: 67 (1973).

Var. **bracteata:**

SR; B; SB; K1–3.—Lower Siam, Malaya, Sumatra.

Tree to 25 m. high, very common in primary Heath forest, peat forest, open Dipterocarp (*Shorea albida*) forest, *padang* forest, *kerangas* forest, *Agathis* forest, *Solenospermum* marsh, seasonal riverine swamp or temporarily submerged flat bog, also in *blukar* and *ladang*, on poor sandy soil or white acid sand overlain by peat, or sandy loam or loam with lime, or giant podsol, or black or clayey soil, or high-level alluvium, up to 240 m. alt.

Usually readily recognizable by the close fuscous puncticulation of the lower surface of the leaves, which often dry a ferrugineous-brown colour. The inflorescences and young growth are shortly ferrugineous-tomentellous; bracts relatively large, boat-shaped. Capsule trilocular, globose, 1·5–2 cm. diam., mostly borne on twigs or small branches.

Sometimes cultivated; fruits eaten by natives of Brunei. The tips of the shoots are said to be used in making *bladeang*.

Var. **crassifolia** (*J. J. Sm.*) *Airy Shaw*, var. & stat. nov.

Baccaurea crassifolia J. J. Sm. in Bull. Jard. Bot. Buitenz. III, 1: 394 (1920); Pax & Hoffm. xv: 66 (1922).

SR; SB; K1, 3.—Sumatra.

Tree to 18 m., in primary shallow peat-swamp forest (sometimes seasonal), or on high-level alluvium, up to 150 m. alt.

An extreme form of *B. bracteata* with more coriaceous and less punctate leaves. It appears to grade into the typical form.

Baccaurea brevipes *Hook. f.*: 372 (1887); Pax & Hoffm. xv: 58 (1922); Ridley: 250 (1924); Corner: 239 (1940); Meijer, 7: 36 (1967); Whitmore: 66 (1973).

SR (SW.).—Malaya, Sumatra.

Tree to 15 m. high, in Mixed Dipterocarp forest on sandy clay or granodiorite-derived soil up to 570 m. alt.

Very close to *B. motleyana* and only distinguishable with certainty by the more manifestly pubescent fruits, which are subsessile or borne on very short (1–2 mm.), upwardly directed pedicels. The leaves tend to be more narrowly cordate and more shortly petiolate than in that species.

BACCAUREA CORDATA Merr. = **B. kunstleri** *King ex Gage*

Baccaurea costulata (*Miq.*) *Muell. Arg.*: 464 (1866); Pax & Hoffm. xv: 68 (1922); S. Moore in Journ. Bot. Brit. & For. 63, Suppl.: 98 (1925).

Mappa costulata Miq., Fl. Ind. Bat., Suppl.: 459 (1861).
Pierardia costulata (Miq.) Muell. Arg. in Flora 47: 469 (1864).
Baccaurea deflexa Muell. Arg.: 462 (1866); J. J. Sm.: 248 (1910) & in Ic. Bogor. 4: 29, t. 310 (1910); Pax & Hoffm. xv: 69 (1922); Merr., Pl. Elmer.: 145 (1929); Backer & Bakh. f.: 454–5 (1963); Meijer, 7: 37 (1967); **synon. nov.**

SB.—Sumatra, Banka.

Tree to 14 m. high, in primary lowland forest on brownish soil up to 45 m. alt.

Allied to *B. minor*, but leaves larger, with more numerous primary nerves, and especially with closely scalariform secondary nerves, which are very prominent on the lower surface. The plant is usually minutely fulvous-puberulous, but is sometimes almost glabrous. The capsule is trilocular, about 1 cm. diam.

The similarity of *B. deflexa* to *B. costulata* was noted already by Mueller (*l.c.*: 463). I have no doubt that the two are conspecific.

BACCAUREA DOLICHOBOTRYS Merr. = **B. hookeri** *Gage*

Baccaurea edulis *Merr.*, Pl. Elmer.: 149 (1929); Meijer, 7: 36 (1967), *in obs.*

SR; SB; K1.—Endemic.

Tree to 33 m. high, in primary Lowland Dipterocarp forest on sandstone or sandy soil, basalt-derived soil or basaltic loam, brown soil or tufa, up to 200 m. alt.

The characteristic feature of this species, by which it can almost invariably be recognized, is the pale, finely tessellate-reticulate venation of the leaves, especially as seen on the upper surface. The leaves usually turn a rather dark brown colour above when dry, and the ultimate nerves then generally show up as a fine, lighter-coloured, subquadrate reticulum. There is sometimes a development of minute tooth-like marginal tufts of hair on the leaves, as in *B. pyriformis*, but far less conspicuously.

According to Merrill, the fruits, which are bilocular, about 2 cm. diam., and indehiscent, were described by the collector Elmer as "the best of the edible Euphorbiaceous fruits here"—i.e. in the Tawao district of Sabah.

BACCAUREA ELMERI Merr. = **B. latifolia** *Hook. f.*

Baccaurea hookeri *Gage* in Rec. Bot. Surv. Ind. 9: 232 (1922); Ridley: 247 (1924); Whitmore: 67 (1973).

B. dolichobotrys Merr., Pl. Elmer.: 147 (1929), **synon. nov.**

SR; SB; KI.—Malaya.

Tree to 40 m. high, in forest on sandstone, or on heavy loam containing limestone, at 20 m. alt.

Leaves ample, almost glabrous, dark brown when dry, rounded or slightly cordate at the base, and often with a shallow indentation opposite each main lateral nerve, producing a kind of very distant shallow crenation. Flowers unknown. Infructescences immensely elongate, up to 60 cm. long, in pendulous fascicles from the older branches, minutely ferrugineous; fruits globose, 1–1·3 cm. diam., trilocular, thin-shelled, tardily dehiscent, minutely puberulous, endocarp densely sericeous within; pedicels 3–5 mm. long, strongly reflexed, thus bearing the capsules erect on the pendulous racemes; seeds flattened, with a sweet orange aril.

Closely related to *B. maingayi*, but apparently differing in the slightly thinner leaves, with laxer venation, which is less closely scalariform beneath and more evident (less immersed) on the upper surface, and in the marginal indentation of the lamina. The infructescences appear to reach greater lengths than those of *B. maingayi*. In the Malayan plant the inner surface of the endocarp is glabrous, but I can detect no other significant difference. The species seems to be everywhere scarce.

I am not convinced that *B. hookeri* is specifically distinct from *B. polyneura* Hook. f., but further material of the latter species (described from Malacca) is needed in order to settle this point.

Baccaurea cf. **javanica** (*Bl.*) *Muell. Arg.*: 465 (1866); Boerl.: 281 (1900); J. J. Sm.: 253 (1910); Merr.: 331 (1921); Holth. & Lam in Blumea 5: 200 (1942); Backer & Bakh. f.: 454–5 (1963); Soejarto in Bot. Mus. Leafl. Harv. Univ. 21: 96–7 (1965); Meijer, 7: 37 (1967); Whitmore: 65 (1973).

Adenocrepis javanica Bl., Bijdr.: 579 (1825).
Hedycarpus javanicus (Bl.) Miq., Fl. Ind. Bat. 1(2): 359 (1859).
Microsepala acuminata Miq., Fl. Ind. Bat., Suppl.: 444 (1861).

Baccaurea acuminata (Miq.) Muell. Arg.: 463 (1866).
B. leucodermis Hook. f. ex Ridley: 244 (1924), **synon. nov.**

SR; SB; K1a.—Malaya, Sumatra, Java, Talaud Is.? (N.B. Specimens collected in Banka in 1949 by *Kostermans & Anta* and distributed as *B. javanica* have strongly papillose-puberulous perianth-segments and are evidently not that species, but are scarcely identifiable with certainty in the absence of fruits.)

Tree to 10 m. high, in primary or secondary forest on brown sandy soil at about 90 m. alt.

Only certainly distinguishable from its immediate allies by the externally glabrous male and female perianth-segments and by the small, thin-walled, bilocular, dehiscent capsule, 7–13 mm. long, surmounted by a very short style and capitate stigma.

The occurrence of *B. javanica* in Borneo must be regarded as doubtful, since no fruiting material has yet been collected. A few gatherings with externally glabrous perianth-segments have been made, but these could well represent glabrous states of *B. racemosa*, etc. *Motley* 579, from Bandjermasin in SE. Borneo, was cited by Mueller (*l.c.*) as *B. javanica*, but a pencilled note by J. D. Hooker on the Kew sheet reads: "*B. javanica* Muell., but I think quite difft. from the Java plant. *J.D.H.*"

Baccaurea kunstleri *King ex Gage* in Rec. Bot. Surv. Ind. 9: 230 (1922); Ridley: 248 (1924); Meijer, 7: 36 (1967); Airy Shaw in Kew Bull. 26: 219 (1972); Whitmore: 65 (1973).

B. cordata Merr., Pl. Elmer.: 147 (1929).

B; SB; K1.—Lower Siam, Malaya, Sumatra, Banka.

Tree to 25 m. high, in primary forest on sandstone, sandy loam or sandy clay alluvium, up to 840 m. alt. on Kinabalu.

The cordate-based leaves, with closely scalariform secondary nervation prominently raised beneath, combined with the rather dense fulvous tomentellum of young parts, petioles, midribs and inflorescences, are usually sufficient for the recognition of *B. kunstleri*. The male inflorescences are short, dense, rarely 5 cm. long, and bare of flowers below. The infructescences are up to 24 cm. long; the capsules trilocular, smooth, velutinous, up to 2 cm. long; seed red, with orange-red aril.

Baccaurea lanceolata (*Miq.*) *Muell. Arg.*: 457 (1866): Hook. f.: 368 (1887); Stapf in Trans. Linn. Soc., Bot. 4: 224 (1894); Boerl.: 280 (1900); J. J. Sm.: 247 (1910); Pax & Hoffm. xv: 60 (1922); Merr.: 331 (1921) & 411 (1923); Ridley: 248 (1924); S. Moore in Journ. Bot. Brit. & For. 63, Suppl.: 98 (1925); Merr., Pl. Elmer: 150 (1929); Pax & Hoffm. in Mitt. Inst. Bot. Hamburg 7: 223 (1931); Hend. in Journ. Malayan Br. Roy. As. Soc. 17: 69 (1939); Corner: 240 (1940); Backer & Bakh. f.: 454–5 (1963); Meijer, 7: 35, tab. (1967); Airy Shaw in Kew Bull. 26: 220 (1972); Whitmore: 65 (1973).

Hedycarpus lanceolatus Miq., Fl. Ind. Bat. 1(2): 359 (1859).
Adenocrepis lanceolata (Miq.) Miq. in Linnaea 32: 82 (1863).

Baccaurea glabriflora Pax & Hoffm. xv: 59 (1922); Merr.: 411 (1923); **synon. nov.**

SR; B; SB; K1, 1a.—Lower Siam and throughout W. Malesia to SW. Philippines (Palawan).

Tree to 27 m. high, in primary or secondary Lowland Dipterocarp forest or *Oncosperma filamentosum* forest, on yellow sandy loam, clay loam, loam with lime, clay-rich alluvium, limestone-derived alluvium, base of limestone cliff, seasonally swampy alluvium, or scree beneath igneous (andesitic) rock-face, up to 900 m. alt. on Kinabalu.

A very distinct species, perhaps distantly related to *B. angulata.* Like it almost completely glabrous, but leaves (and stems) invariably drying green, less oblong in outline, thinner in texture, with closer and more evidently reticulate venation. Perianth-segments (male and female) papillose within only, glabrous outside, the male segments often shortly united; female per. segs. suborbicular, 3–5 mm. diam., much imbricate; ovary globose, 4(–5)-locular; infructescence up to 30 cm. long; fruit globose, unbeaked, yellowish-white when ripe, taste compared to that of 'unripe potatoes'!

Baccaurea lanceolata is almost the only member of the genus able to tolerate limestone soils.

Baccaurea latifolia *King ex Hook. f.*: 373 (1887); Ridley: 249 (1924); Whitmore: 67 (1973).

B. puberula Merr., Pl. Elmer.: 145 (1929); Meijer, 7: 36–37 (1967); **synon. nov.**

B. elmeri Merr., *l.c.*: 146 (1929), **synon. nov.**

SR; SB.—Malaya.

Tree to 24 m. high, in primary Lowland or Mixed Dipterocarp forest, or in swampy *kerangas*, on basalt or shale, or clay-rich or black soil, up to 1650 m. alt.

Among the species with stellate indumentum (§ *Everettiodendron*) readily recognizable by its rather large, long-petioled leaves, up to 25 cm. long, with 9–14 pairs of nerves. Infructescences 12–20 cm. long, on trunk or branches. Fruits globose, 2–3·5 cm. diam., verruculose, puberulous or glabrescent, tardily dehiscent. One collector notes that the aril is pale yellowish, rather than white as in many others.

Baccaurea macrocarpa (*Miq.*) *Muell. Arg.*: 459 (1866); Hook. f.: 375 (1887); Pax & Hoffm. xv: 49 (1922); Meijer, 7: 35 (1967).

Pierardia macrocarpa Miq., Fl. Ind. Bat., Suppl.: 441 (1861).

Baccaurea borneënsis Muell. Arg.: 460 (1866); Boerl.: 280 (1900); Merr.: 330 (1921); **synon. nov.**

Baccaurea griffithii Hook. f.: 371 (1887); Pax & Hoffm. xv: 66 (1922); Ridley: 248 (1924); Merr. in Papers Mich. Acad. Sci. 20: 100 (1935); Corner: 240 (1940); Whitmore: 66 (1973); **synon. nov.**

Baccaurea sp., Merr., Pl. Elmer.: 153 (1929) (*Elmer* 21717, 21240, 21816).

SR; B; SB; K1.—Malaya, Sumatra, Banka.

Tree to 24 m. high, in primary or secondary Mixed Dipterocarp forest on basalt, yellow clay, sandy soil, black soil or alluvium, up to 1440 m. on Kinabalu.

Vegetatively very similar to *B. macrophylla*, but male inflorescence lax, with many-flowered branchlets, papillose-puberulous, ochreous-grey; fruits large, woody, thick-walled, up to 4 cm. diam., borne in short stout racemes on the trunk and branches.

Baccaurea macrophylla (*Muell. Arg.*) *Muell. Arg.*: 460 (1866); Hook. f.: 369 (1887); Boerl.: 281 (1900); Merr.: 331 (1921); Pax & Hoffm. xv: 62 (1922); Ridley: 247 (1924); Airy Shaw in Kew Bull. 26: 220 (1972); Whitmore: 66 (1973).

Pierardia macrophylla Muell. Arg. in Flora 47: 516 (1864).
Baccaurea beccariana Pax & Hoffm. xv: 62 (1922), **synon. nov.**

SR; SB; KI.—Malaya, Sumatra.
Tree to 20 m. high, in primary or secondary forest on sandy soil or yellow clay up to 450 m. alt.

Leaves almost glabrous, smooth, up to 30 cm. long, coriaceous, usually drying a deep reddish-brown, long-petioled; stipules very broad, rounded, caducous; male inflorescences rather dense, cylindric, with 3-flowered branchlets, minutely ferrugineous; infructescence pendulous, up to 30 cm. long, fruits small, globose, 10–15 mm. diam., trilocular, tardily dehiscent, sparsely puberulous, borne on decurved pedicels 5–6 mm. long.
The foliage is almost indistinguishable from that of *B. macrocarpa*, but the inflorescences are quite different, being typical of § *Calyptroön*, whilst the fruits are small and thin-walled.

Baccaurea maingayi *Hook. f.*: 370 (1887); Pax & Hoffm. xv: 63 (1922); Ridley: 246 (1924); Whitmore: 65 (1973).

SR (SW.).—Malaya.
Tree to 18 m. high, in primary Lowland Dipterocarp forest at 90 m. alt.

Close to *B. hookeri*; leaves slightly more coriaceous, very smooth, with immersed venation on the upper surface, but with prominent, closely scalariform secondary veins on the lower surface, and margin entire, without indentations.
So far known in Borneo only from the Arboretum of the Semengoh Forest Reserve near Kuching.

Baccaurea minor *Hook. f.*: 370 (1887); Pax & Hoffm. xv: 64 (1922); Ridley: 245 (1924); Airy Shaw in Kew Bull. 16: 342 (1963); Meijer, 7: 37 (1967); Whitmore: 66 (1973).

Aporosa billitonensis Pax & Hoffm., *l.c.*: 245 (1922); cf. Airy Shaw, *l.c.*
Baccaurea pendula Merr., Pl. Elmer.: 152 (1929), **synon. nov.**

SR; SB; KI.—Malaya, Banka, Billiton.
Tree to 26 m. high, in primary or secondary lowland forest, on sandstone or clay or yellow clayey sand, or greyish soil, up to 1200 m. alt. on Kinabalu.

Leaves rather small, 5–10(–15) cm. long, usually conspicuously and abruptly caudate, practically glabrous; petiole slender, 1–3 cm. long, markedly geniculate at apex; nerves 4–6 pairs. Infructescence up to 25 cm. long; capsule globose, trilocular, 1 cm. diam., minutely ferrugineous-puberulous, endocarp densely sericeous within.

Resembles a small edition of *B. maingayi*, with smaller, caudate, mostly cuneate-based, more shortly petioled leaves. The clearly geniculate petiole, longer acumen, usually elongate infructescence and trilocular capsule distinguish it from the sometimes similar *B. sumatrana*.

Baccaurea motleyana (*Muell. Arg.*) *Muell. Arg.*: 461 (1866); Hook. f.: 371 (1887); Boerl.: 281 (1900); Merr.: 331 (1921); Pax & Hoffm. xv: 53 (1922); Ridley: 250 (1924); Corner: 240 (1940); Backer & Bakh. f.: 454–5 (1963); Soejarto in Bot. Mus. Leafl. Harv. Univ. 21: 73, tt. 9, 10, 12 (1965); Meijer, 7: 36, 37 (1967), *in obs.*; Airy Shaw in Kew Bull. 26: 220 (1972); Whitmore: 66 (1973).

Pierardia motleyana Muell. Arg. in Flora 47: 516 (1864).

SR; B; K1, 1a.—Sumatra, Banka; cult. Malaya, Java.
Tree to 15 m., in forest on yellow sandy clay up to 750 m. alt.

Very close to *B. brevipes*, but distinguishable by its very thinly puberulous fruits, which are borne on much longer (5–10 mm.), recurved or reflexed pedicels, and by its usually longer-petioled and more broadly cordate leaves.

Baccaurea nanihua *Merr.*, Interpr. Rumph. Herb. Amboin.: 315 (1917); Pax & Hoffm. xv: 67 (1922).

Var. **oblongata** *J. J. Sm.* in Bull. Jard. Bot. Buitenz. III, 6: 94 (1924).

K1.—Endemic. (Var. *nanihua* in the Moluccas.)
No ecological information available.

J. J. Smith (*l.c.*) wrote: "I am inclined to believe that this species should be regarded as a form of *B. philippinensis* Merr." I doubt if this is correct: the male inflorescence is much laxer, with much smaller bracts, than in that species. Var. *oblongata* was distinguished by Smith from the type form of *B. nanihua* by its "lax indumentum, the generally narrower acuminate acute leaves, and the conspicuous ♂ disk glands". I feel, however, considerable misgivings as to the value of this variety.

Baccaurea parviflora (*Muell. Arg.*) *Muell. Arg.*: 462 (1866); Hook. f.: 368 (1887); Boerl.: 281 (1900); Pax & Hoffm. xv: 59 (1922); Ridley: 243 (1924); Corner: 241 (1940); Airy Shaw in Kew Bull. 26: 220 (1972); Whitmore: 64 (1973).

Pierardia parviflora Muell. Arg. in Linnaea 32: 82 (1863).
Baccaurea affinis Muell. Arg.: 459 (1866).
B. scortechinii Hook. f.: 368 (1887); Pax & Hoffm. xv: 56 (1922): Ridley: 244 (1924); Corner: 242 (1940).
? *B. odoratissima* Elm. in Leafl. Philipp. Bot. 4: 1276 (1911).
B. singaporica Pax & Hoffm. xv: 54 (1922).
B. rostrata Merr., Pl. Elmer.: 150 (1929); Airy Shaw in Kew Bull. 21: 355 (1968), *in obs.*; **synon. nov.**

SR; SB; K1.—Lower Burma, Malaya.

Tree to 6 m. high, in Mixed Dipterocarp forest on sandy loam or granodiorite-derived soil up to 300 m. alt.

Distinguished from *B. racemosa* and *B. trunciflora* only by its fusiform 2–3-locular fruits, 2–2·5 cm. long by 1 cm. diam., which are sometimes 4–6-angled or shallowly winged.

Apparently rather scarce in Borneo, though common in Malaya.

BACCAUREA PENDULA Merr. = **B. minor** *Hook. f.*

Baccaurea philippinensis (*Merr.*) *Merr.* in Philipp. Journ. Sci. 10, Bot.: 275 (1915); Pax & Hoffm. xv: 68 (1922); Merr.: 411 (1923).

Everettiodendron philippinense Merr. in Philipp. Journ. Sci. 4, Bot.: 279 (1909); Elm., Leafl. Philipp. Bot. 3: 916 (1910).

SB.—Philippines.

Tree to 17 m. high, in primary swamp forest on reddish sandy soil up to 30 m. alt.

Very close to *B. bracteata*, from which it only differs in the complete absence of puncticulations on the lower leaf-surface, and perhaps in the somewhat larger fruit, up to 3 cm. diam. I cannot appreciate the allegedly larger size of the bracts, as given by Pax & Hoffmann.

BACCAUREA PUBERULA Merr. = **B. latifolia** *King ex Hook. f.*

Baccaurea pyriformis *Gage* in Rec. Bot. Surv. Ind. 9: 233 (1922); Ridley: 249 (1924); S. Moore in Journ. Bot. Brit. & For. 63, Suppl.: 98 (1925); Whitmore: 67 (1973).

B. platyphylla Pax & Hoffm. xv: 67 (1922); cf. Ridley, *l.c.* (1924), *in obs.*; **synon. nov.**

B. platyphylloïdes Pax & Hoffm. xv: 68 (1922).

SR; B; SB; K1.—Malaya, Sumatra.

Tree to 21 m. high, in primary or secondary Lowland Dipterocarp forest on sandy loam, yellow sandy clay, Tertiary sandstone, brownish or yellow-brown soil, up to 150 m. alt.

Petioles, midribs beneath, inflorescences and young growth thinly to densely fulvous-tomentellous. Leaves more or less chartaceous, turning a warm chestnut-brown (especially beneath) when dry, often rather brittle, and often bearing a series of minute tooth-like tufts of hair on the margins. Fruit often obovoid or pear-shaped when young, but sometimes globose when mature, 2–2·5 cm. diam.

Baccaurea racemosa (*Reinw.*) *Muell. Arg.*: 461 (1866); J. J. Sm. 249 (1910); Merr.: 331 (1921); Pax & Hoffm. xv: 51 (1922); Merr., Pl. Elmer.: 150 (1929); Backer & Bakh. f.: 455 (1963); Soejarto in Bot. Mus. Leafl. Harv. Univ. 21: 74, tt. 9, 10, 11, 13 (1965); Whitmore: 64 (1973).

Coccomelia racemosa Reinw. apud Bl., Cat. Gewassen Buitenz.: 110 (1823) & in Flora 8(1): 103 (1825) & in Syll. Pl. Ratisb. 2: 5 (1825).
Pierardia racemosa (Reinw.) Bl., Bijdr.: 579 (1825).
? *Baccaurea parviflora* sec. Muell. Arg.: 462 (1866), *pro parte*; Merr.: 331 (1921); Meijer, 7: 37 (1967); *an Pierardia parviflora* Muell. Arg. (1863)?
B. wallichii Hook. f.: 375 (1887); Pax & Hoffm. xv: 53 (1922); Ridley: 245 (1924); **synon. nov.**
? *B. membranacea* Pax & Hoffm. xv: 49 (1922).
? *B. sarawakensis* Pax & Hoffm. xv: 54 (1922).

SR; B; SB; KI.—Malaya, Sumatra, Java.

Shrub or tree to 16 m. high, in primary or secondary Mixed or Lowland Dipterocarp forest or peat-swamp forest, on dry sandstone or deep yellow sand or yellow sandy clay or sandy loam with a little clay, or clay-rich soil, or heavy loam with coral limestone or granodiorite-derived or shale-derived soil, or brown soil or tufa, or riverine alluvium, up to 810 m. alt.

Only distinguishable from *B. trunciflora* and *B. parviflora* (*B. rostrata*) by the globose or ellipsoid, or slightly trigonous, trilocular, thick-walled, red fruit, 2–2·5 cm. diam., which is usually marked with numerous white more or less azolliform pustules. The seeds are covered by a violet-blue aril. The infructescences may reach 30 cm. in length.

Baccaurea membranacea and *B. sarawakensis* were described without fruits, and it is therefore not possible to identify them with certainty. See further note under *B. trunciflora* (below).

Baccaurea reticulata *Hook. f.*: 373 (1887); Pax & Hoffm. xv: 65 (1922); Ridley: 246 (1924); Corner: 241 (1940); Meijer, 7: 36 (1967), *in obs.*; Whitmore: 67 (1973).

SR; B.—Malaya, Sumatra.

Tree to 20 m. high, in primary Lowland Dipterocarp forest or peat-swamp forest on deep yellow sandy loam up to 90 m. alt.

Closely related to *B. bracteata*, but quite devoid of puncticulations on the lower leaf-surface, which moreover is not ferrugineous-brown when dry but ochreous-grey or sometimes a lurid yellow-green; the pubescence also is ochreous-grey, or at least not brightly ferrugineous, and sparser than in *B. bracteata*; the floral bracts tend to be blackish when dry; fruit (according to *Wyatt-Smith* KEP 80107,* from Brunei) much larger, up to 4 cm. diam., trilocular, tardily dehiscent, mostly borne on trunk or branches.

I refer the following Bornean specimens to *B. reticulata*. SARAWAK: Sg. Snibong, *Anderson* 6552; Semengoh Forest Reserve, *Ghazalli* S14976, *Yakup* 8907 (Tree No. 62), *Banyeng & Sibat* S. 25377 (Tree No. 4683). BRUNEI: Bt. Gana Arboretum, *Wyatt-Smith* KEP 80107 (with fruit!); Andulau Forest Reserve, *Jacobs* 5662 (fruits described in field notes, but none with Kew duplicate), *Wood, Smythies & Ashton* SAN 17533.

The female inflorescence and fruit of *B. reticulata* have rarely been collected, though several times referred to in field notes. There are in the Kew Herbarium 21 sheets from Malaya and 7 from Sumatra and Borneo, and with very few exceptions they bear male inflorescences or are sterile. In an en-

* It is, however, very possible that these detached fruits do not belong; they are very like those of *B. macrocarpa*.

velope attached to one of the syntypes, however, *Maingay* 1879, from Malacca, there are, mixed with male inflorescences, a few female flowers and half-grown fruits. These were never described by Hooker, by Pax & Hoffmann or by Ridley (though Hooker made a pencil sketch of a female flower on the sheet), and it is evident that they cannot have been attached to the male specimen. In his MS. notes, however (vol. IV, p. 219), Maingay himself entered a description of them, as follows:—"♀. Calyx 5 partite, villous, persistent. Fruit baccate, fleshy, 3 locular. Loculi biovulate. Styles thick, fleshy, three in number, each deeply bipartite, their inner surfaces stigmatico-papillose."

I am not, however, prepared to accept these female flowers and fruits as those of *B. reticulata* without further evidence. I suspect, in fact, that they may be those of *B. pyriformis*, which occurs in Malacca and which Maingay himself collected (*Maingay* 2656). They were probably picked up off the ground under the tree of *reticulata*, having fallen from a *pyriformis* growing next to it.

In view of the paucity of fruiting material of *B. reticulata*, I reproduce the field notes on these given by two collectors.

"Ramiflorous; fruits dull sordidly orange, hard, splitting into 3 parts, wall inside yellow-whitish; seeds 2–5 in translucent jelly-like sweet-tasting coat.... The juicy seed-coat is well edible." (*Jacobs* 5662, Brunei.)

"Fruits round, marked by 3 lines. Pericarp firm, fleshy, sour, ripening orange flushed red. 3 seeds, white, entirely enclosed in fleshy orange-yellow thickish aril. Ultimately splitting." (*Whitmore* KEP 3818, Pahang, Malaya.)

Baccaurea rostrata Merr. = **B. parviflora** (*Muell. Arg.*) *Muell. Arg.*

Baccaurea stipulata *J. J. Sm.* in Ic. Bogor. 4: 33, t. 311 (1910); Merr.: 331 (1921); Pax & Hoffm. xv: 56 (1922); Meijer, 7: 37 (1967).

B. brevipedicellata Pax & Hoffm. xv: 55 (1922), **synon. nov.**

SR; SB; K1.—Endemic?

Tree to 15 m. high, in primary Mixed Dipterocarp forest, or riparian or swampy or damp open forest, on sandstone or black or brown or white sandy soil, or clay with sand or with scattered volcanic stone, or loam with lime, or occasionally on limestone, up to 1500 m. on Kinabalu.

Distinguished from its close allies *B. racemosa*, *B. parviflora* and *B. trunciflora* by the usually more conspicuous tuft of stipules or bud-scales at the tips of the branches, by the leaves which hardly ever remain green when dry but are mostly plumbeous above and more or less ochreous below, with a kind of indistinct whitish cast or dusty suffusion, by the mostly greater average length of the female inflorescences (20–30 cm.), by the often bright red colour of the rhachis, and by the small more or less ellipsoid fruit, resembling that of *B. racemosa* on a smaller scale.

The closest relative of *B. stipulata* is probably *B. tetrandra* (Baill.) Muell. Arg., endemic in the Philippines. It has a similar pale dusty suffusion to that of *B. stipulata*, but differs in its smaller leaves and stipules and shorter female inflorescences.

Baccaurea sumatrana (*Miq.*) *Muell. Arg.*: 466 (1866); Pax & Hoffm. xv: 63 (1922); S. Moore in Journ. Bot. Brit. & For. 63, Suppl.: 98 (1925);

Merr., Pl. Elmer.: 151 (1929); Meijer, 7: 36 (1967); Whitmore: 65, 66 (1973).
Calyptroön sumatranum Miq., Fl. Ind. Bat., Suppl.: 472 (1861).
Baccaurea kingii Gage in Rec. Bot. Surv. Ind. 9: 231 (1922); Ridley: 245 (1924); **synon. nov.**
B. bivalvis Merr., Pl. Elmer.: 148 (1929), **synon. nov.**

SR; B; SB; K1, 1a.—Malaya, Sumatra, Banka.

Tree to 40 m. high, in primary riparian or Lowland Dipterocarp forest, or in *kerangas* (incl. submontane), on yellow sandy clay, loam with coral limestone, white sand (raised beach), *gading* soil, dark brownish or red-brown soil, or alluvium, up to 1500 m. on Kinabalu.

Leaves mostly more or less spathulate or cuneate-obovate, sometimes oblong-elliptic, 5–12(–16) cm. long, chartaceous, glabrous; petiole slender, 1–4 cm. long, not or slightly geniculate at the apex; inflorescences short, rusty-puberulous or furfuraceous; infructescence up to 6 cm. long; capsule globose to obovoid or pyriform, 7–15 mm. long, bilocular, minutely puberulous, dark brown or black when dry, splitting very readily.

Differs from the related *B. minor* in the not or scarcely geniculate petiole, not or scarcely caudate lamina, short infructescence and bilocular capsule.

Baccaurea trigonocarpa *Merr.*, Pl. Elmer.: 152 (1929); Meijer, 7: 38 (1967) & 10: 232 (1968).

SR; SB; K1.—Endemic.

Tree to 9 m. high, in primary or secondary Mixed Dipterocarp forest or montane *kerangas*, on sandy clay or yellow loam or dark brown soil, up to 800 m. alt.

Leaves membranaceous to chartaceous, usually retaining a tinge of green (at least beneath) when dry, 10–20 cm. long, cuneate at base, shortly acuminate; secondary veins very slender, lax, inconspicuous; midrib beneath with spreading or adpressed short strigose hairs; petiole slender, up to 5 cm. long. Male inflorescence 2–5 cm. long, shortly ferrugineous-tomentellous, mostly bare below, dense-flowered above, bracts very short, ovate, the lower few free, the remainder adnate, almost glabrous at the tips. Infructescences up to 21 cm. long, rhachis papillose-puberulous or shortly ferrugineous-tomentellous; capsules subglobose to pyriform, 10–13 mm. long, (1–2–)3–4-locular, often distinctly tri- or tetra-coccous, always sharply and slenderly 3–4-keeled and often with 3–4 intermediate keels in addition, surface dull and irregularly granular or venulose, sparsely pubescent or glabrescent.

The thin leaves, drying greenish, with strigose midrib beneath, and the sharply keeled capsules, render this a very distinct species.

Baccaurea trunciflora *Merr.*, Pl. Elmer.: 151 (1929).

? *B. parviflora* sec. Muell. Arg.: 462 (1866), *pro parte*; Merr.: 331 (1921); Meijer, 7: 37 (1967); *an Pierardia parviflora* Muell. Arg. (1863)?
? *B. membranacea* Pax & Hoffm. xv: 49 (1922).
? *B. sarawakensis* Pax & Hoffm. xv: 54 (1922).

SR; B; SB; K1.—Endemic.

Tree to 18 m. high, in peat-swamp forest or primary or secondary Lowland Dipterocarp forest or dry Mixed Dipterocarp forest or submontane forest, on sandstone or yellow sandy clay or clay loam or basalt up to 1500 m. alt.

Only certainly distinguishable from *B. racemosa* and *B. parviflora* by the transversely dicoccous or subreniform, bilocular (rarely trilocular), smooth, thin-walled, tardily dehiscent fruit, about 1 cm. long by 1·5 cm. broad, sometimes marked with a shallow longitudinal constriction and with very small whitish pustules. Not infrequently only one ovule develops, and the fruits are then subglobose or ellipsoid, about 1 cm. diam., but their small size and thinly crustaceous pericarp distinguish them from those of *B. racemosa*.

One of the syntype collections of *B. sarawakensis* bears female flowers, in which the ovary is bilocular. This tends to favour an identification with *B. trunciflora*, and would provide an earlier name for the taxon. In the present state of our knowledge of this difficult group of species, however, I am not yet prepared to adopt the name *B. sarawakensis* in preference to that of *B. trunciflora*, the type of which bears mature fruit.

Blachia *Baill.*

Blachia andamanica (*Kurz*) *Hook. f.*: 403 (1887); Boerl.: 284 (1900); Pax & Hoffm. iii: 38 (1911); Merr.: 346 (1921); Gagnep.: 416 (1926); Airy Shaw in Kew Bull. 23: 121 (1969) & 26: 223 (1971); Whitmore: 68 (1973).

Codiaeum andamanicum Kurz, For. Fl. Brit. Burma 2: 405 (1877).
Dimorphocalyx andamanicus (Kurz) Benth. in Benth. & Hook. f., Gen. Pl. 3: 302 (1880).
Blachia philippinensis Merr. in Philipp. Journ. Sci. 4, Bot.: 277 (1909); Pax & Hoffm. v (Euph.-Addit. iii): 285 (1912); Merr.: 455 (1923); **synon. nov.**

K –(*teste* Boerl.; no specimens seen).—Assam, SE. Asia, Malaya (Perak), Philippines, Celebes.
No ecological information available for Borneo.

There is no Bornean material of *Blachia* at Kew, and I have not yet traced the basis of Boerlage's record. The species is evidently rare or extremely local in Borneo. As I noted in 1969 (*l.c.*: 122), the only tangible point distinguishing *B. philippinensis* from *B. andamanica* seems to be the 'subglabrous' ovary, but Philippine gatherings with pubescent and with more or less glabrous ovaries respectively seem otherwise identical and inseparable from *B. andamanica*. A greater range of material must show whether the subglabrous form deserves varietal recognition.

Blumeodendron (*Muell. Arg.*) *Kurz*

1 Petioles short and thick, 10–15 mm. long, 4–5 mm. thick; nervation strongly and closely bullate; lower surface shortly and densely brown-pubescent . . . . **B. bullatum**

1 Petioles much longer and thinner, 3–7 cm. long, 1–3 mm. thick; nervation flat, not bullate; leaves glabrous or almost so:
  2 ♂ inflorescence elongate, falsely racemose; leaves rarely whorled; secondary nerves running almost at right angles to the midrib, rather closely parallel and finely raised beneath . . . . **B. tokbrai**
  2 ♂ inflorescence usually much condensed, but occasionally ± racemose; leaves often whorled:
    3 Leaves usually reddish-brown when dry:
      4 Leaves with a dull and 'shagreened' (i.e. finely roughened) surface, frequently arranged in whorls of 4; drip-tip shorter, broader and less abrupt . . . . **B. kurzii**
      4 Leaves with a smooth and somewhat shining surface, rarely subverticillate; drip-tip longer, slenderer and more abrupt **B. calophyllum**
    3 Leaves yellowish-green when dry:
      5 Leaves narrowly elliptic or oblong, up to 18 × 6·5 cm.; secondary nerves arranged at right angles to the midrib, closely parallel, somewhat as in *B. tokbrai* . . . . **B. borneënse**
      5 Leaves broadly elliptic, up to 29 × 13·5 cm.; secondary nerves less obviously arranged at right angles to midrib and less conspicuously parallel; tertiary nerves forming a fine raised reticulum, especially beneath . . . . **B. concolor**

Blumeodendron borneënse *Pax & Hoffm.* xiv (Euph.-Addit. vi): 14 (1919).

SR; SB.—Endemic.

Tree up to 7·5 m. high, at 270 m. alt.

Leaves narrowly elliptic or oblong, up to 18 × 6·5 cm., often arranged in whorls of 3, remaining greenish when dry; secondary nerves arranged at right angles to the midrib, rather closely parallel, somewhat as in *B. tokbrai*, finely raised; male inflorescence unknown; fruit apparently 2-locular, with slightly flanged sutures.

Described originally from a single gathering from Sarawak (*Beccari* 2976). The Sabah record likewise rests on a single gathering: Temburong District, Kuala Belalong, 27 March 1957, *Smythies, Wood & Ashton* SAN 17086.

Blumeodendron (?) **bullatum** *Airy Shaw* in Kew Bull. 19: 310 (1965) & 27: 86 (1972).

SR.—Endemic.

No ecological information available, but probably growing at low altitudes.

In its leaves with a short, thick petiole, without any pulvinus at base or apex, with a lamina up to 22 cm. long by 10·5 cm. broad, densely brown-pubescent beneath, and with strongly bullate nervation, this species is very aberrant in *Blumeodendron*. The male flowers and inflorescence, however, are typical of the genus. It remains to be seen whether the female flowers and fruits, which are still unknown, will justify the present tentative generic disposition.

Blumeodendron calophyllum *Airy Shaw* in Kew Bull. 19: 309 (1965) & 25: 518 (1971); Whitmore: 70, fig. 2 (1973).

SR; B.—Malaya.

Tree up to 25 m. high, in primary forest in *kerangas* on yellow sandy clay up to 90 m. alt.

Differs from *B. kurzii* in the scattered or only subapproximate arrangement of its leaves, in their very smooth and somewhat shining, not dull and shagreened, surface, and in the somewhat longer, slenderer and more abrupt drip-tip.

Blumeodendron concolor *Gage* in Rec. Bot. Surv. Ind. 9: 244 (1922); Ridley: 281 (1924).

Blumeodendron sp., Merr., Pl. Elmer.: 157 (1929) (*Elmer* 21129).
B. borneënse sec. Whitmore: 70 (1973), *non* Pax & Hoffm.

SB; KI.—N. Malaya (Lankawi Is., Perak).

Tree up to 20 m. high, in primary forest on sandstone or clay loam, or on rocks by brook in deep ravine, ascending to 1500 m. on Kinabalu.

Very similar to *B. borneënse*, but leaves broadly elliptic, up to 29 × 13·5 cm., turning a pale or yellowish-green when dry; the secondary nerves arranged less clearly at right angles to the midrib, and less conspicuously parallel; tertiary nerves forming a distinct raised reticulum. Male inflorescence a condensed cyme; fruit bilocular, with slightly flanged sutures.

The following collections are referred here: SABAH: Tenompok, Kinabalu, April–May 1932, *Clemens* 29614 & 30366; Silabukan For. Res., 7 May 1962, *Pereira* SAN 29752; Tawao, 1922–23, *Elmer* 21129; Quoin Hill, 14 March 1963, *Gibot* SAN 34059. E. INDONESIAN BORNEO: Sangkulirang Distr., Mt. Medadam, 12 Aug. 1957, *Kostermans* 13475; W. Koetai, near Mt. Kemoel, 3 Oct. 1925, *Endert* 3774.

Blumeodendron kurzii (*Hook. f.*) *J. J. Sm.*: 463 (1910); Pax & Hoffm. vii: 48 (1914); Ridley: 281 (1924); Backer & Bakh. f.: 480 (1963); Airy Shaw in Kew Bull. 26: 224 (1971); Whitmore: 70 (1973).

B. tokbrai sensu Kurz in Journ. As. Soc. Bengal 42: 245 (1873) & For. Fl. Brit. Burma 2: 391 (1877), *non Elateriospermum tokbrai* Bl.
Mallotus kurzii Hook. f.: 427 (1887).
Blumeodendron verticillatum sec. Merr., Pl. Elmer.: 157 (1929), *an* Merr. 1920?

SR; B(?); SB; KI.—Andaman Is., Burma, Lower Siam (extreme S.), Malaya, Sumatra, Java, ? Philippines, New Guinea.

Tree up to 50 m. high, in primary or secondary peat swamp or Mixed Dipterocarp forest, on sandstone or yellow sandy clay or dark greyish-brown sandy loam or loam containing lime, at low and medium altitudes, ascending to 2700 m. on Kinabalu.

Leaves very variable in size, shape, and arrangement, but usually recognizable by their distinctly 'shagreened' surface, and by the very inconspicuous secondary nerves, which are immersed or slightly impressed but never raised as in *B. tokbrai*. The leaves are sometimes arranged in a whorl of four, arising from a thickened node. The male inflorescence is usually much condensed, but sometimes shortly racemose. The seeds are said to be eaten by birds and animals.

The status of the related taxa described from the Philippines as *B. philippinense* Merr. & Rolfe, *B. subcaudatum* Merr., *B. subrotundifolium* (Elm.) Merr. and *B. verticillatum* Merr., is quite unclear. Certain small-leaved specimens from Borneo and Malaya seem to correspond to one or other of these, but I can find no clear dividing lines between them and suspect that they are at all events not specifically distinct from *B. kurzii*.

Blumeodendron tokbrai (*Bl.*) *Kurz* in Journ. As. Soc. Bengal 42: 245 (1873) & For. Fl. Brit. Burma 2: 391 (1877), *tantum quoad synon. Blum.*; emend. J. J. Sm.: 463 (1910); Pax & Hoffm. vii: 48 (1914); Merr.: 340 (1921); Backer & Bakh. f.: 479 (1963); Whitmore: 71, fig. 3 (1973).

Elateriospermum tokbrai Bl., Bijdr.: 621 (1825).
Mallotus tokbrai (Bl.) Muell. Arg.: 956 (1866).
Rottlera tokbrai (Bl.) Scheff. in Ann. Mus. Bot. Lugd.-Bat. 4 : 122 (1868–9).
Mallotus ? *vernicosus* Hook. f.: 443 (1887).
? *Elateriospermum paucinervia* [sic!] Elm., Leafl. Philipp. Bot. 2: 484 (1908).
Blumeodendron elateriospermum J. J. Sm. in Bull. Jard. Bot. Buitenz. II, 8: 56 (1912).
? *B. paucinervium* (Elm.) Merr. in Philipp. Journ. Sci. 16: 555 (1920).
B. vernicosum (Hook. f.) Gage in Rec. Bot. Surv. Ind. 9: 244 (1922); Ridley: 282 (1924).

SR; B; SB; Ki.—Malaya, Sumatra, Java, Sangieh & Talaud Is., Moluccas, New Guinea.

Tree up to 24 m. high, in mangrove forest or primary peat swamp (*kerangas*) forest, Lowland or Mixed Dipterocarp forest, primary hill forest, lower or submontane moss forest, on white sand or sandstone (Biban sandstone formation), on deep yellow sandy soil over Tertiary clays, on yellow clay with shallow raw humus, on gabbro, and on blackish, greyish or brownish soil (unspecified), from low altitudes up to 1680 m. on Kinabalu.

Extremely variable, but recognizable from the closely parallel secondary nerves, running at right angles to the midrib and distinctly raised on the lower surface, and from the elongate male inflorescence. The fruit may be 2- or 3-locular, with either grooved or strongly keeled sutures; it is stated to be edible.

Herbarium material usually turns brown on drying, but a form occurs which remains green and in which the leaves are more or less oblong in shape and somewhat larger than usual. This occurs in Sumatra as well as in Borneo. Material was distributed by J. J. Smith with a varietal name attached, but this was apparently never published. Although very distinct in its extreme forms, it is connected with the typical form by intermediates, and at the present time does not seem to warrant a special name.

Borneodendron *Airy Shaw*

Borneodendron aenigmaticum *Airy Shaw* in Kew Bull. 16: 359 (1963) & 25: 504–6 (1971), *in obs.*, & in Hook. Ic. Pl. 37: t. 3633 (1967); Meijer, 7: 24 (1967) & 10: 10, 229 (1968).

SB.—Endemic.

Tree to 39 m. high, in primary hill forest or *Casuarina sumatrana* forest on ultrabasic soil or rock, from sea-level up to 1050 m. alt. on Kinabalu. In the *Casuarina* forest on Pulau Tabawan, Lahad Datu Distr., the habitat is described as follows: 'Acid conditions. Xerophytic vegetation, but many epiphytes. Little undergrowth.'

The only relative of this interesting and taxonomically isolated plant is the genus *Cocconerion* Baill., of New Caledonia, from which it differs in its more abruptly petiolate leaves in whorls of 3 (instead of 6–10); in its nodding terminal male inflorescence of 2–4 three-flowered whorls of subsessile flowers, each flower subtended by a large bract; in the absence of a calyx to the male flower; and in the female flower borne on a stout, rigid, flattened pedicel, with 3 small caducous sepals and a usually bicarpellate gynoecium. It is very characteristic of the ultrabasic areas of Lahad Datu District, but also occurs in the Sandakan and Ranau Districts, for which no geological information is available from the field notes.

Botryophora *Hook. f.*

Botryophora geniculata (*Miq.*) *Beumée ex Airy Shaw* in Kew Bull. 3: 484 (1949) & 14: 375 (1960), *q.v.*, & in Hook. Ic. Pl. 36: t. 3576 (1962); van Steenis in Blumea 12: 15 (1963); Meijer, 7: 29 (1967) & 10: tab. opp. p. 64 (1968); Airy Shaw in Kew Bull. 26: 224 (1972); Whitmore: 72 (1973).

Sterculia geniculata Miq., Fl. Ind. Bat., Suppl.: 164, 400 (1861).
Botryophora kingii Hook. f.: 476 (1888); Pax in Engl. & Prantl, Pflanzenf. III. 5: 116 (1891); Boerl.: 254 (1900); Ridley in Journ. Fed. Malay States Mus. 10: 116 (1920) & Fl. Malay Penins. 3: 282 (1924); Merr. in Papers Mich. Acad. Sci. 20: 101 (1935). [*Non* Gage in Nova Guinea 12: 486 (1917); cf. v. Steenis in Nova Guinea II, 7: 8–9 (1956).]
Ostodes sp., Burkill ex Merr., Pl. Elmer.: 164 (1929).

SB; K1.—Lower Burma, Lower Siam, W. Malesia (exc. Philippines).
Tree to 18 m. high, in primary forest on blackish soil or sandy loam or lime, up to 216 m. alt.

Completely glabrous, or the inflorescences very finely rusty-puberulous. Stems straight, smooth, terete, the leaves arising in crowded groups from slightly swollen or nodulose portions of stem, with long bare portions between; petioles long and very slender, pulvinate at apex; lamina ovate, elliptic or obovate, chartaceous, similar to that of *Mallotus wrayi* King ex Hook. f. Male inflorescence a pendulous panicle, with angular rhachis and short spreading branches bearing numerous subsessile globose flowers; calyx membranous, red-purple; anthers bright yellow, in a globose mass, with subimbricate peltate connectives. Female inflorescence a short simple spike. Fruit a rather large 3–4-coccous capsule, the cocci at first distinctly keeled.

Breynia *J. R. & G. Forst.*

1 Fruit somewhat shining, crowned with a distinct apical shallowly raised ring, not hoary at apex, splitting at maturity; leaves usually thinner and flatter . . . . **B. coronata**

1 Fruit with a dull surface, without an apical raised ring, but slightly hoary at apex, dehiscing very tardily; leaves usually stiffer, with more distinctly reflexed margins . . . . **B. racemosa**

Breynia coronata *Hook. f.*: 330 (1887); Ridley: 218 (1924); Meijer, 7: 38 (1967); Whitmore: 73 (1973).

SR; B; SB.—Malaya; ? Sumatra.

Shrub or small tree to 8 m. high, in primary Mixed Dipterocarp forest, sometimes swampy, or in secondary forest, on yellow or brown sandstone soil or on basalt, up to 210 m. alt.

The only reliable distinction of this species from *B. racemosa* seems to be the conspicuous ring at the apex of the fruit. The other points mentioned in the key are often evident, but are not invariable. A narrow-leaved form (1–1·5 cm. wide) occurs in Sarawak (*Clemens* 21248, 21671, *Purseglove & Shah* P. 4427, *Sibat ak. Luang* S. 21993), in which the styles are connivent or connate into a very short beak, approaching the condition found in *B. rostrata* Merr., of S. China and Indochina. It does not seem to warrant taxonomic recognition.

Used in native medicine (Sabah, 1932, *Goklin* (N. Born. For. Dept.) 2398). A wide variety of native names are recorded. Sarawak: 'changkok manis' (Malay); Sabah: 'tamu tamu' (Brunei), 'tatamoh' (Dusun), 'bamban ambok' (Brunei), 'kubamban-kubamban' (Brunei Benuni), 'hingos hingos' (language?).

Breynia racemosa (*Bl.*) *Muell. Arg.*: 441 (1866); J. J. Sm.: 177 (1910); Merr. in Philipp. Journ. Sci. 11, C. Bot.: 65 (1916) & Enum.: 329 (1921); Pax & Hoffm. in Mitt. Inst. Bot. Hamburg 7: 225 (1931); Backer & Bakh. f.: 465 (1963).

Melanthesa racemosa Bl., Bijdr.: 592 (1825).
Phyllanthus reclinatus Roxb., Fl. Ind. 3: 669 (1832), **synon. nov.**
Melanthesa reclinata (Roxb.) Muell. Arg. in Linnaea 32: 74 (1863).
M. acuminata Muell. Arg. *l.c.*: 74 (1863), **synon. nov.**
M. rhamnoïdes var. *hypoglauca* Muell. Arg., *l.c.*: 73 (1863).
Breynia rhamnoïdes var. *hypoglauca* (Muell. Arg.) Muell. Arg.: 440 (1866).
B. acuminata (Muell. Arg.) Muell. Arg.: 442 (1866); Merr.: 404 (1923); Meijer, 7: 38 (1967).
[*Melanthesopsis fruticosa* sec. Muell. Arg.: 437 (1866), *non Andrachne fruticosa* L.]
Breynia reclinata (Roxb.) Hook. f.: 331 (1887); Ridley: 219 (1924); Airy Shaw in Kew Bull. 26: 226 (1972); Whitmore: 73 (1973).
[*B. fruticosa* sec. Merr.: 329 (1921), quoad *Motley* 557 (!), *non Andrachne fruticosa* L.]
[? *B. rhamnoïdes* sec. Merr.: 329 (1921), quoad *Clemens* 9587 (*non vidi*), an *Phyllanthus rhamnoïdes* Retz.?]

SR; SB; K3.—Lower Siam, Malay Peninsula, Sumatra, Philippines, Java, Sumbawa, Timor.

Shrub or small tree to 9 m. high, in scrub, forest edges, etc., at low altitudes, ascending to 2500 m. on Kinabalu.

Apparently a widespread and very variable taxon, in which I am convinced *B. reclinata* must be included; the Philippine *B. acuminata* also seems indistinguishable. The dull, black fruits, completely lacking a raised apical ring, distinguish *B. racemosa* from *B. coronata*. There is much variation in the size and shape of the female perianth. In an example from Kinabalu (*Chew & Corner* RSNB 4762) it is widely expanded, 7–8 mm. wide at anthesis, with 3 small outer and 3 much larger inner segments, but in other respects the specimen seems identical with the common form.

Bridelia *Willd.*

1 Flowers large, up to 10 mm. in diameter; drupe large, usually ovoid, bilocular, 8–11 mm. long; plant often scandent, softly rufo-tomentose (§ *Stipulares*) . . . . **B. stipularis**

1 Flowers small, up to 6 mm. in diameter:

2 Ovary bilocular (but drupe sometimes 1-seeded); seeds usually plano-convex; petals larger; leaves with a marginal nerve (§ *Scleroneurae*):

3 Leaves chartaceous or submembranaceous, 8–12-nerved, with almost flat margins, the lower surface ± puberulous with elevate reticulate venation . . . . . . **B. tomentosa**

3 Leaves stiffly coriaceous, 7–9-nerved, with strongly revolute margins, the lower surface glabrous and glaucous with conspicuous immersed venation . . . . **B. adusta**

2 Ovary unilocular; seeds with a deep longitudinal groove; petals minute; leaves without a marginal nerve (§ *Cleistanthoïdeae*):

4 Branchlets robust, rough, strongly pustulate-lenticellate; pulvini of inflorescences rather large; whole plant blackening on drying **B. pustulata**

4 Branchlets less robust, not rough or pustulate; pulvini smaller; plant not blackening on drying:

5 Stipules large, up to 12 × 5 mm.; drupe up to 10 mm. long; indumentum strong, rufous . . . . . **B. sosopodonica**

5 Stipules smaller; drupe smaller; indumentum weaker;

6 Drupes distinctly and stoutly pedicelled; undersurface of leaves somewhat glaucescent . . . . **B. glauca**

6 Drupes sessile or subsessile:

7 Leaves rufous-puberulous beneath; flowers 4–4·5 mm. diam.; styles not exserted . **B. griffithii** var. **cinnamomea**

7 Leaves thin, almost glabrous; flowers 2–3 mm. diam.; styles exserted . . . . **B. penangiana**

Bridelia adusta *Airy Shaw* in Kew Bull. *ined.* (1975–6). *Vide* p. 224, *infra.*

SR (NE.).—Endemic.

Tree to 18 m. high, among limestone boulders in primary forest at 840–1050 m. alt.

Very distinctive in its elliptic-oblong, glabrous, stiffly coriaceous, 7–9-nerved leaves, 4–9 cm. long, somewhat shining and finely reticulate-venulose above but smooth and glaucous beneath; margins strongly revolute, with an inconspicuous marginal nerve; stipules narrow-subulate, ferrugineous-pubescent; fruit bilocular but 1-seeded; whole plant taking on a brownish, 'scorched' appearance when dry.

Bridelia glauca *Bl.*, Bijdr.: 597 (1825); Muell. Arg.: 497 (1866); J. J. Sm.: 307 (1910); Jabl. viii: 74 (1915); Merr.: 423 (1923); Holth. & Lam in Blumea 5: 200 (1942); Backer & Bakh. f.: 475 (1963); Meijer, 7: 39 (1967).

SR; SB; K1.—Throughout Malesia (except Malaya) to the Moluccas; Bismarck Archip.

Tree to 15 m. high, or climbing shrub, in mixed peat-swamp forest or primary forest on black rocky soil, yellow sandy loam, sandstone, or sand and limestone, up to 400 m. alt., ascending to 1500 m. on Kinabalu and 1800 m. on Trusmadi.

Recognizable by the smooth, glaucescent (though minutely puberulous) undersurface of the leaves, and the distinctly and rather stoutly pedicelled fruits.

Bridelia griffithii *Hook. f.* var. **cinnamomea** (*Hook. f.*) *Gehrm.* in Engl., Bot. Jahrb. 41, Beibl. 95: 38 (1908).

B. cinnamomea Hook. f.: 273 (1887); Ridley: 185 (1924); Gage in Journ. As. Soc. Beng. 75(5): 490 (1936); Corner: 74 (1973).
B. gehrmannii Jabl. viii: 73 (1915); Merr.: 335 (1921).
B. griffithii sec. Jabl. *l.c.*: 74 (1915), *vix* Hook. f.

SR.—Malaya.

Shrub, probably at low altitudes; no field data available for Borneo.

Var. *cinnamomea* only differs from var. *griffithii*, of Malacca and Johore, in the greater development and rufous colouring of the indumentum.

Bridelia penangiana *Hook. f.*: 272 (1887); Jabl. viii: 75 (1915); Ridley: 185 (1924); Gage in Journ. As. Soc. Beng. 75(5): 492 (1936); Airy Shaw in Kew Bull. 26: 229 (1972); Whitmore: 75 (1973).

B. minutiflora Hook. f.: 273 (1887); Boerl.: 271 (1900); Jabl. viii: 76 (1915); Merr.: 335 (1921) & 423 (1923) & Pl. Elmer.: 155 (1929); Gagnep.: 493 (1926); Holth. & Lam in Blumea 5: 200 (1942); Backer & Bakh. f.: 475 (1963); Meijer, 7: 39 (1967).
B. platyphylla Merr. in Philipp. Journ. Sci. 7, Bot.: 384 (1912).

SR; SB; K1, 1a, 2.—SE. Asia (except, apparently, Siam), and throughout Malesia to N. Queensland and the Solomon Is.

Tree to 21 m. high, in primary or secondary forest in peat swamps, at edge of mangroves, etc., very frequently on river banks (sometimes flooded), on sandy loam or dark brown soil, from sea-level up to 750 m. alt.

The thin, almost glabrous leaves and dense glomerules of very small flowers are characteristic.

Used in Sabah in native medicine, as a cure for headache.

Bridelia pustulata *Hook. f.*: 271 (1887); Gehrm. in Engl., Bot. Jahrb. 41, Beibl. 95: 38 (1908); Jabl. viii: 75 (1915); Ridley: 185 (1924); Airy Shaw in Kew Bull. 23: 67 (1969); Whitmore: 74 (1973).

SB.—Malaya, Sumatra.
Tree of 10 m., in primary forest at 10 m. alt.

Very distinct in its robust, strongly pustulate-lenticellate branches and dense, many-flowered pulvini of small flowers. The whole plant becomes blackish on drying. So far known in Borneo from a single collection only.

Bridelia sosopodonica *Airy Shaw* in Kew Bull. 23: 67 (1969).

SB; K3?—Endemic.
Tree to 9 m. high, in primary or secondary forest on dark brown or black soil at 900–1380 m. alt.; apparently locally frequent on Kinabalu.

Distinguished by its strong rufous indumentum, large stipules (up to 12 × 5 mm.) and rather large fruit (up to 10 mm. long). In the original description of the fruit the measurements should have read 'up to 10 mm. long, 7–8 mm. in diam.'

Bridelia stipularis (*L.*) *Bl.*, Bijdr.: 597 (1825); Muell. Arg.: 499 (1866), *p.p.*; Hook. f.: 270 (1887), *p.p.*: Jabl. viii: 55 (1915); Merr.: 336 (1921) & 424 (1923); Ridley: 183 (1924); Gagnep.: 492 (1926); Gage in Journ. As. Soc. Beng. 75(5): 485 (1936); Corner: 243 (1940); Backer & Bakh. f.: 475 (1963); Meijer, 7: 39 (1967); Airy Shaw in Kew Bull. 26: 230 (1972); Whitmore: 74 (1973).

Clutia stipularis L., Mant.: 127 (1767).

SB.—India, SE. Asia & S. China, and throughout W. Malesia to Timor.
Climbing or occasionally erect shrub to 15 m. high, in primary or secondary forest, swamps, or riversides, on sandy soil, up to 450 m. alt.

Distinguished by the large flowers and large oblong-ovoid fruits. Locally frequent in Sabah, but so far unknown elsewhere in Borneo.
Used in native medicine; said to be poisonous. Fruit edible.

Bridelia tomentosa *Bl.*, Bijdr.: 597 (1825); Muell. Arg.: 501 (1866); Hook. f.: 271 (1887); Jabl. viii: 58 (1915); Merr.: 336 (1921); Ridley: 184 (1924); Gagnep.: 488 (1926); Gage in Journ. As. Soc. Beng. 75(5): 487 (1936); Hend. in Journ. Malayan Br. Roy. As. Soc. 17: 69 (1939); Corner in Gard. Bull. Str. Settlem. 10: 291 (1939) & Ways. Trees: 243 (1940); Airy Shaw in Kew Bull. 26: 231 (1972); Whitmore: 74 (1973).

B. lanceifolia Roxb., Fl. Ind. 3: 737 (1832); Jabl., *l.c.*: 60 (1915).
B. loureirii Hook. & Arn., Bot. Beech. Voy.: 211 (1841), *excl. synon.*
B. tomentosa var. *lanceifolia* (Roxb.) Muell. Arg.: 502 (1866).
B. monoica sec. Merr. in Philipp. Journ. Sci. 13, Bot.: 142 (1918) & Enum.: 423 (1923) & in Trans. Amer. Philos. Soc. II, 24(2): 234 (1935); Backer & Bakh. f.: 475 (1963); *vix Clutia monoica* Lour.
B. glabrifolia Merr.: 423 (1923), *pro nom. nov.* (*nec comb. nov.*, *pace* Merr.!).

K1a.—NE. India, SE. Asia, S. China, Formosa, and thoughout Malesia to N. Australia.
No field data available for Borneo. Known from a single gathering only from Bandjermasin (*Motley* 493).

Leaves small, thin, bluntly subacuminate; branchlets slender; fruits small, globose.

Cephalomappa *Baill.*

1 Petioles usually consistently short, 0·5–1·5 cm. long (rarely up to 2·5 cm.); lamina clearly crenulate; capsule with short (1–2 mm.) conical processes:
 2 Lateral nerves 8–11 pairs; hill plant . . . . **C. malloticarpa**
 2 Lateral nerves 4–6 pairs; peat-swamp plant . . **C. paludicola**
1 Petioles very variable in length, sometimes up to 8 cm. long; lamina crenulate, denticulate or entire, drying greenish or brownish; capsule with short or long processes:
 3 Capsule very densely echinate with slender processes 3–4 mm. long; lamina distinctly crenulate, densely or sparsely minutely stellate-lepidote beneath . . . . **C. lepidotula**
 3 Capsule less densely echinate with short conical processes 1–2 mm. long **C. beccariana**
  4 Leaves thinly chartaceous, drying greenish or greenish-brown, 10–22 cm. long, sparsely and minutely fasciculate-pubescent beneath; margin very shallowly denticulate, sometimes appearing sub-entire; indumentum slightly floccose . . . var. **tenuifolia**
  4 Leaves stiffly chartaceous or subcoriaceous, drying reddish-brown, 5–15(–17) cm. long, entire or with only faint undulations:
   5 Leaves densely fasciculate-pubescent beneath var. **beccariana**
   5′ Leaves rather closely and minutely stellate-lepidote beneath var. **havilandii**
   5″ Leaves (except for the tomentellous nerves) almost glabrous beneath, or with very minute scattered fascicled hairs var. **hosei**

Cephalomappa beccariana *Baill.* in Adansonia 11: 131 (1874); Boerl.: 245 (1900); Pax & Hoffm. ii: 17 (1910); Merr.: 343 (1921); Airy Shaw in Kew Bull. 14: 380 (1960).

Var. **beccariana**; Airy Shaw, *l.c.*

SR (SW. & Central).
Tree to 30 m. high, in primary Lowland or Mixed Dipterocarp forest on clay loam up to 90 m. alt.

Var. **havilandii** *Airy Shaw* in Kew Bull. *l.c.* & 16: 353 (1963).

SR (SW. & Central).
Tree to 21 m. high, on hill slopes up to 270 m. alt.—Fruit said to be edible (*Sibat ak. Luang* S. 23669).

Var. **hosei** *Airy Shaw* in Kew Bull. 14: 380 (1960).

SR (Central & NE.).
Tree to 19 m. high, in primary forest up to 90 m. alt.

Var. **tenuifolia** *Airy Shaw*, var. nov., a ceteris varietatibus foliis chartaceis 10–22 cm. longis siccitate viridulis distincta.

SR (SW.). First Division: Lundu District; Gunong Gading, hill slope, common in Mixed Dipterocarp forest on granodiorite-derived soil, alt.

180 m., 14 Nov. 1963, *Ashton* S. 18041 (holotype, K):—Tree 15 m tall, 60 cm. girth; bole ribbed at base; bark surface greenish-cream, smooth; green fruit. *Ibid.*, Lowland Dipterocarp forest, alt. 75 m., 17 July 1963, *Chew* CWL 570:—Tree 6 m., girth 120 cm., twigs brown; flowers [♂] green. *Ibid.*, 19 July 1963, *Chew* CWL 586:—Tree 9 m., girth 60 cm.; inflorescences [♂] yellow. *Ibid.*, hill slope, path to G. Gading, Tertiary granodiorite, alt. 40 m., 19 July 1963, *Paul Chai* S. 18470:—Tree 7·5 m. tall, 30 cm. girth; flowers [♂] greenish-yellow, with white anthers.

Cephalomappa lepidotula *Airy Shaw* in Kew Bull. 14: 379 (1960); Whitmore: 76 (1973).

SR.—Malaya, Sumatra.

Tree to 15 m. high, on hill slope at 195 m. alt.

Distinguished from the other species by the more finely echinate capsule, with slender processes 3–4 mm. long. Leaves densely or remotely minutely stellate-lepidote beneath.

Cephalomappa malloticarpa *J. J. Sm.* in Bull. Jard. Bot. Buitenz. III, 6: 95 (1924); Airy Shaw in Kew Bull. 14: 380 (1960); Whitmore: 76, fig. 4 (1973).

SR; SB; K1.—Malaya.

Tree to 20 m. high, in primary hill forest, sometimes swampy, on loam or sand, up to 600 m. alt.

Distinguished by the distinctly crenulate leaves, usually greenish when dry, with 8–11 pairs of nerves, and petioles 0·5–1·5 (very rarely up to 2·5) cm. long.

Cephalomappa paludicola *Airy Shaw* in Kew Bull. 14: 380 (1960) & 16: 353 (1963).

SR.—Endemic.

Tree to 18 m. high, locally abundant in primary mixed peat-swamp forest (phasic community 1) with *Shorea albida*, at very low altitudes (up to 30 m.).

Closely related to *C. malloticarpa*, differing chiefly in the leaves with only 4–6 pairs of nerves, and in the peat-swamp habitat. The species is also very close to *C. sinensis* (Chun & How) Kostermans, of S. China; cf. Kew Bull. 16: 354 (1963).

The male flowers are said to be heavily scented and to be covered with small bees and ants at anthesis.

Chaetocarpus *Thw.*

Chaetocarpus castanocarpus (*Roxb.*) *Thw.*, Enum. Pl. Zeyl.: 275 (1861); Muell. Arg.: 1122 (1866) (‘*castaneaecarpus*’); Hook. f.: 460 (1887); Pax & Hoffm. iv: 8 (1912); Ridley: 310 (1924); Gagnep.: 471 (1926); Merr. in Philipp. Journ. Sci. 29: 386 (1926); Corner: 244 (1940) (‘*castaneicarpus*’); Meijer, 7: 28, tab. (1967); Airy Shaw in Kew Bull. 26: 231 (1972); Whitmore: 76 (1973).

Adelia castanicarpa Roxb., Fl. Ind. 3: 848 (1832).
Regnaldia cluytioïdes Baill. in Adansonia 1: 188 (1860); Muell. Arg.: 1257 (1866).
R. myrtioïdes Baill., *l.c.*: 187 (1860).

SR; B; SB; K1, 2.—Ceylon, Assam, SE. Asia, N. Malaya, ? Sumatra, Banka.

Tree to 33 m. high, in primary or secondary Lowland or Mixed Dipterocarp forest, in coastal peat-swamp forest (*kerangas*), on yellow, brown or black sandy soil, on sandy loam or yellow clay, often in low undulating country, once noted on a tufa plateau, up to 300 m. alt.

The coriaceous, glabrous leaves, turning madder-brown when dry, the flowers in axillary clusters, the stamens connate in a column, and especially the densely setose capsules with shining black arillate seeds, make this an unmistakable plant. It is evidently a calcifuge. The genus is related to *Trigonopleura* Hook. f.

Cheilosa *Bl.*

Cheilosa malayana (*Hook. f.*) *Corner ex Airy Shaw* in Kew Bull. 16: 364 (1963) & 20: 49 (1966); Whitmore: 77 (1973).

Baccaurea malayana sec. King ex Hook. f.: 374 (1887); Boerl.: 280 (1900), *in adnot.*; Pax & Hoffm. xv: 70 (1922); Ridley: 247 (1924); Burkill, Dict. Econ. Prod. Mal. Penins. 1: 279 (1935); Corner in Gard. Bull. Str. Settlem. 10: 289 (1939); *non Hedycarpus malayanus* Jack.
Baliospermum malayanum Hook. f.: 463 (1888), *pro sp. nov.*; Ridley: 313 (1924).
Cheilosa homaliifolia Merr. in Philipp. Journ. Sci., Bot. 8: 379 (1913); Pax & Hoffm. xiv: 50 (1919); Merr.: 457 (1923); Pax & Hoffm. in Engl. & Harms, Pflanzenf. ed. 2, 19c: 181 (1931).
Cheilosa sp., Merr.: 346 (1921).
Ch. homaliifolia var. *grandifolia* Merr.: 457 (1923).
Scortechinia malayana (Hook. f.) Ridley, Fl. Malay Penins. 5: 332 (1925), *in obs.*
Ch. montana var. *longifolia* S. Moore in Journ. Bot. Brit. & For. 64, Suppl.: 104 (1925), *e descr.*, **synon. nov.**

SR; SB; K1.—Malaya, Sumatra, Philippines.

Tree to 25 m. high, in primary forest on sandy loam from low altitudes up to 1200 m.

It seems possible that *Ch. malayana* may not be specifically distinct from *Ch. montana* Bl., of Java, the type species. The genus is very close to *Neoscortechinia*, only differing in having 8–10 instead of 5–6 stamens, longer styles, and a thick-walled globose capsule with 3 seeds. *Cheilosa* is of course much the earlier name.

In the field notes to a specimen of *Ch. malayana* from E. Indonesian Borneo (*Kostermans* 4334) the wood is described as 'smelling of fresh sugar-cane'.

Chondrostylis *Boerl.*

Chondrostylis kunstleri (*King ex Hook. f.*) *Airy Shaw* in Kew Bull. 14: 359 (1960) & 16: 345 (1963) & 20: 27, 398 (1966) & 26: 231 (1972); Whitmore: 77 (1973).

Mallotus? kunstleri King ex Hook. f.: 443 (1887).
Kunstlera glumacea King ex Hook. f., *l.c.*, *in obs.*, *nom. event.*
Kunstlerodendron sublanceolata [sic] Ridley: 283 (1924); Pax & Hoffm. in Engl. & Harms, Pflanzenf. ed. 2, 19c: 230 (1931).

SR.—Lower Siam, Malaya, Sumatra.
Tree to 9 m. high, in mixed forest, or in primary Heath forest on humic podsols derived from plateau sandstone formation, or on clayey loam, up to 120 m. alt.

Related to *Agrostistachys* (especially sect. *Sarcoclinium*), from which it differs chiefly in the paniculate inflorescence and absence of petals. The leaves sometimes bear on their upper surface towards the base and on the portion decurrent on the petiole a few black, elliptic, raised macular glands, 0·5–1 mm. long, closely resembling certain scale insects (*Coccidae*).

Claoxylon *Juss.*

1 Mature inflorescences (either ♂ or ♀), at the flowering stage, usually more than 7 cm. long:
 2 Leaves distinctly bullate, thinly chartaceous, not fleshy, slightly shining below; ♂ inflorescence lax, fulvo-puberulous, ♀ unknown **C. subbullatum**
 2 Leaves flat, not bullate:
  3 Stamens very numerous (±100); leaves elliptic or oblong, 16–21 cm. long, obscurely and minutely denticulate; capsule unknown **C. winkleri**
  3 Stamens (where known) fewer than 50:
   4 Loculi of capsule distinctly dorsally keeled; infructescence up to 30 cm. long or more . . . . **C. carinatum**
   4 Loculi of capsule (where known) rounded, not keeled, but sutures sometimes raised to form a kind of double keel at dehiscence; infructescence not exceeding 15 cm.:
    5 Indumentum softly and shortly tomentellous, not adpressed:
     6 Petioles 5–14 cm. long; leaves usually manifestly crenate-dentate; plant often purplish; capsules usually dehiscing by raised sutures . . . . **C. indicum**
     6 Petioles less than 5 cm. long; leaves very shortly repand-dentate; plant not purplish; capsules unknown **C. insigne**
    5 Indumentum sparse and adpressed, or almost absent:
     7 Usually a coarse plant; leaves 12–40 cm. long; ♂ inflorescence robust; stamens 35–50 . . . . . . **C. longifolium**
     7 More slender plants; leaves 5–17 cm. long; ♂ inflorescence very slender; stamens 17–30;

8 Leaves up to 17 × 7 cm., petiole up to 8 cm. long **C. praetermissum**

8 Leaves up to 9 × 4 cm., petiole up to 2·5 cm. long **C. pseudo-insulanum**

1 Mature inflorescences (♂ or ♀), at the flowering stage, less than 7 cm. long:

9 Indumentum rather harsh, spreading, tawny-fulvous; leaves long-acuminate; capsules egg-yellow when dry . **C. hirsutellum**

9 Indumentum not harsh and spreading:

10 Indumentum strongly velutinous; leaves up to 35 × 13·5 cm., not acuminate; stamens ±100 . . . . **C. velutinum**

10 Indumentum much weaker or absent; leaves smaller:

11 Leaves with strong closely parallel transverse tertiary venation:

12 Ovary and capsule with strong, rough, shaggy indumentum **C. stapfianum**

12 Ovary and capsule almost glabrous . . **C. hosei**

11 Leaves without, or with less strong, transverse tertiary venation:

13 Plant distinctly purplish-tinged when dry, almost glabrous; leaves cuneate-based, manifestly crenate-dentate; stamens ±30 . . . . **C. brachyandrum**

13 Plant not purplish-tinged when dry:

14 Leaves narrow, elongate, 12–28 cm. long, 2·5–5·5 cm. wide, long-cuneate-attenuate below, distantly dentate towards apex; capsule trilocular . . **C. attenuatum**

14 Leaves less elongate, 5–15 cm. long:

15 Leaves 4–8·5 cm. wide; ovary and capsule trilocular **C. kinabaluense**

15 Leaves 1–2·5(–4·7) cm. wide; ovary and capsule bilocular . . . . **C. salicinum**

Claoxylon attenuatum *Airy Shaw* in Kew Bull. 29: 314 (1974).

SB.—Endemic.
Slender shrub or small tree to 3 m. high, probably in forest by river, at 1500–1740 m. alt.

Distinguished by the very elongate, membranous, glabrous leaves, up to 28 cm. long and 5·5 cm. broad, with a distantly crenate-dentate or sharply dentate margin in the distal portion, entire and long-cuneate-attenuate below, and by the abbreviated fulvous-pubescent inflorescences, somewhat resembling those of *C. salicinum.*

Claoxylon brachyandrum *Pax & Hoffm.* vii: 115 (1914); Merr.: 430 (1923).

C. rubescens sec. Muell. Arg.: 788 (1866), *p.p.*; Pax & Hoffm. vii: 118 (1914), *pro majore parte*; Hutch. in Journ. Linn. Soc., Bot. 42: 135 (1914); Merr.: 338 (1921); *non* Miq.

SB.—Philippines; New Guinea (var.?).
Tree to 4·5 m. high, in forest or swampy land up to 300 m. alt.

Almost glabrous, distinctly purplish-tinged when dry; leaves cuneate-based, manifestly crenate-dentate; inflorescences up to 6 cm. long; stamens about 30; capsule 2–3-locular, 4–7 mm. in diam.

Leaves edible; also used medicinally (as purgative).

Claoxylon carinatum *Airy Shaw* in Kew Bull. 20: 401 (1966).

KI.—Endemic.

Shrub or small tree, on brook bank in deep ravine or on rock wall along river at 700–1100 m. alt.

Very similar to *C. longifolium*, but with greatly elongate infructescences, up to 30 cm. or more, with smooth minutely adpressed-puberulous capsules, each loculus with a conspicuous dorsal keel.

Claoxylon hirsutellum *Airy Shaw* in Kew Bull. 20: 402 (1966).

KI.—Endemic.

Treelet of 2 m., in *Agathis* forest on waterlogged, sandy acid soil at 500 m. alt.

Perhaps nearest to *C. velutinum*, but very distinct in its much smaller, narrowly elliptic, gradually long-acuminate leaves and harsh, spreading, tawny-fulvous indumentum.

Claoxylon hosei (*Merr.*) *Airy Shaw* in Kew Bull. 14: 391 (1960).

Coelodepas hosei Merr. in Philipp. Journ. Sci. 11, C. Bot.: 66 (1916) & Enum.: 338 (1921); Pax & Hoffm. xiv (Euph.-Addit. vi): 22 (1919).

SR (NE.).—Endemic.

No ecological information; known from a single collection only, in female flower, from the Baram region of the Fourth Division of Sarawak.

In the numerous elongate spreading filaments of the deeply laciniate stigmatic branches this species differs from all others known to me, except *C. kinabaluense* (which see). The female inflorescence is only 2–4 cm. long, floriferous almost to the base; the male inflorescence is unfortunately unknown. Apart from the inflorescence the plant is practically glabrous. The leaves are oblong-elliptic, minutely and rather distantly denticulate.

Claoxylon indicum (*Reinw. ex Bl.*) *Hassk.*, Cat. Pl. Hort. Bogor. Alter: 235 (1844); Muell. Arg.: 782 (1866); Hook. f.: 410 (1887); J. J. Sm.: 369 (1910); Pax & Hoffm. vii: 108 (1914); Ridley: 271 (1924); Gagnep.: 422 (1926); Corner in Gard. Bull. Str. Settlem. 10: 292 (1939) & Ways. Trees: 245 (1940); Airy Shaw in Kew Bull. 26: 233 (1972); Whitmore: 78 (1973).

Erytrochilus indicus Reinw. ex Bl., Bijdr.: 615 (1825).
E. mollis Bl., *l.c.* (1825).
E. minor Bl., *l.c.*: 616 (1825).
Croton halecum Roxb., Fl. Ind. 3: 683 (1832).
Claoxylon parviflorum Hook. & Arn., Bot. Beech. Voy.: 212 (1841).
C. macrophyllum Hassk., Pl. Jav. Rar.: 251 (1848).
C. minus (Bl.) Hassk., *l.c.* (1848).
C. molle (Bl.) Miq., Fl. Ind. Bat. 1(2): 386 (1859).
C. polot sec. Merr., Interpr. Rumph. Herb. Amboin.: 200 (1917), *in obs.*, & Enum.: 337 (1921); Backer & Bakh. f.: 480 (1963); *vix Croton polot* Burm. f.

K1, 1a.—India, SE. Asia & S. China, and throughout Malesia to New Guinea.
Tree to 11 m. high, in forest on heavy loam containing lime at 10 m. alt.

Softly and shortly (sometimes rather sparsely) grey- or ochraceous-tomentellous; leaves broad, usually dentate or serrate, cordate to cuneate at the base; male inflorescences very elongate, many-flowered; capsule shortly and softly grey-tomentose, cocci dehiscing by distinctly raised sutures.

Claoxylon insigne *Airy Shaw* in Kew Bull. 29: 313 (1974).

SR.—Endemic.
Shrub of 90 cm., in secondary forest on hillside at 870 m. alt.

Differs from *C. longifolium* in the densely tomentellous indumentum of stem, petioles and midribs, in the shortly repand-denticulate leaf-margin, in the completely glabrous inner surface of the male sepals and in the very small juxta-staminal glands, each bearing a solitary apical hair.

Claoxylon kinabaluense *Airy Shaw* in Kew Bull. 20: 401 (1966).

SB.—Endemic.
Tree to 4·5 m. high, in primary forest on hillsides and ridges, etc., on black soil, at 900–1800 m. alt.

Close to *C. stapfianum*, differing in the more numerous (12–14) flowers of the female inflorescence and in the very short smooth indumentum of the capsules. In the flowering stage the stigmatic branches of the female flower are deeply laciniate, with fine elongate spreading segments almost exactly as in *C. hosei*; these fine segments are lost in the fruiting stage. It is just possible that *C. kinabaluense* could ultimately be reducible to *C. hosei*.

Claoxylon longifolium (*Bl.*) *Endl. ex Hassk.*, Cat. Pl. Hort. Bogor. Alter: 235 (1844); Muell. Arg.: 781 (1866), *p.p.*; Hook. f.: 411 (1887); J. J. Sm.: 366 (1910); Pax & Hoffm. vii: 117 (1914); Ridley: 272 (1924); Gagnep.: 421 (1926); Hend. in Journ. Malayan Br. Roy. As. Soc. 17: 69 (1939) (var. *brachystachys* Hook. f.); Corner: 245 (1940); Holth. & Lam in Blumea 5: 200 (1942); Backer & Bakh. f.: 481 (1963); Airy Shaw in Kew Bull. 26: 234 (1972); Whitmore: 79 (1973).

Erytrochilus longifolius Bl., Bijdr.: 616 (1825).
Claoxylon papyraceum Airy Shaw in Kew Bull. 23: 77 (1969), *quoad typum.*

Var. **longifolium**: capsula matura laevissima siccitate viridi, coccis rotundatis dense adpresse sericeo-puberulis.

SR; SB; K3.—NE. India, SE. Asia, and throughout W. Malesia (exc. Philippines?); New Guinea?
Shrub or small tree to 9 m. high, in primary or secondary forest on granodiorite or sedimentary derived soils, or black or red-brown soil, from low levels up to 1200–1500 m. on Kinabalu.

Var. **rugifrux** *Airy Shaw*, var. nov., capsula matura valde rugosa siccitate ochraceo-aurantiaca, indumento brevissimo magis patulo nec adpresse sericeo.

SR.—Third Division: Gat, Upper Rejang River, forested slopes, 1929, *J. & M. S. Clemens* 21681:—Tree 4·5 m. *Ibid.*, 1929, *J. & M. S. Clemens* 21682:—Tree 9 m. (♂ flower, probably this).

Ki.—W. Koetai, No. 34/35, near Long Petah, forest on stony river bank, alt. 450 m., 7 Sept. 1925, *Endert* 4078:—Small tree 5 m. high. Central Kutei, Pedohon River near Tabang, in clay soil along river, alt. . . ., 26 April 1955, *Kostermans* 10602:—Treelet 5 m. tall; fls. [♀] green. Sg. Mentawir region, near Balikpapan, sandy clay soil in moist valley, alt. . . ., 26 July 1954, *Kostermans* 9832 (holotype, K):—Tree 10 m. tall, diam. 20 cm.; bark smooth, brown, 0·5 mm. thick; living bark 5 mm., pale; wood white; fr. green.

I suspect that this form may deserve specific rank, but at present I can find no distinguishing character beyond the very different-looking fruit. There is a superficial resemblance to *C. elegans* Airy Shaw, of Sumatra, but in that species the fruit is attenuate below and borne on a slender pedicel 7–9 mm. long.

Claoxylon praetermissum *Airy Shaw* in Kew Bull. 23: 78 (1969).

SB.—Endemic.

Small tree 3–9(–15) m. high, in primary forest on brownish or black sandy soil, mostly on low level land, up to 120 m. alt.

Thinly puberulous or almost glabrous; leaves ovate or narrowly elliptic, up to 17 × 7 cm., thinly membranous, with slender petiole; male inflorescence very slender and elongate, up to 19 cm. long; male flower very small, with 17–25 stamens; capsule smooth, densely adpressed-puberulous.

Perhaps not distinct from *C. pseudo-insulanum*, of which no material has been seen; apparently only differs (*e descr.*) in the larger leaves, longer petioles and longer male inflorescence.

Claoxylon pseudo-insulanum *Pax & Hoffm.* vii: 113 (1914); Merr.: 338 (1921).

Ki (?).—Endemic.

No ecological information.

Apparently only differs from *C. praetermissum* in the smaller dimensions of leaves, petioles and male inflorescence. No material available.

Claoxylon salicinum *Airy Shaw* in Kew Bull. 21: 376 (1968); Meijer, 10: 231 (1968).

SB.—Endemic.

Shrub or small tree up to 2·5 m. tall, in wet primary or secondary jungle or in more open situations, at 1500–1800 m. alt.

Leaves narrowly elliptic-oblanceolate, up to 15 cm. long, 1–2·5 (rarely up to 4·7) cm. broad, almost entire, dull and somewhat fleshy in appearance; inflorescences (both male and female) short, 1–4 cm. long; stamens about 10; capsule dicoccous, 10–12 mm. wide, densely shortly fulvo-sericeous; aril red. Apparently locally frequent on and near Kinabalu.

Claoxylon stapfianum *Airy Shaw* in Kew Bull. 20: 400 (1966).

C. pauciflorum Stapf in Trans. Linn. Soc. Lond. II, 4: 225 (1894); Pax & Hoffm. vii: 107 (1914); Merr.: 337 (1921); *non* Muell. Arg. (1864).

SR (N.); **SB.**—Endemic.
Slender, shrubby tree to 6·5 m. tall, in primary and secondary forest on limestone, among boulders, on blackish soil, at 1000–2100 m. alt.

Characterized by the abbreviated few-flowered inflorescences, especially the female, by the prominent closely parallel transverse secondary venation, and by the rough almost shaggy indumentum of the young capsules.

Claoxylon subbullatum *Airy Shaw* in Kew Bull. 21: 375 (1968); Meijer, 10: 231 (1968).

SB.—Endemic.
Slender tree 2 m. high, near caves at 1800 m. alt.

Very distinctive in its thinly chartaceous, non-fleshy leaves, slightly shining on the lower surface, and almost glabrous except for the nerves, with a narrowly revolute minutely denticulate margin, and markedly bullate venation. The male inflorescences are up to 11 cm. long, lax, conspicuously fulvo-puberulous; stamens about 17; female inflorescence and fruit unknown.

Claoxylon velutinum *J. J. Sm.* in Bull. Jard. Bot. Buitenz. III, 1: 395, t. 44 (1920); Pax & Hoffm. xvii (Euph.–Addit. vii): 182 (1924).

K1.—Endemic.
No ecological information available.

Characterized by the possession of very numerous (about 100) stamens in the male flower, as in *C. winkleri*, but differing from that species in the stronger, densely velutinous indumentum, the leaves without a long acumen, the biseriate warts or glands at the apex of the petiole, the shorter male inflorescence and the very dense and abbreviated female inflorescence. From *C. hirsutellum* (of which the male flower is at present unknown) it differs in its much larger leaves, up to 35 × 13·5 cm., with an almost rounded non-acuminate apex, and in its softer and less harsh indumentum.

Claoxylon winkleri *Pax & Hoffm.* vii: 106 (1914); Winkler in Engl., Bot. Jahrb. 50, Suppl.: 207 (1914), *in obs.*; Merr.: 338 (1921).

K1a.—Endemic.
No ecological information available.

Noteworthy especially for the very numerous (about 100) stamens. The minor nerves are transversely arranged, rather closely parallel and quite prominent below.

Cleidion *Bl.*

Cleidion javanicum *Bl.*, Bijdr.: 613 (1825); Muell. Arg.: 987 (1866); Hook. f.: 444 (1887); Pax & Hoffm. vii: 290 (1914); Ridley: 296 (1924); Gagnep.: 450 (1926); Backer & Bakh. f.: 487 (1963).

? *Acalypha spiciflora* Burm. f., Fl. Ind.: 203 (sphalm. '303') (1768).

Lasiostylis salicifolia Presl, Bot. Bemerk.: 149 (1849).
Rottlera urandra Dalz. in Hook. Journ. Bot. & Kew Garden Misc. 3: 229 (1851).
? *Tetraglossa indica* Bedd. in Madras Journ. Sci. II, 22: 70 (1861).
Macaranga tamiana K. Schum. in Notizbl. Bot. Gart. Berlin 1: 52 (1895).
Cleidion spiciflorum sec. Merr., Interpr. Rumph. Herb. Amboin.: 322 (1917), *in obs.*, & Enum.: 439 (1923); Airy Shaw in Kew Bull. 26: 234 (1972); Whitmore: 79 (1973); *an Acalypha spiciflora* Burm. f.?
Cephalomappa sp.?, Hend. in Journ. Malayan Br. Roy. As Soc. 17: 69 (1939)

SR (C.); **SB.**—India, SE. Asia and S. China, and throughout Malesia to the Bismarcks and Solomon Is. and N. Queensland.

Tree to 11 m. high, in primary or riverine forest on black soil up to 150 m. alt.

Apparently scarce or very local in Borneo; only two collections seen. In Siam, Malaya and Java it is sometimes associated with limestone.

Cleistanthus *Hook. f. ex Planch.*

1 Capsule distinctly (3–15 mm.) pedicellate (but not stipitate):
  2 Leaves large, almost oblong, (13–)17–30 cm. long, 5–10 cm. broad, coriaceous, sometimes pruinose below; flowers densely ferrugineous-tomentose; fruiting pedicel up to 7 mm. long **C. pseudopodocarpus** var. **leptopus**
  2 Leaves smaller, 5–15 cm. long; flowers not ferrugineous-tomentose:
    4 Leaves 2–6 cm. broad; flowers borne on short leafless or small-leaved branchlets, the bracts and calyx often blackish when dry; dried material sometimes slightly aromatic **C. praetermissus**
    4 Leaves 1–3·5 cm. broad; plant not aromatic when dry:
      5 Leaves chartaceous or subcoriaceous, very smooth above, glaucous-pruinose beneath; nerves very obscure; flowers borne on greatly abbreviated brachyblasts . . . . **C. erycibifolius**
      5 Leaves thinly chartaceous, with finely reticulate nervation on both surfaces, not glaucous beneath, usually madder-brown (especially beneath) when dry; flowers borne on slender pedicels in axillary fascicles . . . . **C. pedicellatus**
1 Capsule not or very shortly (1–2 mm.) pedicellate:
  6 Capsule stipitate (i.e. with a gynophore):
    7 Leaves densely or sparsely finely adpressedly coppery-sericeous beneath, very variable in size and shape; ovary glabrous **C. myrianthus**
    7 Leaves not coppery-sericeous beneath; ovary ± pubescent:
      8 Calyx externally ferrugineous-tomentellous:
        9 Flower-glomerules axillary; capsule long-ferrugineous-lanate, at least when young . . . . **C. pyrrhocarpus**
        9 Flower-glomerules borne on short ± leafless branchlets; capsule thinly or shortly pilose . . . . . . **C. bakonensis**
      8 Calyx not ferrugineous-tomentellous:
        10 Flower-glomerules borne in sharply distinct slender leafless inflorescences; leaves glabrous, up to 14 × 6 cm., with only 3–6 pairs of nerves; flowers glabrous . . **C. podopyxis**

10 Flower-glomerules axillary or sometimes in ill-defined ± small-leaved lateral inflorescences:
11 Leaves 12–30 cm. long:
12 Leaves chartaceous, brownish and dull when dry, with numerous parallel ascending primary nerves and closely parallel transverse secondaries; stipules large, 5–7 mm. long; capsule ± fleshy, up to 2·5 cm. diam. **C. megacarpus**
12 Leaves coriaceous, greenish and shining when dry, with ± few somewhat bullately impressed primary nerves and strong elevate reticulate secondaries; stipules small; capsule not fleshy, up to 2 cm. diam. **C. macrophyllus**
11 Leaves 5–18 cm. long:
13 Leaves not *Bridelia*-like, glabrous or almost so, distinctly decurrent into the petiole, with 4–5 pairs of primary nerves and closely and finely reticulate secondary venation on both surfaces . . . **C. decurrens**
13 Leaves *Bridelia*-like, finely adpressedly rufous-puberulous beneath, not distinctly decurrent into the petiole, with 6–8 pairs of parallel primary nerves and closely parallel non-reticulate secondary venation **C. vestitus**
6 Capsule not stipitate:
14 Leaves pubescent or puberulous or adpressed-pilose, at least on the midrib beneath:
15 Leaves pubescent below throughout, not only on the midrib:
16 Nerves usually strongly bullately impressed; pubescence of branchlets strong, rufous-brown; leaves 15–28 cm. long, ± rufous-villous on the margin; styles pilose . **C. paxii**
16 Nerves not or less strongly bullate; pubescence of branchlets less strong; leaves not villous on margin; styles glabrous:
17 Nerves 8–15 pairs; leaves up to 30 cm. long **C. hirsutulus**
17 Nerves 5–8 pairs; leaves up to 9 cm. long **C. maingayi**
15 Leaves pubescent on the midrib only (or sometimes also on a few basal nerves):
18 Pubescence (especially of flower-glomerules) white **C. celebicus**
18 Pubescence ochraceous or tawny:
19 Leaves 3–15 cm. long:
20 Nerves 5–8 pairs; ovary pubescent **C. brideliifolius**
20 Nerves 9–12 pairs; ovary glabrous . **C. glabratus**
19 Leaves 13–20 cm. long; nerves 10–13 pairs:
21 Branchlets and stipules rufous-tomentellous **C. beccarianus**
21 Branchlets and stipules glabrous or sparsely puberulous **C. acuminatissimus**

14 Leaves quite glabrous:
22 Leaves contracted at the base into a short pseudo-petiole, glaucous beneath; flower-glomerules axillary **C. contractus**
22 Leaves not contracted into a pseudo-petiole:
23 Leaves strongly glaucous beneath:
24 Stipules large, striate, 5–12 mm. long; leaves chartaceous **C. striatus**
24 Stipules small, not conspicuously striate, 3–5 mm. long:
25 Leaves ± oblong, coriaceous, 8–20 cm. long **C. elongatus**
25 Leaves elliptic-ovate, chartaceous, 4–13 cm. long **C. baramicus**
Cf. also **C. glabratus**
23 Leaves not or only weakly glaucous beneath:
26 Stipules large, striate, 5–12 mm. long; leaves chartaceous **C. striatus**
26 Stipules small, 1–3 mm. long*:
27 Leaves relatively long and narrow, 15–27 × 3–6·5 cm.:
28 Floral bracts leafy, 1–2·5 cm. long; leaves chartaceous . . . . **C. venosus**
28 Floral bracts scarious, 2–3 mm. long; leaves coriaceous . . . . **C. coriaceus**
27 Leaves less long and narrow, not exceeding 17 × 6·5 cm.:
29 Flower-glomerules usually axillary; leaves thinly chartaceous, usually wrinkling somewhat in drying, variable in shape . . **C. glaber**
29 Flower-glomerules usually borne on leafless or small-leaved lateral inflorescences; leaves more firmly coriaceous, usually remaining flat on drying:
30 Calyx of ♀ flower 6–7 mm. long; lateral nerves 6–8 pairs . . . . . **C. winkleri**
30 Calyx of ♀ flower 3–4 mm. long; lateral nerves 4–6 pairs:
31 Leaves larger, usually more gradually acuminate, with more distinct venation **C. sumatranus**
31 Leaves smaller, usually conspicuously and abruptly caudate, with inconspicuous venation . . . . **C. gracilis**

N.B. *C. sarawakensis* is not included in the above key; *vide* p. 86, below.

Cleistanthus acuminatissimus *Merr.*, Pl. Elmer.: 155 (1929); Meijer, 7: 41 (1967).

SB.—Endemic.

Shrub or small tree in densely forested ravines, probably at low alt.

* Cf. also **C. pseudopodocarpus** var. **pseudopodocarpus,** with leaves up to 30 × 10 cm. and densely ferrugineous-tomentose flowers.

Closely related to *C. beccarianus*, of which it could be treated as a subspecies or variety, but apparently differing consistently in its glabrous or sparsely puberulous branchlets and stipules; the stipules are also longer and finer.

Cleistanthus bakonensis *Airy Shaw* in Kew Bull. 21: 369 (1968).

SR; SB.—Endemic.

Tree to 12 m. high, in primary Heath or *kerangas* forest, or Mixed Dipterocarp forest, on sandstone boulders or rocks or white sand, up to 270 m. alt.

In the small section *Ferruginosi*, characterized by the quickly glabrescent branchlets, the leaves up to 18 cm. long, glaucescent beneath and with densely elevate-reticulate nervation, and the flowers borne on short, mostly leafless, rufous-sericeous branchlets 1·5–4·5 cm. long; the capsule is thinly pilose, very shortly (1–2 mm.) pedicellate, and usually shortly stipitate also.

Flowers noted as weakly fragrant.

Cleistanthus baramicus *Jabl.* viii: 24 (1915); Merr.: 334 (1921).

? *C. glaucus* Jabl. viii: 25 (1915); Ridley: 188 (1924).

SR (C. & NE.)**; SB; K1.**—Malaya?

Tree to 15 m. high, in open primary or secondary Mixed Dipterocarp forest, on sandstone hills, chocolate-coloured sandy soil, basalt ridges, once noted from a small forested area in a sandy heath region near the sea, up to 800 m. alt.

Distinguished from the closely related *C. elongatus*, of SW. Sarawak, by the smaller, thinner (more or less chartaceous), elliptic-ovate rather than oblong leaves. A very small-leaved specimen from the Ulu Balleh in the 3rd Division of Sarawak (*Othman bin Haron* S. 29215), with leaves only 1–3 cm. long, may possibly represent an extreme condition of this species.

Cleistanthus beccarianus *Jabl.* viii: 19 (1915); Merr.: 334 (1921).

SR.—Endemic.

Tree of 9 m. in understorey of forest on limestone hill at 600 m. alt.

Branchlets rufous-tomentellous when young; leaves membranous or chartaceous, caudate-acuminate, base rounded or cordate, glabrous except for the midrib beneath, lower surface slightly glaucous or plumbeous when dry; stipules subulate, rufous-puberulous. The rufous indumentum of branchlets and stipules distinguishes this species from the closely allied *C. acuminatissimus*.

Cleistanthus brideliifolius *C. B. Rob.* in Philipp. Journ. Sci. 3, Bot.: 191 (1908) & 6: 323 (1911); Jabl. viii: 25 (1915); Merr.: 419 (1923).

C. pallidus var. *subcordatus* J. J. Sm.: 304 (1910).
C. subcordatus (J. J. Sm.) Jabl. viii: 22 (1915); Backer & Bakh. f.: 474 (1963).

SR; SB; K1.—Philippines, Malaya (?), Sumatra, Java.

Tree to 12 m. high, in primary or secondary Mixed Dipterocarp forest, mostly on limestone or coral limestone rocks, sometimes on basalt, or on yellowish clay, or in stony ground near a stream, up to 600 m. alt.

Branchlets slender, very shortly tomentellous or puberulous; leaves lanceolate to elliptic, 3–7(–13) cm. long, 1–2(–4·5) cm. wide, rounded or very shortly subcordate at the base, mostly distinctly caudate at apex (cauda obtuse), chartaceous, glabrous except for midrib; nerves 7–8 pairs; flowers in small axillary glomerules, but sometimes tending to be borne on special bare or small-leaved branchlets; calyx externally tomentellous, but female calyx sometimes thinly puberulous or subglabrous; capsule pubescent.

Flowers said to be fragrant.

Jablonszky's inclusion of this species in his key under the heading A.b. α.II.1, 'Limbus apice acutus, non acuminatus', is misleading, since the leaves are often long-acuminate or long-caudate.

Cleistanthus celebicus *Jabl.* viii: 22 (1915).

SB.—Celebes.

Small tree to 6 m. high, in primary forest on steep limestone cliff or on rocky ridge-top in black soil up to 470 m. alt.

Branches slender, at first thinly puberulous; leaves thinly chartaceous, elliptic to oblong, glabrous except for midrib beneath, broadly cuneate to rounded or very slightly cordate at base, shortly caudate-acuminate at apex, green when dry; stipules subulate, 2–3 mm. long; flower-glomerules axillary, densely whitish-tomentellous.

Near *C. hirsutulus*, but differing in the much sparser (or obsolete) pale indumentum of branchlets and midribs, in the smaller stipules, and especially in the densely whitish-tomentellous flower-glomerules. I refer to this species two Sabah collections: *Ah Wing* SAN 38103, from Gomantong Hill Forest Reserve in the Sandakan District, and *Meijer* SAN 37945, from Madai in Lahad Datu Distr. Ah Wing's collection especially matches the Celebes type (*Riedel s.n.*!) closely.

Cleistanthus contractus *Airy Shaw* in Kew Bull. 25: 511 (1971); Whitmore: 80 (1973).

SB.—Malaya (Johore).

Tree to 10 m. high, in primary forest on brown sandy soil up to 300 m. alt.

Branchlets almost glabrous; leaves elliptic, 9–17 cm. long, 7–12 cm. wide, conspicuously contracted at the base into a short pseudo-petiole, often with a semi-annular enlargement where it joins the true petiole, chartaceous, smooth and glabrous above, glaucous beneath and sometimes minutely fuscous-puberulous, with prominulous nervation. Flower-glomerules axillary. Calyx almost glabrous.

Very similar superficially to *C. glandulosus* Jabl., of Malaya, but differing in the midrib raised above, in the much shorter glabrous stipules, and in the glabrous or thinly pilose, not densely white-tomentose calyx.

Cleistanthus coriaceus *Airy Shaw* in Kew Bull. 21: 362 (1968).

SR.—Endemic.

Tree to 12 m. high, in primary Lowland or Mixed Dipterocarp forest on clay loam or sandstone or yellow leached sandy soil (Begrih formation) up to 150 m. alt.

Closely related to *C. sumatranus*, but differing consistently in its larger and proportionately narrower leaves, up to 23 cm. long, and in their much thicker and more coriaceous texture. Also close to *C. venosus*, but differing in the minute, non-foliaceous floral bracts and coriaceous leaves.

Cleistanthus decurrens *Hook. f.*: 278 (1887); Jabl. viii: 33 (1915); Ridley: 191 (1924); Hend. in Journ. Malayan Br. Roy. As. Soc. 17: 69 (1939); Whitmore: 81 (1973).

C. decipiens sec. Elmer, Leafl. Philipp. Bot. 4: 1284 (1911), Merr.: 420 (1923), *p.p.*, quoad *Elmer* 12888, 12988; *non* C. B. Rob.

C. mattangensis Jabl. viii: 33 (1915); Merr.: 334 (1921); **synon. nov.**

SR.—Malaya, SW. Philippines (Palawan).

Tree of 4·5 m., in hill forest at 60 m. alt.

Leaves elliptic, 6–18 cm. long, 2·5–6 cm. wide, glabrous or almost so, with closely and finely elevate-reticulate venation on both surfaces, plumbeous or greenish above when dry, pale brownish beneath, conspicuously decurrent into the short petiole; young parts brown-rufous-sericeous, soon glabrescent; flower-glomerules axillary or on special small-leaved or leafless branchlets; bracts membranous, ferrugineous-pilose; flowers sessile; capsule variously (1–7 mm.) stipitate, glabrous except for the pilose base.

Apparently everywhere a scarce species. In Malaya it occurs on limestone.

Cleistanthus elongatus *Jabl.* viii: 23 (1915); Merr.: 334 (1921).

SR (SW.).—Endemic.

Small tree 10 m. high on *kerapah* at sea-level.

Almost completely glabrous; leaves oblong or lanceolate or oblanceolate or elongate-elliptic, stiffly coriaceous, up to 20 cm. long, shortly acuminate, glaucous beneath.

Closely related to *C. baramicus*, of C. & NE. Sarawak and Sabah, but distinguished by the larger, more oblong and more coriaceous leaves.

Cleistanthus erycibifolius *Airy Shaw* in Kew Bull. 20: 389 (1966) & 27: 76 (1972).

KI.—Malaya.

Tree to 20 m. high, locally very common in primary forest on sandy soil or sandy loam, once noted from a tufa plateau, up to 100 m. alt.

A very characteristic species, without close relatives in Borneo, but perhaps related to *C. isabellinus* Elm., of the Philippines, from which it differs in the perfectly glabrous underside of the leaves and in the long-pedicelled but non-stipitate capsule. The relatively small leaves (up to 11 × 4 cm.), with their pinkish-glaucous undersurface and reflexed or almost revolute margins, are very distinctive. Material with inflorescences is still lacking.

Cleistanthus glaber *Airy Shaw* [ex Meijer, 7: 40 (1967), *anglice*, *in clavi*, &] in Kew Bull. 21: 366 (1968), *latine*; Meijer, 10: 230 (1968); Whitmore: 80, 81 (1973).

SR; SB.—Malaya, Philippines (Mindanao).

Tree to 14 m. high, in primary Mixed Dipterocarp forest, on sandstone, brown or black sandy soil, or shale, up to 90 m. alt.

Near *C. baramicus* and *C. elongatus*, but differing in the green or only slightly glaucous leaf-undersurface, the minute stipules, scarcely 1 mm. long, and the externally white-pilose calyx. The leaves are extraordinarily variable in shape and size—mostly elliptic, but occasionally shortly ovate or elongate-oblong-lanceolate (and then up to 21 cm. long).

Cleistanthus glabratus *Jabl.* viii: 40 (1915); Merr.: 334 (1921).

SR (SW.).—Endemic.

Tree to 12 m. high, in primary forest by streams on sandy clay soil at low altitudes.

A little-known species. Young parts and petioles thinly adpressed-rufous-pubescent. Leaves elliptic to oblong, up to 14·5 × 4·5 cm., chartaceous to coriaceous, almost completely glabrous, but thinly adpressedly rufous-pilose towards the base of the midrib beneath, smooth and slightly shining above, slightly glaucous beneath; lateral nerves 9–12 pairs, rather steeply ascending, very slender; minor nerves very numerous, closely scalariform; petiole 5–10 mm. long. Flowers in axillary glomerules or on short small-leaved lateral branchlets. Calyx glabrous outside; ovary glabrous. Capsule not or scarcely stipitate.

The only collection of *C. glabratus* that I have seen, other than the type (*Haviland* 805—the number was omitted by Jablonszky—from the Rejang River in the 3rd Division of Sarawak), is a fruiting specimen, *Mamit* S. 29847, from Stabut, Padawan, in the 1st Division. On account of its glabrous ovary Jablonszky included the species in his section *Nanopetalum* (*C. myrianthus*, etc.), but Mamit's collection shows that it differs from all others in that group in its practically sessile capsule. In this character, and in the tendency of the flowers to be produced on special short lateral branchlets, it agrees more closely with Sect. *Leiopyxis*, but there the glabrous ovary is anomalous. I suspect that there is in *Cleistanthus* a complex situation of permutations and combinations in which natural groupings may be difficult to establish.

Cleistanthus gracilis *Hook. f.*: 277 (1887); Jabl. viii: 15 (1915); Ridley: 190 (1924) (incl. var. *parvifolia* Ridl.); Gage in Journ. As. Soc. Bengal 75(5): 506 (1936); Hend. in Journ. Malayan Br. Roy. As. Soc. 17: 70 (1939); Meijer, 7: 40 (1967); Airy Shaw in Kew Bull. 26: 236 (1971); Whitmore: 80, 82 (1973).

C. dasyphyllus F. N. Williams in Bull. Herb. Boiss. II, 5: 31 (1905); Jabl., *l.c.*: 18 (1915).

? *C. blancoi* Rolfe f. *dubius* Jabl., *l.c.*: 13 (1915); Merr.: 334 (1921).

SR; SB; K1.—Penins. Siam, N. Malaya, Anamba Is., ? Philippines.

Tree to 18 m. high, in primary lowland coastal forest or *kerangas* or Mixed Dipterocarp forest, once among sandstone boulders on a plateau sandstone outcrop, up to 570 m. alt.

Close to *C. sumatranus* (Miq.) Muell. Arg. (*C. laevigatus* Jabl.), but differing in the slender branches and small ovate or elliptic leaves, which are often conspicuously and abruptly caudate, with the nerves obscure or very inconspicuous and scarcely raised on the dull lower surface.

Cleistanthus hirsutulus *Hook. f.*: 278 (1887); Jabl. viii: 26 (1915); Ridley: 192 (1924); Gage in Journ. As. Soc. Beng. 75(5): 511 (1936); Hend. in Journ. Malayan Br. Roy. As Soc. 17: 70 (1939); Airy Shaw in Kew Bull. 26: 237 (1972); Whitmore: 81, 82 (1973).

C. siamensis Craib in Bull. Misc. Inf. Kew 1913: 71 (1913); Jabl. viii: 23 (1915); Ridley: 190 (1924); Gage in Journ. As. Soc. Beng. 75(5): 507 (1936).
C. penangensis Jabl., *l.c.*: 21 (1915).
C. cochinchinae Jabl., *l.c.* (1915).
Paracleisthus siamensis (Craib) Gagnep. in Bull. Soc. Bot. France 70: 497 (1923) & in Lecomte, 5: 497 (1926).

SR (NE.); **SB.**—Siam, Indochina, Malaya, Sumatra.
Tree to 18 m. high, in primary forest on yellowish clay, clay mixed with sand, sandstone and lime, brownish or black stony soil, stony loam, basalt, or litter overlying and between limestone rocks, up to 1065 m. alt.

Superficially similar to *C. brideliifolius*, but leaves mostly longer, often glaucous beneath and nerves 12–14(–18) pairs. Pubescence variable, mostly rather strong.

Cleistanthus macrophyllus *Hook. f.*: 278 (1887); Jabl. viii: 35 (1915); Ridley: 192 (1924); Meijer, 7: 40 (1967); Whitmore: 81 (1973).

SR; SB; KI.—Malaya.
Tree to 6 m. high, in primary forest on greyish or black rocky or stony soil, once noted on limestone, up to 420 m. alt.

Recognizable by its often robust branches, large, coriaceous, smooth, shining, almost completely glabrous leaves, 15–30 cm. long, 5–9 cm. wide, often remaining greenish when dry, the primary nerves somewhat bullately impressed, the lesser nerves strongly reticulate below. Flowers in axillary glomerules. Capsule 1·5–2 cm. diam., blackish, slightly venose or verruculose, glabrous, stoutly stipitate, stipes 1–8 mm. long.
The dividing line between *C. macrophyllus* and certain large-leaved forms of *C. myrianthus* is sometimes by no means clear, depending ultimately upon the presence (*myrianthus*) or absence (*macrophyllus*) of fine sericeous hairs on the underside of the leaves, and upon the presence (*macrophyllus*) or absence (*myrianthus*) of a dense ferrugineous indumentum on the ovary. As female flowers seem to be rarely collected, and as the capsule of *C. macrophyllus* seems to lose every trace of pubescence at a very early stage, it would perhaps be better to subordinate *C. macrophyllus* to *C. myrianthus* (which is already very variable) with infra-specific rank.

Cleistanthus maingayi *Hook. f.*: 280 (1887); Jabl. viii: 25 (1915); Ridley: 193 (1924); Whitmore: 82 (1973).

C. borneënsis Jabl., *l.c.*: 25 (1915); Merr.: 334 (1921); **synon. nov.**

SR (SW.).—Malaya.
No field notes available for Borneo.

Differs from *C. brideliifolius* by the denser ferrugineous indumentum of the young branchlets and by the evident though thin ferrugineous pilosity of the lower leaf-surface. The leaves are in fact often more *Bridelia*-like (cf. *B. cinnamomea* Hook. f.) than are those of *C. brideliifolius*.

Cleistanthus megacarpus *C. B. Rob.* in Philipp. Journ. Sci. 6, Bot.: 232 (1911); Jabl. viii: 31 (1915); Merr.: 420 (1923); Holth. & Lam in Blumea 5: 200 (1942); Meijer, 7: 39 (1967).

SR (NE.)**; SB; K1.**—Philippines, Sangie and Talaud Is.

Tree to 12 m. high, in primary or secondary forest, sometimes swampy or near rivers, on black or blackish soil, sand and limestone, or loam with coral limestone rocks, up to 400 m. alt.

A characteristic species, recognizable by its large, thinly chartaceous leaves, with numerous ascending primary nerves and closely parallel transverse secondaries, elongate subulate stipules, 5–7 mm. long, and large, somewhat fleshy capsules, up to 2·5 cm. diam. The flowers are borne in axillary glomerules.

Cleistanthus myrianthus (*Hassk.*) *Kurz*, For. Fl. Brit. Burma 2: 370 (1877); Hook. f.: 275 (1887); Boerl.: 271 (1900); J. J. Sm.: 297 (1910); Jabl. viii: 37 (1915); Merr.: 334 (1921) & 421 (1923); Ridley: 194 (1924); S. Moore in Journ. Bot. 63, Suppl.: 93 (1925); Gagnep.: 482 (1926); Gage in Journ. As. Soc. Beng. 75(5): 515 (1935); Holth. & Lam in Blumea 5: 200 (1942); Backer & Bakh. f.: 474 (1963); Meijer, 7: 41 (1967); Airy Shaw in Kew Bull. 26: 237 (1972); Whitmore: 82 (1973).

Nanopetalum myrianthum Hassk. in Verh. Kon. Akad. Wetensch. Amsterd. 24: 140 (1855).
Cleistanthus cupreus Vidal, Rev. Pl. Vasc. Filip.: 235 (1886), **synon. nov.**
C. pseudo-canescens Elm., Leafl. Philipp. Bot. 3: 910 (1910).
C. myrianthus subsp. *cupreus* (Vidal) Jabl., *l.c.*: 39 (1915); Merr.: 421 (1923).
C. pseudomyrianthus Jabl., *l.c.*: 41 (1915); Ridley: 194 (1924), *teste* Whitmore, *l.c.* (1973) (typum non vidi).

SR; SB; K1, 1a, 3(?).—SE. Asia and throughout Malaysia to Solomon Is.

Tree to 16 m. high, in primary Mixed Dipterocarp forest (sometimes periodically inundated), mostly on limestone, sometimes on brownish sandy soil; also noted on yellowish clay soil, basalt, granodiorite-derived soil, and clay-rich pale pink-grey soil between dacite boulders, up to 400 m. alt.

An exceedingly variable entity. 'Typical' *C. myrianthus* (from Java, where it is very scarce) has chartaceous to coriaceous, small to medium-sized leaves, rarely exceeding 20 × 5 cm., with flat (not bullate) primary nervation, and a thin but dense covering of fine, adpressed, coppery hairs beneath. Plants of this type occur not infrequently in Borneo.

In Borneo there occurs also a form with more stiffly coriaceous leaves, sometimes reaching 30 × 10 cm., with bullately impressed primary nervation, tertiary nerves very prominent below, and a much thinner or almost obsolete covering of fine adpressed hairs. This form sometimes approaches very closely to *C. macrophyllus*, which is completely glabrous. There seems to be an almost complete intergrading between the two species (placed by Jablonszky in different sections!), but the extremes look very different. The group requires study in the field.

An extreme form of *C. myrianthus* in the other direction is perhaps represented by a single collection (*Othman bin Haron* S. 29089) from Ulu Balleh in the Third Division of Sarawak. The leaves are narrowly oblong-lanceolate, only 5–8·5 cm. long and 1–2·2 cm. wide, very long- and slenderly acuminate, with reflexed or revolute margins, and finely coppery below. Further material is needed in order to show whether this plant deserves taxonomic recognition. —I have not yet seen material of forma *ovalis* Jabl., *l.c.*: 39, from W. Indonesian Borneo (**K3**).

Cleistanthus paxii *Jabl.* viii: 27, fig. 4 (1915); Merr.: 335 (1921); Meijer, 7: 41 (1967).

SR (SW.).—Endemic.

Shrub or small tree to 4 m. high, in primary Lowland Dipterocarp forest on sandstone substratum up to 500 m. alt.

Branchlets 3–4 mm. thick, with a dense rufous or chocolate-coloured tomentum; leaves oblong, elliptic or narrowly obovate, 15–28 cm. long, 4–9 cm. wide, shortly and slenderly caudate, chartaceous or scarcely coriaceous, glabrous above, rusty-pilose below; lateral nerves 8–12 pairs, slenderly prominent below, often bullately impressed above; flowers rather large (about 7 mm. diam.), styles pilose.

Resembles a very robust version of *C. hirsutulus*, but with larger flowers, a darker indumentum and usually bullately impressed nerves. The flowers are said to have a weak sourish fragrance.

Cleistanthus pedicellatus *Hook. f.*: 281 (1887); Jabl. viii: 44 (1915); Ridley: 187 (1924); Whitmore: 80 (1973).

C. integer C. B. Rob. in Philipp. Journ. Sci. 3, Bot.: 196 (1908); Elm., Leafl. Philipp. Bot. 3: 908 (1910); Jabl. viii: 45, fig. 6, *E–G* (1915); Merr.: 420 (1923); **synon. nov.**

C. quadrifidus C. B. Rob., *l.c.*: 197 (1908); Jabl. viii: 44 (1915); Merr.: 421 (1923); Airy Shaw in Kew Bull. 23: 64 (1969); **synon. nov.**

C. dichotomus J. J. Sm. in Nova Guinea 8(4): 786, t. 136 (1912); Jabl. viii: 43 (1915); **synon. nov.**

SB (Pulo Gaya only).—Malaya, Philippines, New Guinea.

Shrub or small tree on hill; no further ecological information for Borneo.

A very isolated species, unmistakable from its relatively small, glabrous, closely reticulate leaves, drying a madder-brown colour, and especially from its long- (5–15 mm.) and slenderly pedicelled flowers and capsules. The Pulo Gaya plant has unusually small leaves, up to 7 × 2·5 cm., but these are matched by a specimen from Peñablanca, Cagayan Province, Luzon, *Ramos* Bur. Sci. 76705.

I am now convinced that the synonyms cited above only represent minor variants of this widely scattered taxon (cf. Kew Bull. *l.c. supra*).

Cleistanthus podopyxis *Airy Shaw* [ex Meijer, 7: 39 (1967), *anglice, in clavi*, &] in Kew Bull. 21: 363 (1968), *latine*; Meijer, 10: 230 (1968).

SB; K1.—Endemic.

Tree to 12 m. high, in primary or secondary forest on black or clayey soil, or sandstone, or coral limestone rocks, up to 600 m. alt.

In Jablonszky's classification this seems clearly referable to Sect. *Leiopyxis*, the flowers being borne on special leafless branchlets. From almost all the species of that section, however (cf. also *C. praetermissus*, below), it differs in having a manifestly stipitate (but non-pedicellate) capsule. In other respects the species shows no remarkable features. The leaves are mostly thinly chartaceous, but sometimes subcoriaceous, 6–14 × 2–6 cm., quite glabrous, and with only 3–6 pairs of rather prominent nerves. The inflorescences are elongate and rather slender; the flowers are quite glabrous, and are said to be fragrant.

Cleistanthus praetermissus *Gage* in Bull. Misc. Inf. Kew 1914: 240 (1914); Ridley: 193 (1924); Gage in Journ. Roy. As. Soc. Beng. 75(5): 511 (1936); Meijer, 10: 230 (1968); Airy Shaw in Kew Bull. 21: 372 (1968) & 26: 238 (1972); Whitmore: 81 (1973).

C. hosei Pax & Hoffm. ex Meijer, 7: 40 (1967), *anglice, in clavi*, **synon. nov.**

SB.—Lower Siam, Malaya.
Tree to 18 m. high, in primary forest on black sandy soil up to 150 m. alt.

Cleistanthus praetermissus is unusual in Sect. *Leiopyxis* in having a definitely, though shortly (2–4) mm.), pedicellate (but non-stipitate) capsule. From *C. podopyxis* it differs in its mostly flatter leaves, usually greyish or glaucescent, beneath, with rather more numerous and less prominent primary nerves and in its much shorter and less slender inflorescences, with much more evident bracts, often blackish rather than brownish when dry. The dried material often gives off a faint but distinct aromatic odour.

Cleistanthus pseudopodocarpus *Jabl.* viii: 29 (1915); Merr.: 335 (1921); Meijer, 7: 41 (1967); Whitmore: 83 (1973).

? *C. rufescens* Jabl., *l.c.*: 30 (1915), *e descr.*
C. sp., Merr., Pl. Elmer.: 155 (1929) (*Elmer* 21565).

Var. **pseudopodocarpus**: fruiting pedicel 0–2 mm. long.

SR; SB.—Malaya, ? Sumatra.
Tree to 18 m. high, in primary Lowland or Mixed Dipterocarp forest, or in damp forested depressions, on brown or blackish sandy soil up to 90 m. alt.

Closely related to *C. podocarpus* Hook. f., of Malaya, but differing in the rounded or contracted, not cordate, base of the leaves.

Var. **leptopus** *Airy Shaw* in Kew Bull. 27: 76 (1972).

SR.—Endemic.
Treelet a few m. high, in primary Dipterocarp forest on sandstone substratum at 100–500 m. alt.

Differs from the type form in the elongate fruiting pedicel, up to 7 mm. long.

Cleistanthus pubens *Airy Shaw* in Kew Bull. 21: 365 (1968); Meijer, 10: 230 (1968).

SR; SB; K1, 1a.–Endemic.
Shrub or tree to 30 m. high, in primary Mixed Dipterocarp forest on sandstone or sandy loam or friable granodiorite-derived soil up to 1100 m. alt.

Related to *C. sumatranus*, but readily distinguished by the usually evident development of pubescence on the branchlets and on the nerves beneath, and by the usually thinner texture of the leaves, with more numerous nerves (6–8 pairs).

Cleistanthus pyrrhocarpus *Airy Shaw* in Kew Bull. 21: 370 (1968).

SR; B.—Endemic.
Tree to 15 m. high, in primary Lowland or Mixed Dipterocarp forest on dry soil or sandstone or deep yellow sandy soil over Tertiary clays up to 700 m. alt.

In the section *Ferruginosi*, closely related to *C. nitidus* Hook. f. (so far only known from Penang in Malaya), but the leaves are up to 5 cm. wide and cuneate (not subcordate) at the base, and the petiole is much longer (5–8 mm.) and finally glabrous. The sometimes almost shaggily ferrugineous-tomentose capsule is very distinctive.
A native collector in the Miri District of Sarawak has noted the fruit as edible.

Cleistanthus sarawakensis *Jabl.* viii: 34 (1915); Merr.: 335 (1921).

SR.—Endemic.
No ecological information available.

I have not seen the type of this (*Beccari* 2469), nor any other material so named. The original description suggests a plant somewhat approaching *C. contractus*, but it differs from that species in the midrib not prominent above, in the considerably fewer (about 6) nerves, and perhaps in the 'exiguous' stipules, those of *C. contractus* being 1·5–2·5 mm. long. Forms of *C. glaber* might also be considered, but that species apparently never has a glaucescent leaf-underside as *C. sarawakensis*. The latter character, together with the description of the lamina as 'basi in petiolum contractus, ad finem petioli et limbi incrassatus', actually narrows down the possibilities considerably, almost limiting them to *C. contractus*, *C. glaber*, and *C. hirsutipetalus* Gage, of Malaya, which has not yet been collected in Borneo. The flower-glomerules are described as 'spicati'. Only an inspection of authentic material can settle the identity of *C. sarawakensis*.

Cleistanthus striatus *Airy Shaw* in Kew Bull. 23: 63 (1969).

SR; SB.—Endemic.
Tree to 18 m. high, in primary forest on yellowish-brown sandy soil, or sandy clay, up to 300 m. alt.

Rather similar to *C. hirsutulus*, but almost completely glabrous, the leaves more or less glaucous beneath, with large, conspicuous, subulate, striate stipules, 5–12 mm. long.

Cleistanthus sumatranus (*Miq.*) *Muell. Arg.*: 504 (1866); Jabl. viii: 13, fig. 2, *A–E* (1915); Merr.: 335 (1921) & Philipp. Journ. Sci. 29: 381 (1926); Backer & Bakh. f.: 474 (1963); Meijer, 7: 40 (1967); Airy Shaw in Kew Bull. 26: 238 (1972); Whitmore: 82 (1973).

Leiopyxis sumatrana Miq., Fl. Ind. Bat., Suppl.: 446 (1861).
? *Cleistanthus blancoi* Rolfe in Journ. Linn. Soc., Bot. 21: 315 (1884); Hallier in Meded. Rijks Herb. 1910: 7 (1911); Jabl. viii: 13 (1915); Merr.: 419 (1923).
C. heterophyllus Hook. f.: 276 (1887); Jabl. viii: 14 (1915); Ridley: 190 (1924); Gage in Journ. As. Soc. Beng. 75(5): 505 (1936).
C. laevis Hook. f.: 277 (1887); Jabl. *l.c.*: 13 (1915).
? *C. vidalii* C. B. Rob. in Philipp. Journ. Sci. 3, Bot.: 193 (1908); Jabl. *l.c.*: 15 (1915); Merr.: 422 (1923).
C. laevigatus Jabl. viii: 12 (1915); Merr.: 334 (1921); Meijer, 7: 40 (1967); Airy Shaw in Kew Bull. 21: 363 (1968), *in obs.*; **synon. nov.**
C. oligophlebius Merr. in Philipp. Journ. Sci. 13, Bot.: 80 (1918) & Enum.: 335 (1921) & Pl. Elmer.: 154 (1929), **synon. nov.**
C. saichikii Merr. in Philipp. Journ. Sci. 23: 248 (24 July 1923).
Paracleisthus subgracilis Gagnep. in Bull. Soc. Bot. France 70: 500 (Aug. 1923) & in Lecomte: 500 (1926).
Cleistanthus euphlebius Merr. in Papers Mich. Acad. Sci. 24: 78 (1939), **synon. nov.**

SR; SB; KI.—Siam, Indochina, Hainan, and throughout Malesia (incl. Philipp.?) to Moluccas, Tenimber and Aru Is.

Shrub or tree to 15 m. high, in primary or secondary Mixed Lowland Dipterocarp forest (sometimes dry), on sandstone or clay loam or yellow or brown soil, or on a tufa plateau, up to 240 m. alt.

Wood hard, used for sword handles (Sabah).

An exceedingly polymorphic species, whose boundaries, especially as regards *C. gracilis*, are still far from clear. The larger, usually more gradually acuminate leaves and more distinct venation seem to constitute the most reliable characters for its separation from that entity. I believe it is necessary to add *C. laevigatus*, *C. oligophlebius* and *C. euphlebius* to the already long synonymy given in Kew Bull. 26: 238 (1971).

Cleistanthus venosus *C. B. Rob.* in Philipp. Journ. Sci. 3, Bot.: 192 (1908); Jabl. viii: 12 (1915); Merr.: 335 (1921) & 422 (1923).

C. elmeri Merr., Pl. Elmer.: 154 (1929); Meijer, 7: 40 (1967); **synon. nov.**

SB.—Malaya, Philippines (Mindanao).

Tree to 12 m. high, in primary hill forest and on river banks, on dark brown or sandy soil, up to 90 m. alt.

Closely related to *C. winkleri*, with which it shares the rather large flowers and long narrow sepals, but differing in that the sepals and petals are externally long-pilose.

Cleistanthus vestitus *Jabl.* viii: 32 (1915); Merr.: 335 (1921); Whitmore: 83 (1973).

C. vestitus f. *perakensis* Jabl., *l.c.*

C. barrosii Merr. in Philipp. Journ. Sci. 20: 400 (1922) & Enum.: 419 (1923), **synon. nov.**
C. perakensis (Jabl.) Gage ex Ridley: 15 (1924).
C. sp., Merr., Pl. Elmer.: 155 (1929) (*Elmer* 21567).

SR; SB; K1(?), **3.**—Malaya, Philippines.
Shrub-like tree along margins of clearings; no further data.

Leaves *Bridelia*-like (cf. *B. cinnamomea* Hook. f.), 5–12 cm. long, chartaceous, finely adpressedly rufous-puberulous beneath, nerves rather steeply ascending and markedly parallel; flowers in axillary glomerules, surrounded by densely ferrugineous-tomentose bracts.

Cleistanthus winkleri *Jabl.* viii: 12 (1915); Merr.: 335 (1921).

SR; B.—Endemic.
Tree to 6 m. high, in primary Heath forest on Quaternary white sand terrace or yellow sandy loam up to 45 m. alt.

Near *C. venosus,* differing in its glabrous sepals and petals.

Cnesmone *Bl.*

Cnesmone javanica *Bl.*, Bijdr.: 630 (1825) (sphalm. '*Cnesmosa*'), corr. Bl., Fl. Jav. Praef. p. vi (1828); Muell. Arg.: 926 (1866); Scheff. in Ann. Mus. Bot. Lugd.-Bat. 4: 122 (1868); Hook. f.: 466 (1888); Boerl.: 262 (1900); J. J. Sm.: 514 (1910); Pax & Hoffm. ix–xi: 102 (1919); Merr.: 343 (1921); Ridley: 306 (1924); Gagnep.: 385 (1926); Merr., Pl. Elmer.: 161 (1929); Backer & Bakh. f.: 490 (1963) (*Cnesmosa*); Airy Shaw in Kew Bull. 26: 240 (1972); Whitmore: 83 (1973).

Tragia hastata Reinw. ex Hassk., Pl. Jav. Rar.: 245 (1868).
? *Cenesmon tonkinense* Gagnep. in Bull. Soc. Bot. France 71: 869 (1924) & in Lecomte: 389 (1926), *e descr.*
? *Cnesmone tonkinensis* (Gagnep.) Croiz. in Journ. Arn. Arb. 22: 429 (1941).

SB; K1a.—Assam, SE. Asia, W. Malesia (exc. Philipp.).
Climber, more or less herbaceous, in thickets and forests, mostly at low altitudes; no precise field data available for Borneo.

Strongly hirsute and urticating; leaves closely denticulate or sometimes subentire, deeply cordate, with a wide sinus; stipules large, conspicuous, often brownish when dry.
Only three specimens seen from Borneo: Sabah (Tawao, *Elmer* 20663); SE. Indonesian Borneo (Bandjermasin, *Motley* 199; Lumowia to Kumam, approx. 2° S., 116° E., *Hub. Winkler* 2913).

Codiaeum *A. Juss.*

1 Stamens 20–30; disk-glands of ♂ flower 5; ovary glabrous; styles short and thick . . . . C. VARIEGATUM
1 Stamens ±50; disk-glands of ♂ flower 10–15(?); ovary pubescent; styles slender, 4–5 mm. long . . . . **C. affine**

Codiaeum affine *Merr.* in Philipp. Journ. Sci. 29: 385 (1926).

SB (Banggi).—Endemic.
Shrub of 2 m. near the seashore.

According to Merrill (*l.c.*) this species is related to the group of seven species endemic in the Philippines, especially *C. luzonicum* Merr. and *C. palawanense* Elm. He makes no reference to petals or disk-glands in his description; the number of glands given in the above key as '10–15' is that of the related *C. luzonicum.* I confess to feeling some scepticism as to the value of many of the alleged 'species' in this genus. There is need of a careful review of all the taxa, based upon abundant materials, in order to gain some idea of the populations.

Codiaeum variegatum (*L.*) *Bl.*, Bijdr.: 606 (1825); Muell. Arg.: 1119 (1866); Hook. f.: 399 (1887); Boerl.: 284 (1900); J. J. Sm.: 22 (1910); Pax & Hoffm. iii: 23 (1911); Merr.: 454 (1923); Gagnep.: 408 (1926); Backer & Bakh. f.: 493 (1963); Whitmore: 84 (1973), *in obs.*

Croton variegatus L., Sp. Pl. ed. 3: 1424 (1764).

Cultivated shrub, probably native in E. Malesia, N. Queensland and the Pacific.

"Curiously, no representative of *Codiaeum* is recorded from Borneo proper, although numerous cultivated forms of *C. variegatum* Blume certainly occur there."—Merr. in *Philipp. Journ. Sci.* 29: 386 (1926), *in obs.* I have so far seen no material of *C. variegatum* from Borneo, and include the species here solely on the strength of Merrill's statement.

Croton *L.*

1 Leaves densely silvery or coppery or white-lepidote beneath (cf. also *C. krabas*):
  2 Leaves distinctly pseudo-verticillate, separated by long bare portions of stem, rounded or narrowly cordate at base, less acute at apex; stipules smaller; inflorescence short, mostly under 7 cm. (rarely reaching 13 cm.); capsule small . . . . **C. cascarilloïdes**
  2 Leaves not distinctly pseudo-verticillate, not cordate at base, acutely acuminate at apex; stipules linear, elongate; inflorescence elongate (to ±15 cm.); capsule large. . . . . . . **C. argyratus**
1 Leaves densely or sparsely stellate-pubescent or hirsute beneath, or glabrous:
  3 Leaves very shortly and densely whitish-tomentellous or sublepidote beneath, finely, closely and irregularly dentate, subchartaceous **C. krabas**
  3 Leaves green beneath, hirsute, sparsely lepidote or glabrous:
    4 Leaves densely or sparsely stellate-hirsute beneath, at least on the nerves; stellate hairs with relatively few, longer, slender rays:
      5 Leaves mostly less than twice as long as broad; plant often scandent **C. caudatus**
      5 Leaves mostly more than twice as long as broad; plant not scandent:

6 Capsule strongly muricate, larger; leaves relatively broader, thinly chartaceous, thinly stellate-pubescent beneath (at least on the nerves) . . . . **C. singularis**

6 Capsule not muricate, smaller; leaves relatively narrower:

7 Leaves membranous or thinly chartaceous, thinly stellate-pubescent beneath, up to 30 × 9 cm. **C. erythrostachys**

7 Leaves firmly chartaceous, almost glabrous except on the nerves, up to 15 × 6 cm. . . . . . **C. borneënsis**

4 Leaves glabrous or almost glabrous beneath, or with small, scattered stellate hairs with relatively numerous short rays:

8 Rheophytic shrubs, with narrowly oblong or narrowly elliptic leaves up to 1·6 cm. broad:

9 Leaves broader, narrowly elliptic, quite entire; stellate hairs (except on ovary and capsule) almost lacking **C. rheophyticus**

9 Leaves narrower, narrowly oblong, shallowly dentate, with scattered brownish stellate hairs beneath . **C. ensifolius**

8 Plants not rheophytic; leaves ovate or elliptic or obovate, mostly more than 2 cm. broad:

10 Leaves strongly trinerved at base, membranous, elliptic to ovate; capsule up to 2 cm. long . . . . **C. tiglium**

10 Leaves not or weakly trinerved; capsule smaller:

11 Leaves obovate, strongly cuneate-based and conspicuously crenate at the apex, coriaceous, glabrous **C. heterocarpus**

11 Leaves less conspicuously cuneate-based, not or less conspicuously crenate at the apex:

12 Leaves rigidly coriaceous, with shining upper surface; plant of peat swamp or heath forest . **C. coriifolius**

12 Leaves less rigidly coriaceous to membranous, not shining above:

13 Leaves coriaceous, more or less brownish when dry, very variable in shape . . . . **C. oblongus**

13 Leaves chartaceous to membranous, green or ochraceous when dry:

14 Petioles less than 2 cm. long; leaves 2–4 cm. wide, with flat entire margins, green when dry **C.** aff. **hookeri**

14 Petioles very variable, up to 6 cm. long; leaves up to 9 cm. wide, with ± dentate reflexed margin, green to ochraceous when dry **C. griffithii**

Croton argyratus *Bl.*, Bijdr.: 602 (1825); Muell. Arg.: 526 (1866) (incl. var. *hypoleucus* Muell. Arg.); Hook. f.: 385 (1887); Merr.: 336 (1921) (incl. var. *hypoleucus*) & 425 (1923); Ridley: 260 (1924); Gagnep.: 277 (1925); Merr. in Philipp. Journ. Sci. 29: 381 (1926); Hend. in Journ. Malayan Br. Roy. As. Soc. 17: 70 (1939); Corner: 247 (1940); Backer & Bakh. f.: 476 (1963); Meijer, 7: 41 (1967); Airy Shaw in Kew Bull. 26: 243 (1972); Whitmore: 85 (1973).

C. budopensis Gagnep. in Bull. Soc. Bot. France 58: 549 (1922) & in Lecomte: 281 (1925).
C. maieuticus Gagnep.: 284 (1925).
C. tawaoënsis Croiz. in Journ. Arn. Arb. 23: 497 (1942), **synon. nov.**
? *C. avellaneus* Croiz. in Journ. Arn. Arb. 23: 498 (1942), *e descr.*

SR; B; SB; K1, 1a, 3.—SE. Asia, and throughout W. Malesia to Moluccas and Bali.

Tree to 24 m. high, in primary Mixed or Lowland Dipterocarp forest on yellow sandy clay over Tertiary sandstone, yellow sandy loam, heavy wet loam with coral limestone, brownish or black fertile soil, or basalt, from very low altitudes up to 1500 m. on Kinabalu.

Used in native medicine (unspecified, Sabah); timber used for firewood.

An exceedingly variable complex, perhaps ultimately to be divided up into smaller units. I am at present unable to grasp the essential features of the 'manifestly larger and coarser' female perianth of *C. avellaneus* Croiz., but no type material is available at Kew, and I cannot see how *Clemens* 20099 and *Elmer* 21201 (duplicates at Kew), stated by Croizat to be 'probably' or 'almost certainly' that species, differ from the great mass of material under *C. argyratus*. Type material of *C. tawaoënsis* is available, however, and the rough indumentum of the capsule appears to be exactly repeated in collections of presumably 'typical' *C. argyratus* from Java.

Croton borneënsis *J. J. Sm.* in Ic. Bogor. 4: 45, t. 314 (1910); Merr.: 336 (1921).

SR; K1.—Endemic.

Shrub or small tree to 3 m. high, in poor open Dipterocarp forest on poor sandy loam soil with standing water, up to 50 m. alt.

Very near *C. erythrostachys*, scarcely differing except in the stiffer leaves and in the weaker and less rufous indumentum.

Croton cascarilloïdes *Raeusch.*, Nomencl., ed. 3: 280 (1797); Merr. in Lingnan Sci. Journ. 13: 60 (1934) & in Trans. Amer. Philos. Soc. II, 24(2): 234 (1935); Croiz. in Journ. Arn. Arb. 23: 46 (1942); Airy Shaw in Kew Bull. 16: 344 (1963) & 26: 244 (1972); Meijer, 7: 41 (1967), *in obs.*; Whitmore: 84 (1973).

C. punctatus Lour., Fl. Cochinch.: 581 (1790); Muell. Arg.: 565 (1866); Gagnep.: 290 (1925); *non* Jacq. (1787).
C. cumingii Muell. Arg. in Linnaea 34: 101 (1865) & in DC.: 566 (1866); Merr.: 426 (1923); Ridley: 261 (1924); Gagnep.: 264 (1925); Hend. in Journ. Malayan Br. Roy. As Soc. 17: 30, 70 (1939).
C. pierrei Gagnep. in Bull. Soc. Bot. France 58: 558 (1922) & in Lecomte: 265 (1925).

SB.—SE. Asia, Ryu-Kyu Is., W. Malesia (exc. Java).

Shrub of 2 m., common in undergrowth of forest on ridge at 450 m. alt.

Readily distinguished from *C. argyratus* (*s.l.*) by its abbreviated inflorescences, small capsules, and pseudo-verticillate leaves separated by long bare portions of stem. Only two specimens seen from Sabah; evidently only locally common.

Croton caudatus *Geisel.*, Croton. Monogr.: 73 (1807); Muell. Arg.: 599 (1866); Hook. f.: 388 (1887); J. J. Sm.: 352 (1910); Merr.: 336 (1921) & 425 (1923); Gagnep.: 286 (1925); Merr. in Philipp. Journ. Sci. 29: 381 (1926); Backer & Bakh. f.: 477 (1963); Meijer, 7: 41 (1967); Airy Shaw in Kew Bull. 26: 245 (1972); Whitmore: 85 (1973).

C. caudatus var. *malaccanus* Hook. f.: 389 (1887); Ridley: 259 (1924).
C. caudatus var. *harmandii* Gagnep., *l.c.* (1925).

SR; SB.—E. Himalayas to Ceylon & S. China, SE. Asia, and throughout W. Malesia.
Tree, shrub or woody climber to 27 m. high, in primary Mixed Dipterocarp forest (sometimes swampy) on black sandy soil or shale-derived soil up to 930 m. alt.

Sparsely stellate-scabrid; leaves narrowly to broadly ovate, rather coarsely toothed, more or less membranous, brittle when dry, nerves often incised above; capsule globose, bluntly 3- or 6-angled; seeds with scattered stellate hairs.

Croton coriifolius *Airy Shaw* in Kew Bull. 29: 311 (1974).

SR.—Endemic.
Shrub or tree to 9 m. high, locally abundant in peat-swamp forest or heath woodland (*kerangas* or *kerapah* forest) up to 120 m. alt.

Possibly a sclerophyllous ecotype of *C. oblongus*, but very distinct in its rigidly coriaceous shining leaves.

Croton ensifolius *Merr.* in Philipp. Journ. Sci. 11, C. Bot.: 66 (1916) & Enum.: 337 (1921); Airy Shaw in Kew Bull. 27: 78 (1972); Whitmore: 85 (1973).

C. viminalis Becc., Nelle Foreste di Borneo: 524 (1902), *nomen*; Merr.: 337 (1921), *nomen*; Meijer, 7: 41 (1967), *anglice, in clavi* ('vinimalis'); *non* Griseb.

SR.—Malaya, W. New Guinea.
Rheophytic shrub to 1·5 m. high, commonest shrub on silty soils near sandstone boulders between flood levels near rapids at about 50 m. alt.

Close to *C. rheophyticus*, and to *C. lanceilimbus* Merr. (Philippines), differing in its narrower, shallowly serrulate leaves, with scattered brownish stellate hairs on the lower surface.

Croton erythrostachys *Hook. f.*: 391 (1887); Ridley: 260 (1924); Whitmore: 85 (1973).

C. calcicola Ridley in Bull. Misc. Inf. Kew 1923: 360 (1923) & Fl. Malay Penins. 3: 262 (1924), **synon. nov.**
C. adumbratus Croiz. in Journ. Arn. Arb. 23: 495 (1942), *pro parte*, saltem quoad *Griffith* 4777 in Herb. Kew.

SR.—Malaya.
Shrub of 1·2 m. on hillside; no further ecological data.

Differs from *C. borneënsis* in its thinner leaves and much stronger and usually rufous indumentum, and from *C. singularis* in its non-muricate capsule.

Croton griffithii *Hook. f.*: 392 (1887); Ridley: 261 (1924); Airy Shaw in Kew Bull. 26: 246 (1972).

C. laevifolius sec. Corner in Gard. Bull. Str. Settlem. 10: 294 (1939); Whitmore: 85 (1973), *p.p.*, *an* Bl.?

SR; SB; K1.—Lower Siam, Malaya, Sangië & Talaud Is.

Tree to 12 m. high, in primary submontane forest on brown sandy soil, or on limestone (Bau series), or on heavy loam with coral limestone, up to 1050 m. alt.

Differs from *C. oblongus* (*C. laevifolius*) in its thinner leaves, drying a green or ochraceous colour, usually glabrous, but sometimes thinly stellate-pubescent. See note in Kew Bull., *l.c. supra.*

Croton heterocarpus *Muell. Arg.*: 621 (1866); Hallier in Meded. Rijks Herb. 1910: 7 (1911); Merr.: 337 (1921) & 426 (1923) & in Philipp. Journ. Sci. 29: 382 (1926); Meijer, 7: 42 (1967); Whitmore: 85 (1973).

C. ardisioïdes Hook. f.: 393 (1887); cf. Hallier, *l.c.*
C. heteropetalum (sphalm.) Ridley: 262 (1924).

SR; SB; K1, ?1a, 3.—Sumatra, Malaya, SW. Philippines (Palawan).

Shrub or tree to 7·5 m. high, in primary forest near coast, in or near mangrove swamps, on tidal waterways or on tidal river banks, occasionally in primary forest on hillsides, up to 15 m. alt.

The obovate, glabrous, strongly cuneate-based leaves, usually conspicuously crenate in the upper half, are very characteristic for this species.

Said to be a host plant for the lac insect. Root used in native medicine (unspecified).

Croton aff. **hookeri** *Croiz.* in Journ. Arn. Arb. 21: 498 (1940); Airy Shaw in Kew Bull. 26: 246 (1971).

SB.—*C. hookeri* (*C. laevifolius* sec. Hook. f.: 391 (1887), *non* Bl.) in Assam, Siam & SW. China.

Shrub of unknown stature at 300 m. alt.; no ecological data.

A collection from the Tawau District, Mt. Andrassy, E. side, 25 March 1962, *Meijer* SAN 29435, closely approaches *C. hookeri* in its thin, dull, flat, bluntly acuminate, very shortly petioled leaves, remaining green when dry, but differs in its almost complete lack of stellate scales, the almost entire margin of the leaves, and the sessile, not slenderly stalked, glands at the apex of the petiole. A further collection, from Mt. Kinabalu (Bembangan River, alt. 1500 m., 23 Feb. 1964, *Chew & Corner* RSNB 4480), is very similar but has rather thicker, very shortly caudate leaves. The material scarcely warrants the description of a new species.

Croton krabas *Gagnep.* in Bull. Soc. Bot. France 58: 555 (1922) & in Lecomte: 286 (1925); Airy Shaw in Kew Bull. 26: 247 (1972).

SB.—Siam, Indochina.

Shrub to 4 m. high, on hill slope at 3 m. alt.

A very distinctive species, with broadly cordate-ovate, closely and irregularly crenate-dentate leaves, closely whitish-stellate-lepidote beneath (but not shining as in *C. argyratus*), with the nerves incised above. From a half-open capsule on the only specimen seen (Bod Gaya I., E. of Timbun Mata, Semporna, 1934, *Orolfo* N.B. For. Dept. 3792), it seems that the seeds are glabrous, whereas in the continental form they bear numerous stellate hairs. The species is related to *C. caudatus*.

Croton oblongus *Burm. f.*, Fl. Ind.: 205, t. 64, fig. 2 (1768); Merr. in Philipp. Journ. Sci. 19, Bot: 361 (1921) & Pl. Elmer.: 156 (1929); Backer & Bakh. f.: 476 (1963); Meijer, 7: 42 (1967).

C. laevifolius Bl., Bijdr.: 603 (1825); Muell. Arg.: 619 (1866); J. J. Sm.: 341 (1910); Merr.: 337 (1921); Corner in Gard. Bull. Str. Settlem. 10: 294 (1939); Whitmore: 85 (1973) [*non* sec. Hook. f.: 391 (1887); cf. Croiz. in Journ. Arn. Arb. 21: 498 (1940)].
C. korthalsii Muell. Arg.: 527 (1866).
C. confusus Gage in Rec. Bot. Surv. Ind. 9: 237 (1922); Ridley: 261 (1924).
? *C. oreoborneicus* Croiz. in Journ. Arn. Arb. 23: 496 (1942), *e descr.*

SR; B; SB; K1.—Throughout W. Malesia (exc. Philippines?).

Tree to 21 m., in primary Mixed Dipterocarp forest, *Coccoceras* forest, *Agathis* forest on waterlogged sandy acid soil, black or brownish soil, sandy clay loam, yellow sandy soil, shale-derived soil, igneous (andesitic) derived soil, basalt or dacite, once noted protruding from a limestone cliff, up to 1200 m.

Used for firewood because easy to split.

Neither the type of *C. oreoborneicus* Croiz., nor any of the other collections cited by Croizat as 'to all appearances' belonging to that species, are available in the Kew herbarium.

Croton rheophyticus *Airy Shaw* in Kew Bull. 27: 78 (1972).

SB.—Endemic.

Rheophytic shrub of 60 cm., by river on ultrabasic stone bed at 15 m. alt.

Differs from *C. ensifolius* by the broader, more elliptic, entire leaves, and by the paucity or absence of stellate hairs on all parts except the ovary and capsule.

Croton singularis *Airy Shaw* in Kew Bull. 21: 374 (1968).

SR (C.); **K1.**—Endemic.

Tree to 15 m. high, locally common on river bank or in primary Mixed Dipterocarp forest, on basalt ridge or igneous (andesitic) derived soils, at 250–840 m. alt.

Differs from *C. erythrostachys* and *C. borneënsis* in its strongly muricate capsule.

Croton tiglium *L.*, Sp. Pl.: 1004 (1753); Muell. Arg.: 600 (1866); Hook. f.: 393 (1887); Merr.: 337 (1921) & 427 (1923); Ridley: 262 (1924); Gagnep.:

285 (1925); Merr. in Philipp. Journ. Sci. 29: 382 (1926); Corner: 248 (1940); Backer & Bakh. f.: 477 (1963); Airy Shaw in Kew Bull. 26: 250 (1972); Whitmore: 84 (1973), *in obs.*

SR; SB; K1, 1a.—India, Ceylon, SE. Asia, China, and throughout Malesia.

Shrub or small tree to 3 m. high, ascending to 900 m. on Kinabalu; perhaps everywhere planted.

Seeds pounded and used as a fish poison ('tubai' fishing).

Dicoelia *Benth.*

Dicoelia beccariana *Benth.* in Hook. Ic. Pl. 13: 70, t. 1289 (1879); Pax in Engl. & Prantl, Pflanzenf. III. 5: 27 (1890); Boerl.: 209, 272 (1900); Merr.: 330 (1921); Pax & Hoffm. xv: 17 (1922) & in Engl. & Harms, Pflanzenf. ed. 2, 19c: 46 (1931); Croiz. in Journ. Arn. Arb. 23: 38 (1942); Airy Shaw in Kew Bull. 27: 3 (1972); Whitmore: 86 (1973).

D. affinis J. J. Sm. in Bull. Jard. Bot. Buitenz. III, 1: 392 (1920), **synon. nov.**

SR; K3.—Malaya, Sumatra, Banka.

Shrub or small tree to 9 m. high, in primary *kerangas* forest at 90–150 m. alt.

Leaves ample, elliptic-obovate or oblong-obovate, coriaceous, with a dull surface, almost glabrous, or occasionally persistently puberulous beneath, nerves widely spaced. Inflorescences axillary, elongate, puberulous, bearing fascicles of male flowers, in many of which a solitary female flower may be present. Flowers petaliferous, the male petals shortly cymbiform and hooded, bearing two deep excavations on the upper surface enclosing the anthers in bud, reflexed at anthesis; anther-thecae large, rounded and gaping widely at dehiscence; female petals simple, erect; ovary bearing three elongate simple divaricate styles with small capitate stigmas. Fruit a tricoccous capsule.

A very isolated genus, of quite obscure affinities. *D. affinis* J. J. Sm. seems to represent no more than a local variant.

Dimorphocalyx *Thw.*

1 Petiole exceeding 1 cm. in length, usually more than 1·5 cm.:
  2 Fruiting calyx not accrescent; capsule lamellate-tuberculate, glabrous; leaves manifestly toothed . . . . **D. muricatus**
  2 Fruiting calyx accrescent; capsule smooth . . . . . **D. luzoniensis**
    3 [Capsule glabrous; leaves entire or almost so; Philippines var. **luzoniensis**]
    3 Capsule adpressed-pubescent; leaves denticulate towards apex; robust plant with stout petioles . . . . var. **trichocarpus**
1 Petiole not exceeding 1·5 cm. in length, usually less than 1 cm.:
  4 Fruiting calyx not accrescent; capsule conic-muricate, glabrous; leaves toothed . . . . **D. denticulatus**

4 Fruiting calyx accrescent; capsule smooth:
  5 Leaves toothed; capsule adpressed-pubescent . . **D. murinus**
  5 Leaves entire or obscurely sinuate; capsule glabrous or rarely pubescent . . . . **D. malayanus**

Dimorphocalyx denticulatus *Merr.* in Philipp. Journ. Sci. 4, Bot.: 278 (1909); Pax & Hoffm. iv: 285 (1912); Merr.: 455 (1923); Airy Shaw in Kew Bull. 26: 252 (1972), *in obs.*

Ostodes pauciflorus (sphalm. '*faucifl.*') Merr. in Philipp. Journ. Sci. 11, C. Bot.: 72 (1916), & Enum.: 345 (1921).
Dimorphocalyx pauciflorus (Merr.) Airy Shaw in Kew Bull. 20: 413 (1966).

SR.—S. Philippines (Mindanao).
Shrub or small tree to 6 m. high, in primary Lowland Dipterocarp forest up to 150 m. alt.

Distinguished from *D. muricatus*, which also has a non-accrescent calyx and muricate capsules, by its very short-petioled (2–7 mm.) and much smaller (up to 18 × 6·5 cm.) leaves, which are sometimes only 2·5 cm. broad, by its very few-flowered inflorescences, and by the finer, more conical murication of the capsule.

Dimorphocalyx luzoniensis *Merr.* in Philipp. Journ. Sci. 5, Bot.: 192 (1910); Pax & Hoffm. iv: 284 (1912); Merr.: 455 (1923); Whitmore: 87 (1973). [*Non* sec. Airy Shaw in Kew Bull. 26: 251 (1972); vide *D. malayanus* Hook. f.]

Var. **trichocarpus** *Airy Shaw*, var. nov., habitu robustiore, foliis apicem versus denticulatis, et praesertim capsula dense adpresse pubescente distinctus.

SR.—First Division: Bidi, Bau, on limestone rocks, with intervening igneous derived soil, at base of limestone hill, alt. 90 m., 12 April 1965, *Anderson* S. 20974 (K, type):—Tree 37 cm. girth; bark smooth, grey.

Further collections may show this plant to be deserving of specific rank. The limestone habitat should be noted. Typical *D. luzoniensis*, from the northern Philippines, is a less robust plant with entire or obscurely sinuate leaves and a glabrous capsule. It differs from *D. murinus* and *D. malayanus* in its longer petioles, and from *D. muricatus* in its smooth capsule and accrescent calyx. No data regarding the substratum in the Philippines appear to be available.

Dimorphocalyx malayanus *Hook. f.*: 404 (1887); Pax & Hoffm. iii: 33 (1911); Merr.: 346 (1921); Ridley: 266 (1924); Stern in Am. Journ. Bot. 54: 668 (1967); Whitmore: 86 (1973).

D. kunstleri King ex Hook. f.: 405 (1887); Pax & Hoffm. iii: 32 (1911); Ridley: 266 (1924); **synon. nov.**
D. beccarii Gagnep. in Bull. Soc. Bot. France 71: 621 (1924), **synon. nov.**
D. luzoniensis sec. Airy Shaw in Kew Bull. 26: 251 (1971), *non* Merr.

SR.—Malaya.
No ecological information available for Borneo.

The combination of small, almost entire, short-petioled leaves, accrescent fruiting calyx, and smooth, usually glabrous capsule, readily distinguishes *D. malayanus*.

Dimorphocalyx beccarii was based upon the same two collections (*Beccari* 2214 & 2215) as those seen by Hooker in Herb. Kew. and annotated by him as *D. malayanus*, though not specifically cited in the *Flora of British India*. The species appears never to have been re-collected in Borneo.

Dimorphocalyx muricatus (*Hook. f.*) *Airy Shaw* in Kew Bull. 20: 412 (1966) & 26: 252 (1972); Whitmore: 87 (1973).

Ostodes muricatus Hook. f.: 401 (1887); Pax & Hoffm. iii: 21 (1911); Ridley: 269 (1924).
Dimorphyocalyx [!] sp.?, Merr., Pl. Elmer.: 163 (1929).

SR; B; SB.—Malaya, Sumatra.

Tree to 14 m. high, in primary Lowland or Mixed Dipterocarp forest in low undulating country on sandstone or yellow sandy clay up to 540 m. alt.

Distinguished from *D. denticulatus*, the only other Bornean species with muricate fruits and a non-accrescent calyx, by the larger and especially broader, long-petioled leaves, and by the broader more or less lamellate murication of the capsule.

The variety *minor* (Hook. f.) Airy Shaw (Kew Bull. 20: 412 (1966); Whitmore: 87 (1973)) occurs in Sarawak, but is probably not worth distinguishing.

Dimorphocalyx murinus *Elm.*, Leafl. Philipp. Bot. 4: 1285 (1911); Pax & Hoffm. vii: 404 (1914); Merr.: 455 (1923); Whitmore: 87 (1973).

SR; SB.—S. Malaya; Philippines (Palawan).

Tree to 15 m. high, in primary or disturbed Mixed Dipterocarp forest, on sandy clay, blackish sandy soil, loam or basalt, up to 400 m. alt.

Distinguished by the very shortly petioled, glandular-toothed leaves, accrescent fruiting calyx and smooth pubescent capsule.

Drypetes *Vahl*

1 Stipules pectinate, up to 1·5 cm. long; fruit densely long-villous, 2·5–3 cm. diam., very shortly (1–2 mm.) pedicelled . . . . **D. eriocarpa**
1 Stipules entire; fruit rarely long-villous (cf. *D. impressinervis*):
  2 Fruit very shortly (1–4 mm.) and stoutly pedicelled:
    3 Leaves chartaceous, dull; fruit 1 cm. diam., villous **D. impressinervis**
    3 Leaves coriaceous, shining; fruit 2–3 cm. diam., glabrous or almost so:
      4 Petioles stout, 1–7 mm. long; leaves mostly larger **D. crassipes**
      4 Petioles more slender, 5–15 mm. long; leaves mostly smaller **D. sibuyanensis**
  2 Fruit borne on longer and slenderer pedicels:
    5 Leaves very narrow, 10–18 mm. wide . . . . **D. rheophila**

5 Leaves more than 18 mm. wide:
6 Lateral nerves unusually numerous (10–15 pairs) and closely arranged in proportion to the size of the leaf (8–17 cm. long); leaves usually shallowly indentate-crenate (cf. also *D. stylosa*) **D. polyneura**
6 Lateral nerves less numerous and less closely arranged, relative to the size of the leaf:
7 Petiole very short (1–4 mm.) for the size of the leaf (up to 25 cm. long); leaf greenish or ochraceous when dry; fruit small, rounded-subcubic or transverse, 12 mm. diam. **D. curtisii**
7 Petiole longer in proportion to size of leaf:
8 Leaves distinctly thin in texture, chartaceous rather than coriaceous:
9 Leaves shallowly crenate-serrate; fruit broadly fusiform; nerves not bullately impressed . **D. roxburghii** *var.*
9 Leaves entire; fruit not fusiform; nerves often subbullately impressed:
10 Branchlets, midribs beneath and pedicels bearing a very short, spreading, golden or ochraceous indumentum; ♂ flowers golden-sericeous . . . **D. ochrothrix**
10 Indumentum, if present, not golden or ochraceous:
11 Fruit somewhat flattened, plum-like; leaves with a dull surface; nerves 6–8 pairs . **D. prunifera**
11 Fruit globose; leaves smooth and rather shining:
12 Leaf-base cuneate; nerves 4–5 pairs; fruit 1·5–2 cm. diam. . . . . . . **D. castilloii**
12 Leaf-base rounded or unilaterally cordate; nerves 5–7 pairs; fruit 2–3 cm. diam. **D. polyalthioïdes**
8 Leaves coriaceous; nerves not bullately impressed:
13 Pedicel and tepals of ♂ flowers finely and densely golden-puberulous; leaves distinctly crenate-dentate; ovary glabrous . . . . **D. stylosa**
13 Pedicel and tepals of ♂ flowers not golden-puberulous:
14 Leaves distinctly and closely reticulate-veined:
15 Leaves 12–20 cm. long; fruit ovoid, plum-like, 1–1·5 cm. long . . . **D. maquilingensis**
15 Leaves less than 10 cm. long; fruit globose, pedicel less than 1 cm. long . . **D. hainanensis**
14 Leaves not closely reticulate-veined:
16 Leaves (at least some, and at least towards apex) dentate or crenate:
17 Fruit ovoid, 2–2·5 cm. long, somewhat flattened, narrowed to a subacute apex, ochraceous when dry; pedicel 1–1·5 cm. long **D. neglecta**
17 Fruit shortly oblong or sometimes ovoid, 10–13 mm. long, mostly obtuse at apex, softly grey-brown when dry; pedicel 5–9 mm. long . . . . **D. littoralis**

16 Leaves entire:
18 Leaves 2–12(–19) cm. long:
19 Petiole 7–17 mm. long; leaves very smooth and shining:
20 Fruits subglobose, densely ochraceous-tomentellous . . **D. aëtoxyloïdes**
20 Fruits oblong-ellipsoid, olive-like, glabrous . . . . **D. kikir**
19 Petiole 2–10 mm. long:
21 Nerves strong and prominent:
22 Fruit ovoid, 2–2·5 cm. long, somewhat flattened, narrowed to a subacute apex, ochraceous when dry; pedicel 1–1·5 cm. long **D. neglecta**
22 Fruit shortly oblong or sometimes ovoid, 10–13 mm. long, mostly obtuse at apex, softly grey-brown when dry; pedicel 5–9 mm. long **D. littoralis**
21 Nerves weaker and less prominent:
23 Leaves mostly gradually narrowed to the apex; fruit transversely bilobed, about 1 cm. diam., adpressed-pubescent . **D. rhakodiskos**
23 Leaves abruptly and conspicuously caudate; fruit globose, up to 2 cm. diam., shortly ochraceous-farinose **D. microphylla**
18 Leaves 15–60 cm. long:
24 Branches and leaves smooth but dull, suffused with a slight glaucous-grey bloom; petioles up to 2·5 cm. long; leaves with symmetrical base and narrow dark brown margin; fruits asymmetrically fusiform, 4–5 cm. long; stipules subulate, convolute, up to 2·5 cm. long **D. fusiformis**
24 Not as above; leaves mostly with asymmetrical base and without discolorous margin; fruits globose; stipules small, not convolute . . . **D. longifolia** *et aff.* **(D. caesia, D. macrostigma, D. megacarpa, D. pendula)**

Drypetes aëtoxyloïdes *Airy Shaw* in Kew Bull. 23: 58 (1969).

SB.—Endemic.

Tree of 10–12 m., in primary or secondary hillside forest on brown or yellowish soil at 100–1350 m. alt.

Leaves elliptic, 8–11 cm. long, 3·5–6 cm. broad, coriaceous, smooth and shining on both surfaces, glabrous; petiole slender, 5–17 mm. long, slightly glaucous; fruits subglobose, densely ochraceous-tomentellous, pedicel 5–16 mm. long. The leaves are quite reminiscent of *Aëtoxylon sympetalum* (v. Steen. & Domke) Airy Shaw, in the *Thymelaeaceae*.

The type collection originated from a rather high altitude in the Kinabalu National Park. Two collections (*Aban Gibot* SAN 65924 & 65936) from the Tenom District of Sabah, at only 60 m. alt., approach it quite closely, but differ in their thinner leaves, somewhat recalling those of *D. castilloii*, with the midrib raised rather than impressed above (but compare remarks on *D. longifolia*, below). The fruiting pedicels are sometimes almost 2 cm. long. Further material will be needed in order to show whether the two forms are distinct or connected by intermediates.

Drypetes caesia *Airy Shaw* in Kew Bull. 20: 387 (1966); Meijer, 7: 44 (1967).

SB; K1 ?—Endemic.

Tree to 12 m. high, in primary forest on *gading* soil at unknown alt.

Very close to *D. longifolia*, but differing in the matt greyish-subglaucous undersurface of the leaves, somewhat recalling the leaves of *Ryparosa* species. The greyish appearance is seen under a magnification of ×20 or more to be due to minute criss-crossed fibrils of wax (?) lying on the surface. The half-dozen or so specimens so far known from Sabah all originated in the Tawau District.

Drypetes castilloii (*Merr.*) *Merr.*, Pl. Elmer.: 140 (1929); Meijer, 7: 43 (1967).

Cyclostemon castilloii Merr. in Philipp. Journ. Sci. 21: 522 (1922).

SB.—Endemic.

Tree to 11 m. high, in primary forest on brown or black sandstone soil up to 40 m. alt.

Leaves mostly broadly elliptic, thinly chartaceous, glabrous, usually rather smooth and shining above, nerves few (4–5 pairs), somewhat bullately impressed above; fruit globose, 1·5–2 cm. diam., almost glabrous, pedicel 5–7 mm. long.

Fruit tasteless and inedible.

Certain specimens seem to some extent intermediate between this and *D. aëtoxyloïdes*; see under that species.

Drypetes crassipes *Pax & Hoffm.* xv: 241 (1922); Meijer, 7: 42 (1967).

SR; SB; K1.—Malaya.

Tree to 24 m. high, in primary Lowland Dipterocarp forest on sandstone or clay loam up to 1000 m. alt.

A robust species, almost glabrous; leaves medium to large, entire, coriaceous, strongly nerved, mostly dark brown when dry, petioles remarkably short, 1–7 mm. long; fruit subglobose, 3 cm. diam., truncate or retuse at base and apex, minutely granular, minutely adpressed-puberulous and finely brown-pulverulent, very shortly (1–2 mm.) and stoutly pedicelled.

Drypetes curtisii (*Hook. f.*) *Pax & Hoffm.* xv: 250 (1922); Ridley: 223 (1924); Meijer, 7: 42 (1967); Airy Shaw in Kew Bull. 26: 254 (1972); Whitmore: 88 (1973).

Cyclostemon curtisii Hook. f.: 343 (1887).

SB; K1.—Malaya.

Shrub or tree to 6 m. high, in primary or secondary forest (sometimes periodically inundated), on brown sandstone soil or loam with sandstone, or on alluvium, up to 150 m. alt.

The very shortly petioled, elliptic leaves, which are usually greenish (sometimes ochraceous) when dry, and the small rounded-subcubic fruits, somewhat resembling those of *D. subcubica* (J. J. Sm.) Pax & Hoffm. (Java and Lesser Sunda Is.), are the distinctive features of this everywhere scarce species.

Drypetes eriocarpa *Airy Shaw* in Kew Bull. 21: 360 (1968) & 29: 297 (1974).

SR.—Endemic.

Tree to 18 m. high, in primary forest on clayey soil up to 150 m. alt.

Differs from all other Bornean species in its elongate, subulate, caducous, filiform-pectinate stipules, 1·5 cm. long, and in its striking subsessile, subglobose, densely and softly long-villous, almost shaggy-lanuginose fruits, 2·5–3 cm. diam., described by one collector as 'rusty brown covered with silvery hairs'. Branchlets and nerves fulvous-tomentellous. Closely related to *D. nervosa* (Hook. f.) Pax & Hoffm., of Malaya.

Drypetes fusiformis *Airy Shaw* in Kew Bull. 19: 306 (1965) & 23: 62 (1969).

SB; K1.—Endemic.

Tree to 35 m. high, in primary forest on black or clay soil, or loam with coral limestone, up to 100 m. alt.

Leaves oblong or elliptic-oblong, 25–35 cm. long, 8–10 cm. wide, glabrous, slightly glaucous beneath; petiole robust, 1–2 cm. long; stipules subulate, convolute when young, up to 2·5 cm. long, very acute, caducous, leaving prominent scars. Male flowers in dense axillary fascicles, very shortly pedicelled. Fruits fusiform, 4–5 cm. long, somewhat asymmetrical, almost beaked apically, densely minutely puberulous or tomentellous.

A very distinct species, not closely related to any other in Borneo, but near *D. convoluta* Airy Shaw of the Philippines, which has even longer stipules but much smaller fruits. The leaves of *D. fusiformis* are often narrowly purplish-brown-edged beneath, recalling the similar condition in *Gonystylus keithii* Airy Shaw (*Thymelaeaceae*).

Drypetes hainanensis *Merr.* in Journ. Arn. Arb. 6: 134 (1925) & in Lingnan Sci. Journ. 5: 107 (1927); Airy Shaw in Kew Bull. 26: 254 (1972); *var.*

SB.—Siam; (Indochina?); Hainan.
Small tree 6 m. high, in forest at 6 m. alt.

A single specimen from Sabah (Mostyn Distr., Timbun Mata, *Lantoh* SAN 68053) must, I think, be referred to this small-leaved species, which is already quite variable in Siam. The Bornean gathering bears female flowers only. The leaves are elliptic or ovate, up to 9·5 cm. long and 4 cm. wide, gradually though shortly acuminate, mostly obtuse at the apex, with prominulous, closely reticulate venation, dark greenish-brown when dry.

Drypetes impressinervis *Airy Shaw* in Kew Bull. 29: 298 (1974).

SR (SW.).—Endemic.
Small tree 3 m. high, by riverside, probably at low altitude.

Leaves elliptic, chartaceous, 14–19 cm. long, 5–6·5 cm. wide, abruptly and rather shortly (1–2 cm.) caudate, entire, almost glabrous, the nerves very slender, sharply prominulous below and somewhat bullately impressed above, anastomosing conspicuously 4–7 mm. from the margin. Flowers not seen. Fruit (immature) depressed subglobose, bilocular, laxly ochraceous-pilose, with a short conical style and small stigmas; pedicel only 1–2 mm. long.

The leaves are remarkably similar to those of *D. prunifera*, in § *Drypetes*. The fruit resembles that of *D. ochrothrix*, in the same group, but has a longer pilose indumentum, and its bilocular condition would place the species technically in § *Sphragidia*. The sectional limits in the genus are obviously in need of review.

Drypetes kikir *Airy Shaw* in Kew Bull. 23: 60 (1969); cf. Meijer, 7: 43 (1967); Whitmore: 88, 89 (1973).

SB; Ki.—Malaya.
Tree to 25(–31) m. high, locally very common in primary or secondary forest on sandstone or sandy loam or black sandy soil, or on a tufa plateau, up to 100(–900) m. alt.

Vegetatively extremely similar to *D. laevis* (Miq.) Pax & Hoffm., of Sumatra (leaves of *D. kikir* slightly more shining, with rather less evident and less spreading nerves), but with a glabrous unilocular ovary and entirely different, oblong-ellipsoid, olive-like fruits, about 1·5 × 1 cm. in size. *D. laevis* has a bilocular, tomentellous ovary and large globose fruits, 3–4 cm. in diameter. Cf. remarks under *D. impressinervis*, above. (Whitmore: 89 (1973) records *D. laevis* from Borneo, but I think this is an error.)

The epithet *kikir* is taken from the native Malayan name, 'kayu kikir', under which the tree is commonly known in E. Indonesian Borneo. The word 'kikir' means a file or rasp, and refers to the very characteristic bark, which is variously described in field notes as 'densely pustular', 'roughish, brown, scaling off in numerous quadrangular, 2 mm. wide, 1 mm. thick, pieces', 'rather rough, cracked in minute pieces, giving it the appearance of a file', 'dark brown, rough, like a rasp, consisting of about 3 mm. wide, 2 mm.

thick, rectangular pieces', 'dark brown, hard, broken in regular pieces of 4 × 4 mm., 2 mm. thick'. A scanty orange sap was once noted.

Drypetes littoralis (*C. B. Rob.*) *Merr.* in Philipp. Journ. Sci. 29: 380 (1926).

Cyclostemon littoralis C. B. Rob. in Philipp. Journ. Sci. 3, Bot.: 198 (1908); Merr.: 407 (1923); Holth. & Lam in Blumea 5: 201 (1942).
C. iwahigensis Elm., Leafl. Philipp. Bot. 4: 1278 (1911), *p.p.*; Merr.: 406 (1923).
C. mindorensis Merr. in Philipp. Journ. Sci. 9: 479 (1914) & Enum.: 408 (1923), **synon. nov.**
Drypetes mindorensis (Merr.) Pax & Hoffm. vii: 274 (1922).

SB.—Philippines, Java, Celebes (?).
Tree to 9 m. high, on sea-beaches in sandy loam up to 3 m. alt.

A characteristic coastal species, though apparently also occurring in inland hill forests at low altitudes. It is very probably, as Pax & Hoffmann thought (*l.c.*: 238), only a form of the imperfectly known *D. cumingii* (Baill. ex Muell. Arg.) Pax & Hoffm., but it was retained as distinct by Merrill, on the ground of the more entire leaves. Distinguished from the superficially similar *D. neglecta* by the characteristic small, shortly oblong or ovoid, minutely adpressed-puberulous fruits, 10–13 mm. long, which assume a soft dull grey-brown colour when dry; pedicels 5–9 mm. long.—*D. mindorensis*, on the basis of several collections, seems quite indistinguishable from *D. littoralis*.

Drypetes longifolia (*Bl.*) *Pax & Hoffm.* xv: 245 (1922); Ridley: 222 (1924); S. Moore in Journ. Bot. Brit. & For. 63, Suppl.: 97 (1925); Corner in Gard. Bull. Str. Settlem. 10: 296 (1939), *in obs.*; Backer & Bakh. f.: 473 (1963); Airy Shaw in Kew Bull. 23: 55 (1969) & 26: 256 (1972); Whitmore: 88, 89 (1973).

Cyclostemon longifolius Bl., Bijdr.: 598 (1825); Hook. f.: 341 (1887), *pro parte*; J. J. Sm.: 206 (1910).
C. macrophyllus Bl., *l.c.* (1825); J. J. Sm.: 204 (1910); Merr.: 329 (1921).
C. macrophyllus var. *malaccensis* Hook. f.: 341 (1887).
? *C. megacarpus* Merr. in Philipp. Journ. Sci. 7: 387 (1912) & Enum.: 407 (1923), *cum synon.*
Drypetes macrophylla (Bl.) Pax & Hoffm., *l.c.*: 247 (1922); Merr., Pl. Elmer.: 140 (1929); Backer & Bakh. f.: *l.c.* (1963); Meijer, 7: 42 (1967).
? *D. megacarpa* (Merr.) Pax & Hoffm., *l.c.*: 248 (1922).
D. myrmecophila Merr., Pl. Elmer.: 141 (1929), **synon. nov.**

SR; SB; K1.—?Burma, ?Andamans, Lower Siam, and throughout Malesia.
Tree to 23 m. high, in primary Mixed Dipterocarp forest on black or brown sandy soil, yellow sandy clay, sand and limestone, limestone, alluvial soil at foot of limestone hill, or Tertiary granodiorite, up to 800 m. alt., ascending to 4000 m. on Kinabalu.

Leaves elongate, oblong, glabrous (rarely shortly pubescent—Sabah, Lahad Datu distr., *Muin Chai* SAN 25600); flowers and fruits cauliflorous; fruits globose, up to 2·5 cm. diam. Stems often myrmecophilous. Flowers stated to have a 'nauseating fragrance'—possibly fly-pollinated.

The status of *D. megacarpa* is puzzling. The only clear distinction from *D. longifolia* seems to be the fact that the midrib is raised on the upper surface instead of impressed (cf. Kew Bull. 23: 56 (1969)). In many genera this is a valid and reliable character for specific distinction, but here there seem to be no other associated characters, and forms with raised or channelled midribs seem to be otherwise identical. Provisionally I am treating them as conspecific. Some other large-leaved Philippines species (*D. bordenii* (Merr.) Pax & Hoffm., *D. grandifolia* (C. B. Rob.) Pax & Hoffm., etc.) should probably be reduced to *D. longifolia*.

Drypetes macrostigma *J. J. Sm.* in Bull. Jard. Bot. Buitenz. III, 6: 85 (1924).

K(**1**?).—Endemic.

Small tree; no further information, unless *Kostermans* 6645 & 10405, from E. Borneo, can be identified as this species: the field notes to these indicate a tree of 20–25 m., growing in sandy loam at 30–60 m. alt.

An insufficiently known species, apparently indistinguishable from *D. longifolia* except by the broad, convex, reflexed stigmas, 4 mm. long, 7 mm. wide, and the larger fruit, 3 cm. diam. According to an isotype at Kew the leaves are somewhat more shining beneath, with rather more reticulate nervation, than those of *D. longifolia.* A specimen from Sarawak (Mt. Penrissen, *Jacobs* 5072) has leaves tending to show these features, and also large reflexed stigmas, but the tepals are almost completely glabrous, instead of densely subsericeous as in *D. macrostigma.* Two Sabah specimens (*Simbut* A 1617, *Bakar* SAN 24995) show the leaf-characters, as do the two *Kostermans* gatherings mentioned above, but *K.* 10405 has glabrous tepals.

Drypetes maquilingensis (*Merr.*) *Pax & Hoffm.* xv: 240 (1922).

Cyclostemon maquilingensis Merr. in Philipp. Journ. Sci. 9, Bot.: 477 (1914) & Enum.: 407 (1923).

Drypetes sp., 'probably undescribed', Merr. in Philipp. Journ. Sci. 29: 380 (1926).

SB (Banggi I.).—Philippines.

In forests; no further ecological information available.

Completely glabrous (except flowers); leaves finely and conspicuously reticulate-nerved, 12–20 cm. long, somewhat oblique at the base, gradually narrowed to the obtuse apex; male and female flowers rather long-pedicelled (5–15) mm.); ovary ochraceous-pulverulent, style conspicuous, very broadly flabellate-infundibuliform, 4 mm. wide, margins reflexed, coarsely fimbriate-lobulate; fruit ovoid, plum-like, 2–3 cm. long, 1–2 cm. diam., pedicel slender, 1–1·5 cm. long.

So far known in Borneo only from a solitary fruiting collection (*Castro & Melegrito* 1563) from Banguey (Banggi) Island, agreeing perfectly with Philippine material, though Merrill in 1926 (*l.c. supra*) failed to recognize it.

Drypetes megacarpa (*Merr.*) *Pax & Hoffm.* xv: 248 (1922); Whitmore: 89 (1973), *in obs.*

Cyclostemon megacarpus Merr. in Philipp. Journ. Sci. 9, Bot.: 479 (1914) & Enum.: 407 (1923).

See note under *D. longifolia*, above.

Drypetes microphylla (*Merr.*) *Pax & Hoffm.* xv: 237 (1922); Meijer, 7: 43, tab. (1967); Whitmore: 88 (1973).

Cyclostemon microphyllus Merr. in Gov. Lab. Publ. (Philipp.) 17: 27 (1904) & in Philipp. Journ. Sci. 1, Suppl.: 76 (1906) & Enum.: 407 (1923).

SR; SB; K1.—Malaya, Philippines.

Tree up to 18 m. high, in primary Mixed Dipterocarp forest, sometimes marshy, on basalt or sandy soil or coral limestone up to 720 m. alt.

Leaves elliptic or oblong-elliptic, 5–11 cm. long, 2–4 cm. wide, rather abruptly and conspicuously caudate (cauda usually obtuse), glabrous, smooth; nerves prominent below, inconspicuous above; male flowers very short-pedicelled, tepals 1·5–2 mm. diam., densely papillose-puberulous within and partly so without; stamens 10–15, very short, around a central hairy disk; fruit globose, up to 2 cm. diam., shortly and evanescently ochraceous-farinose.

Drypetes neglecta (*Koord.*) *Pax & Hoffm.* xv: 242 (1922); Backer & Bakh. f.: 473 (1963).

Cyclostemon neglectus Koord. in Koord.-Schum., Syst. Verz. I Abt. § 1: 21 (1912).

SB; K1.—Sangië & Talaud Is., Java, Lesser Sunda Is.

Tree to 27 m. high, in primary or disturbed forest on black soil or clay or sandstone or limestone, up to 300 m. alt.

Very similar to *D. littoralis*, but leaves mostly larger (up to 20 cm. long) and more shining, and fruit much larger, ovoid, 2–2·5 cm. long, 1–1·3 cm. wide, somewhat flattened, narrowed to a subacute apex, finely adpressed-puberulous and ochraceous-furfuraceous, pedicel 1–1·5 cm. long.

Drypetes ochrothrix *Airy Shaw* in Kew Bull. 21: 361 (1968) & 25: 501 (1971) & 26: 256 (1971); Meijer, 10: 230 (1968).

SB.—Lower Siam.

Tree to 12 m., in primary forest on brown sandstone soil at 90 m. alt.

Readily distinguished from its closest ally, *D. impressinervis*, by the very short, spreading, ochraceous or golden indumentum of the branchlets, midribs beneath, and pedicels, and by the golden-sericeous male flowers. From the more distantly related *D. prunifera* it differs further in the smaller leaves and much smaller fruit.

Drypetes pendula *Ridley* in Bull. Misc. Inf. Kew 1923: 365 (Dec. 1923) & Fl. Malay Penins. 3: 222 (1924); Corner: 248 (1940); Airy Shaw in Kew Bull. 26: 256 (1971); Whitmore: 87, 89, fig. 5 (1973).

Cyclostemon longifolius sec. Hook. f.: 341 (1887), *p.p.*, *non* Bl.
Drypetes gigantifolia J. J. Sm. in Bull. Jard. Bot. Buitenz. III, 6: 89 (July 1924), **synon. nov.**

SB; K1.—S. Penins. Siam, Malaya.
Tree of 12 m., in forest at 1200 m. alt. (on Kinabalu).

Near *D. longifolia*, but branches very robust and leaves up to 60 cm. long, very coriaceous, with an unequally cordate base. Fruit globose, 2·5 cm. diam., trilocular.

Drypetes polyalthioïdes *Airy Shaw* in Kew Bull. 29: 299 (1974).

K1.—Endemic.
Tree to 14 m. high, in forest on sandstone or loam with lime, up to 250 m. alt.

Related to *D. castilloii*, but distinguished by its shorter-petioled (3–8 mm.), much rounder-based, sometimes unilaterally cordate leaves, and larger, finely reticulate, almost glabrous fruits, on a 12–20 mm. long pedicel.

Drypetes polyneura *Airy Shaw* in Kew Bull. 20: 388 (1966); Meijer, 7: 43 (1967); Airy Shaw in Kew Bull. 27: 75 (1972) & in Hook. Ic. Pl. 38: t. 3710 (1974).

SB; K1.—Banka.
Tree to 35 m. high, in primary hill forest (sometimes with bamboos, sometimes periodically inundated), on sandstone or brown soil up to 360 m. alt.

The numerous nerves (10–15 pairs) in proportion to the moderate size of the leaves (8–17 cm. long), together with the usually conspicuously though shallowly indentate-crenate (occasionally subentire) leaf-margin, distinguish this species from all others in Borneo. All nerves (major and minor) are distinctly raised on both surfaces. The stipules are narrowly linear, up to 1 cm. long, but very quickly caducous. Except for the flowers and fruit the plant is quite glabrous.

Drypetes prunifera *Airy Shaw* in Kew Bull. 19: 304 (1965); Meijer, 7: 44 (1967) (sphalm. *D. prunifera* 'Merr.').

SB.—Endemic.
Tree to 12 m. high, in primary forest on black or brown soil up to 120 m. alt.

Distinguished from the other entire-leaved species of subsect. *Anaua* (*D. impressinervis*, *D. ochrothrix*) by its much larger, glabrous, somewhat flattened, plum-like fruit, up to 2·5 × 2 × 1 cm.

Drypetes rhakodiskos (*Hassk.* 1855) *Airy Shaw* in Kew Bull. 19: 300 (1965) & 27: 76 (1972); Whitmore: 88 (1973).

Cyclostemon mucronatus Bl., Bijdr.: 598 (1825); *non Drypetes mucronata* Griseb. (1865).
Hemicyclia? rhakodiskos Hassk. in Versl. & Meded. Kon. Akad. Wetensch., Afd. Natuurk., 4: 140 (1855) & in Bot. Zeit. 14: 802 (1856).
H.? rhacodiscus Hassk., Hort. Bog. Descr. s. Retziae ed. nov. 1: 43 (1858).
Pierardia rhacodiscus (Hassk.) in Bull. Soc. Bot. France 6: 713 (1859).
Cyclostemon rhacodiscus (Hassk.) Muell. Arg.: 484 (1866).

C. palawanensis Merr. in Philipp. Journ. Sci., C. Bot., 9: 480 (1914) & Enum.: 408 (1923).
Drypetes mucronata (Bl.) Pax & Hoffm. xv: 238 (1922); Meeuse & Adelb. in Backer, Bekn. Fl. Java IVC: 65–68 (1943); *non* Griseb. (1865).
D. palawanensis (Merr.) Pax & Hoffm. xv: 249 (1922).
D. palawanensis var. *parvifolia* Merr., Pl. Elmer.: 141 (1929).
D. rhacodiscus (Hassk. 1858) Bakh. f. in Blumea 12: 62 (1963); Backer & Bakh. f.: 472 (1963).
D. "rhakodiskus" Meijer, 7: 43 (1967).

SB; KI.—Scattered throughout W. Malesia.
Shrub or small tree to 7·5 m. high, in primary or secondary forest on loam with limestone, ascending to 1260 m. on Kinabalu.

Leaves ovate, elliptic or oblong-lanceolate, 3–12 cm. long. 2–5 cm. wide, mostly gradually (occasionally more abruptly) shortly and bluntly acuminate, glabrous or sparsely pubescent beneath; fruit slenderly (up to 12 mm.) pedicelled, transversely bilobed, about 10 mm. wide, adpressed-pubescent.

Drypetes rheophila *Airy Shaw* in Kew Bull. 19: 302 (1965).

SR.—Endemic.
Small shrub, locally common on river bank above tide level.

Perhaps a very distinct ecotype of *D. rhakodiskos*, from which it only differs in its very narrow oblong-lanceolate leaves, not exceeding 18 mm. in width, coupled with the riverside habitat.

Drypetes roxburghii (*Wall.*) *Hurusawa* in Journ. Fac. Sci. Univ. Tokyo, Sect. III Bot., 6: 337 (1954); Airy Shaw in Kew Bull. 26: 257 (1972).

Putranjiva roxburghii Wall., Tent. Fl. Nap.: 61 (1826); Muell. Arg.: 443 (1866); Hook. f.: 336 (1887); J. J. Sm.: 223 (1910); Beille: 658 (1927); v. Steenis in Blumea 8: 515 (1957); Backer & Bakh. f.: 465 (1963).
Nageia putranjiva Roxb., Fl. Ind. 3: 766 (1832).
Putranjiva sphaerocarpa Muell. Arg.: 443 (1866).
P. amblyocarpa Muell. Arg.: 444 (1866).

Var. **timorensis** (*Bl.*) *Airy Shaw*, comb. nov.

Pycnosandra timorensis Bl., Mus. Bot. Lugd.-Bat. 2: 192 (1856); cf. v. Steenis, *l.c. supra*.
Dodecastemon teysmanni var. *timorensis* (Bl.) Miq., Fl. Ind. Bat. 1(2): 360 (1859).
Cyclostemon teysmanni var. *timorensis* (Bl.) Muell. Arg.: 481 (1866).
Drypetes timorensis (Bl.) Pax & Hoffm. xv: 278 (1922).

A var. *roxburghii* stylis alte bipartitis laciniis anguste linearibus (nec late flabellatis), drupa (an semper?) brevius pedicellata recedit.

SB.—Lesser Sunda Is.
SABAH. Port Myburgh [Sandakan], 1895, *Creagh s.n.* Labuan Haji, plain, alt. 7·5 m., 25 July 1933, *Polonio* (N. Born. For. Dept.) 3229:—Tree 9 m. high, 30 cm. girth; fruit green.
SUMBAWA (W.). Pernak, Olat Seli, ±12 km. S. of Sumbawa Besar,

monsoon forest, slope, alt. 100 m., 20 May 1961, *Kuswata* 250:—Tree ±30 m. high, 30 cm. diam.; bark brown, up to 3 mm. thick, lenticellate; living bark light brown, paler inside, 7·5 mm. thick; fruit green, somewhat brownish.

TIMOR. Without details of locality, date or collector, *teste* Blume, *l.c.* (type, L).

Blume (*l.c.*), in his generic description of *Pycnosandra*, described the stigmas as '3–4, maxima, patentissima, bipartita, *laciniis filiformibvs*, arcuato-divergentibus, verrucosis'. The above Bornean and Sumbawan specimens show just this character of filiform style-segments, but seem otherwise (apart from the shorter fruiting pedicels) to be indistinguishable from typical *D. roxburghii*. Further material from E. Malesia is desirable.

Drypetes sibuyanensis (*Elm.*) *Pax & Hoffm.* xv: 247 (1922); Airy Shaw in Kew Bull. 23: 59 (1969).

Cyclostemon sibuyanensis Elm., Leafl. Philipp. Bot. 3: 914 (1910); Merr.: 408 (1923).

SR; B.—Sumatra, Philippines.

Tree to 36 m. high, in peat swamps, primary mixed swamp forest, forest intermediate between Heath and Dipterocarp forest, or submontane *kerangas*, on Tertiary sandstone over deep yellow clay, up to 300 m. alt.

Leaves coriaceous, elliptic to oblong-elliptic, 7–15 cm. long, base distinctly narrowed into the petiole, abruptly shortly caudate or sometimes cuspidate, smooth and usually rather shining above, petiole 5–15 mm. long; flowers very shortly pedicellate or subsessile, stamens 12–14, ovary densely whitish-pubescent; fruit hard, smooth, subglobose or shortly ellipsoid or shortly oblong, up to 3 × 2·5 cm., subtruncate at the base, minutely papillose-puberulous, very shortly (2–4 mm.) and stoutly pedicellate. The very shortly pedicelled fruit recalls the similar situation in *D. crassipes*, which differs in its larger but more shortly petiolate leaves.

Drypetes stylosa *Airy Shaw* in Kew Bull. 29: 301 (1974).

SR (C.).—Endemic.

Tree of 24 m., on hill slope at unknown alt.

Pedicels and tepals with fine, dense, golden pubescence as in *D. ochrothrix*, but leaves larger, much more coriaceous, conspicuously though shallowly crenate and with more numerous nerves, and the ovary quite glabrous, with a conspicuous slender style 1 mm. long.

Elateriospermum *Bl.*

Elateriospermum tapos *Bl.*, Bijdr.: 621 (1825); Muell. Arg.: 1131 (1866); Benth. in Hook. Ic. Pl. 13: 73, t. 1294 (1879); Hook. f.: 382 (1887); Boerl.: 227 (1900); J. J. Sm.: 572 (1910); Pax in Engler IV. 147 (Hft. 42): 17 (1910); Hubert Winkler in Engl., Bot. Jahrb. 50, Suppl.: 206 (1914), *in obs.*; Merr.: 344 (1921); Ridley: 252 (1924); Corner: 249 (1940); Backer & Bakh. f.: 497 (1963); Airy Shaw in Kew Bull. 16: 371 (1963) & 26: 258 (1972); Meijer, 2: 17 (1964) & 7: 25 (1967); Medway in Biol. Journ. Linn. Soc. 4: 131 (1972); Whitmore: 91 (1973).

SR; B; SB; K1.—Lower Siam, Malaya, Sumatra, Java.

Tree to 39 m. high, in primary Lowland Dipterocarp or Mixed Dipterocarp forest on yellow loam, clay loam, yellow sandy clay, yellow clay with sandstone, sandstone screes, shale or basalt, up to 600 m. alt.

Leaves almost oblong, glabrous, very smooth and shining, with strongly reticulate venation.

Yields a bluish-white latex. Fruit red, drupaceous, 2–3-locular, said to resemble that of Pará rubber (*Hevea*) in shape and size; eaten by monkeys.

Endospermum *Benth.*

1 Inflorescences simple or almost so . . . . **E. diadenum**
1 Inflorescences paniculiform . . . . **E. peltatum**

Endospermum diadenum (*Miq.*) *Airy Shaw* in Kew Bull. 14: 395 (1960); Schaeffer in Blumea 19: 186 (1971); Airy Shaw in Kew Bull. 26: 258 (1972).

Melanolepis? diadena Miq., Fl. Ind. Bat., Suppl.: 455 (1861).
Endospermum malaccense Benth. ex Muell. Arg. in Flora 47: 469 (1864) & in DC.: 1132 (1866); Hook. f.: 458 (1887); Pax & Hoffm. iv: 34 (1912); Ridley: 305 (1924); Corner in Gard. Bull. Str. Settlem. 10: 296–8 (1939) & Ways. Trees: 251 (1940); Whitmore: 93, fig. 7 (1973).
E. borneënse Benth. ex Muell. Arg., *ll.cc.* (1864 & 1866); Pax & Hoffm. iv: 35 (1912); Merr.: 346 (1921); Corner, *l.c.* 10: 298 (1939).
Mallotus diadenus (Miq.) Muell. Arg.: 959 (1866); Pax & Hoffm. vii: 206 (1914), *in adnot.*
Rottlera diadena (Miq.) Scheff. in Ann. Mus. Bot. Lugd.-Bat. 4: 125 (1868–9).
Endospermum chinense? sec. Hook. f.: 458 (1887); Ridley: 306 (1924); *non* Benth.
E. ovalifolium Pax & Hoffm. iv: 34 (1912); Ridley: 305 (1924); Corner, *l.c.* 10: 297 (1939).
E. beccarianum Pax & Hoffm. iv: 35 (1912); Merr.: 346 (1921); Corner, *l.c.* 10: 298 (1939).
E. chinense var. *malayanum* Pax & Hoffm. iv: 36 (1912), *p.p.*, *typo excluso.*
E. malayanum Chatterjee in Kew Bull. 4: 564 (1950).

SR; B; SB; K1, 2, 3.—Lower Siam, Malaya, Sumatra, Banka, Billiton, Natuna Is.

Tree to 36 m. high, in primary and secondary Mixed Dipterocarp or freshwater swamp forest in low undulating country, in *kerangas*, on white sand, yellow sandy loam or clay, or black stony soil, up to 1750 m. alt. on Kinabalu.

The flowers are said to be fragrant or aromatic.

Endospermum peltatum *Merr.* in Govt. Lab. Publ. (Philipp.) 35: 35 (1906); Pax & Hoffm. iv: 37 (1912); Merr.: 457 (1923); Schaeffer in Blumea 19: 188 (1971); Airy Shaw in Kew Bull. 26: 259 (1972).

SR; B; SB; K1.—Andamans, Lower Siam, Malaya, Philippines, NE. Celebes.

Tree to 45 m. high, in primary and secondary forest, frequently a prominent emergent tree, often in small groups associated with *Octomeles sumatrana*, on silty sandy clay loam, alluvium (rarely flooded), clay-rich soil, black or brown sandy soil, sandstone, or heavy loam containing coral limestone, up to 1000 m. alt.

The seeds are said to be edible and to contain a good oil.

[*Endospermum eglandulosum* Pax & Hoffm. vii (Euph.–Addit. v): 418 (1914), Merr.: 346 (1921), described from Sarawak, is **Sterculia macrophylla** *Vent.*, *teste* Schaeffer, *l.c.*: 191 (1971).]

Erismanthus *Wall. ex Muell. Arg.*

Erismanthus obliquus *Wall. ex Muell. Arg.*: 1138 (1866); Hook. f.: 405 (1887); Boerl.: 236 (1900); Pax & Hoffm. iii: 34 (1911); Merr.: 346 (1921); Ridley: 271 (1924); Gagnep.: 464 (1926); Airy Shaw in Kew Bull. 26: 260 (1972); Whitmore: 95 (1973).

K1.—Lower Siam, Malaya, Sumatra.
Tree to 7 m. high, on low ridge at 30 m. alt.

Apparently very scarce in Borneo: one collection of *Kostermans* (21118) from E. Indonesian Borneo. Hooker's (*l.c.*) alleged record for 'Borneo' was erroneous: the *Beccari* specimens cited were from Sumatra, and '632' should have been 652.

It should be noted that the inflorescence of *Erismanthus* is essentially bisexual. Although solitary axillary male inflorescences and solitary axillary female flowers occur very frequently, bisexual inflorescences, consisting of a male 'cone' or raceme accompanied by a single, basal, long-pedicelled female flower from the same axil, are by no means uncommon. This brings *Erismanthus* into line with its relatives *Syndyophyllum* and (to a lesser extent) *Moultonianthus*.

Euphorbia *L.*

by A. Radcliffe-Smith

1 Succulent shrub with angled stems (Subgen. *Diacanthium*). **E.** cf. **lacei**
1 Non-succulent herbs with terete stems:
  2 Leaves alternate, at least below (Subgen. *Poinsettia*) E. HETEROPHYLLA
  2 Leaves all opposite (Subgen. *Chamaesyce*):
    3 Perennial herbs or subshrubs (§ *Sclerophyllae*) . . . . . **E. atoto**
    3 Annual herbs:
      4 Capsule ripening within the cyathium, causing it to split down the side (§ *Chamaesyce*) . . . . **E. thymifolia**
      4 Capsule ripening outside the cyathium on an elongated, decurved pedicel (§ *Hypericifoliae*):
        5 Leaves broadly rounded at apex . . . **E. hypericifolia**
        5 Leaves tapering at the apex . . . . E. HIRTA

Euphorbia atoto *Forst. f.*, Prodr.: 36 (1786); Boiss. in DC.: 12 (1862); Hook. f.: 248 (1887); Merr.: 348 (1921) & 461 (1923); Ridley: 181 (1924); Gagnep.: 245 (1925); Merr. in Philipp. Journ. Sci. 29: 387 (1926); Backer & Bakh. f.: 503 (1963); A. R. Smith in Kew Bull. 26: 263 (1972).

Chamaesyce atoto (Forst. f.) Croiz. in Degener, Fl. Hawaii, Fam. 190, Chamaesyce, leafl. 4 (1936), *in obs.*

SB.—Old World tropics from S. India to Polynesia and from the Ryu-Kyu Is. to Queensland.

A typical strandline plant found chiefly on sandy beaches.

Glabrous, glaucous herb or subshrub up to 60 cm. high.

Euphorbia heterophylla *L.*, Sp. Pl.: 453 (1753); Boiss. in DC.: 72 (1862); Merr.: 462 (1923); Ridley: 181 (1924); Gagnep.: 244 (1925); Backer & Bakh. f.: 502 (1963); A. R. Smith in Kew Bull. 26: 264 (1972).

E. geniculata Ort., Hort. Matr. Dec.: 18 (1797); Boiss. in DC.: 72 (1862).
E. prunifolia Jacq., Hort. Schoenbr. 3: 15, t. 277 (1798); Hook. f.: 266 (1887), *in adnot.*; Backer & Bakh. f.: 502 (1963).
Poinsettia heterophylla (L.) Klotzsch & Garcke ex Klotzsch in Monatsber. Akad. Berl. 1859: 253 (1859); Dressler in Ann. Miss. Bot. Gard. 48: 329 (1961).
P. geniculata (Ort.) Klotzsch & Garcke ex Klotzsch, *l.c.* (1859).

SB.—Pantropical. Origin in C. America.

Naturalized; in waste places, and in open grassland on rocky soil up to 20 m. alt.

Herb to *c.* 1 m. in height, often woody at the base, with panduriform leaves, the upper ones red at the base, and densely clustered cyathia.

Euphorbia hirta *L.*, Sp. Pl. 454 (1753); Merr.: 348 (1921) & 462 (1923); Ridley: 181 (1924); Merr. in Philipp. Journ. Sci. 29: 387 (1926); Pax & Hoffm. in Mitt. Inst. Bot. Hamburg 7: 230 (1931); Hend. in Journ. Malayan Br. Roy. As. Soc. 17: 70 (1939); Backer & Bakh. f.: 504 (1963); A. R. Smith in Kew Bull. 26: 264 (1972).
E. pilulifera L., *l.c.* (1753); Boiss. in DC.: 21 (1862); Hook. f.: 250 (1887); Gagnep.: 245 (1925).
Chamaesyce pilulifera (L.) Small, Fl. SE. U.S.: 708 (1903).
Ch. hirta (L.) Millsp. in Publ. Field Columb. Mus., Bot. 2: 303 (1909).

SR; SB; K1a, 3.—Pantropical weed, originating in C. America.

Herb up to 90 cm. high, but more commonly much less; young stems densely covered with yellow multicellular hairs.

Euphorbia hypericifolia *L.*, Sp. Pl.: 454 (1753); Boiss. in DC.: 23 (1862); Hook. f.: 249 (1887); Hutch. in Journ. Linn. Soc., Bot. 42: 133 (1914); Merr.: 348 (1921); Backer & Bakh. f.: 504 (1963); A. R. Smith in Kew Bull. 26: 265 (1972).

Chamaesyce hypericifolia (L.) Millsp. in Publ. Field Columb. Mus., Bot. 2: 303 (1909).

E. indica sec. Gagnep.: 248 (1925), *non* L.

K3.—Pantropical weed.

Erect-ascending, sparingly pubescent or subglabrous herb to 30 cm., with markedly asymmetrical opposite leaves.

Euphorbia cf. **lacei** *Craib* in Bull. Misc. Inf. Kew 1911: 456 (1911), & in Aberdeen Univ. Stud. No. 57: 182 (1912); Gagnep.: 255 (1925); A. R. Smith in Kew Bull. 26: 265 (1972).

SB.—Mt. Sidungol, 18 July 1938, *Keith* 9337; Segarong For. Res., Semporna, 26 Aug. 1938, *Keith* 9415. (Typical *E. lacei* in Burma & Siam.)

Succulent shrub to 5 m., on rocky limestone hillsides up to 20 m. alt.
The Bornean plant is almost indistinguishable from *E. lacei*, and may well be conspecific therewith.

Euphorbia thymifolia *L.*, Sp. Pl.: 454 (1753); Boiss. in DC.: 47 (1862); Hook. f.: 252 (1887); Bernard in Ic. Bogor. 4(1): 49, t. 315 (1910); Merr.: 348 (1921) & 464 (1923); Ridley: 182 (1924); Gagnep.: 246 (1925); Backer & Bakh. f.: 503 (1963); A. R. Smith in Kew Bull. 26: 267 (1972).

Chamaesyce thymifolia (L.) Millsp. in Publ. Field Mus. Nat. Hist. Chicago, Bot. 2: 412 (1916).

SR; SB; K1.—Widespread in the Old World Tropics.
A prostrate weed of open, waste places.

This species is often confused with *E. prostrata* Ait., but the latter has the capsule ripening outside the cyathium, and furthermore the hairs on the capsule are confined to the keels, and not evenly spread over the three faces as in *E. thymifolia*.

Excoecaria *L.*

1 Leaves opposite:
  2 Leaves manifestly crenate, with closer nerves; capsule 2–3 cm. diam. **E. bantamensis**
  2 Leaves obscurely denticulate, with fewer nerves; capsule about 1 cm. diam. . . . . **E. borneënsis**
1 Leaves alternate:
  3 Capsule globose, almost unlobed, up to 3 cm. diam., black when dry; leaves crenulate . . . . **E. indica**
  3 Capsule tricoccous, lobed, about 1 cm. diam., not black when dry:
    4 Leaves coriaceous, shining, up to 20 cm. long, with prominent nerves; inflorescences lax, up to 10 cm. long . **E. philippinensis** var.
    4 Leaves chartaceous, not shining, up to 10 cm. long, nerves not prominent; inflorescences dense, 1–5(–7·5) cm. long **E. agallocha**

Excoecaria agallocha *L.*, Syst., ed. 10, 2: 1288 (1759) & Sp. Pl., ed. 2: 1451 (1763); Muell. Arg.: 1220 (1866); Hook. f.: 472 (1888); J. J. Sm.: 616 (1910); Pax & Hoffm. v: 165 (1912) (incl. vars.); Merr.: 347 (1921) &

458 (1923); Ridley: 314 (1924); Gagnep.: 407 (1926); Merr. in Trans. Amer. Philos. Soc. II, 24(2): 240 (1935); Corner: 254 (1940); Backer & Bakh. f.: 499 (1963); Meijer, 7: 44 (1967); Airy Shaw in Kew Bull. 26: 268 (1972); Whitmore: 96 (1973).

Stillingia agallocha (L.) Baill., Ét. Gén. Euphorb.: 518 (1858).

SR; B; SB.—S. India & Ceylon to Formosa & Ryu-Kyu Is., and throughout Malesia to N. Australia & Pacific.

Tree to 9 m. high, in mangrove swamps, occasionally by tidal rivers or along sandy seashores, once noted on undeveloped soil on sandstone rocks, once in primary forest at 5 m. alt.

The relatively small (5–10 cm. long), alternate, entire to shallowly crenulate leaves, densely spicate male inflorescence and small tricoccous capsule are usually sufficient for the recognition of this well-known species.

Excoecaria bantamensis *Muell. Arg.* in Linnaea 32: 124 (1863) & in DC.: 1219 (1866); J. J. Sm.: 610 (1910); Pax & Hoffm. v: 161 (1912); Backer & Bakh. f.: 499 (1963); Airy Shaw in Kew Bull. 26: 269 (1971).

E. macrophylla J. J. Sm.: 611 (1910); Pax & Hoffm. v: 162 (1912); Merr. in Philipp. Journ. Sci. 16: 577 (1920) & Enum.: 347 (1921) & 458 (1923). [*Nec* Merr. in Philipp. Journ. Sci. 29: 387 (1926) = *Syndyophyllum excelsum* Lauterb. & K. Schum., *q.v.*]

K3.—Siam (?), Philippines, Java.

Recorded for Borneo (as *E. macrophylla*) by J. J. Smith, *l.c.*: 612, on the basis of a collection of *Hallier* which I have not seen. No field data available.

Distinguished from the only other opposite-leaved Bornean species, *E. borneënsis* Pax & Hoffm., by the manifestly crenate leaves and by the much larger capsule, 2–3 cm. in diameter. It is more closely related to *E. oppositifolia* Griff., of SE. Asia and Malaya, which differs in its even larger capsule, 4–5 cm. in diameter, and leaves sometimes reaching 30 cm. in length.

Excoecaria borneënsis *Pax & Hoffm.* vii (Euphorb.-Addit. v): 422 (1914); Merr.: 347 (1921).

E. bicolor var. *viridis* sec. Merr. in Philipp. Journ. Sci. 11, C. Bot.: 285 (1916), *non* Pax & Hoffm.

E. mirandae Merr. in Philipp. Journ. Sci. 16: 576 (1920) & Enum.: 458 (1923), *e descr.*, **synon. nov.**

E. cochinchinensis var. *viridis* sec. Meijer, 7: 44 (1967), *non* (Pax & Hoffm.) Merr.

SR; SB; K1.—Sumatra; Philippines; Sangië & Talaud Is.; Moluccas.

Shrub of 1 m., rarely to 9 m., in primary forest on or near limestone rock, on black or dark brown soil, up to 345 m. alt.

Very close to *E. cochinchinensis* Lour. var. *viridis* (Pax & Hoffm.) Merr., but differing consistently in the subentire or obscurely denticulate leaves. In *E. cochinchinensis* the leaves are closely and sharply, though shallowly, denticulate-serrate. I can find no significant difference in the points mentioned by Pax & Hoffmann (longer, slender, gland-bearing petiole, differently shaped limb, different arrangement of flowers).

In Sarawak (Fourth Division) the plant is said to be used in cases of rheumatism, by warming the leaves over the fire and applying them to the body.

Excoecaria indica (*Willd.*) *Muell. Arg.* in Linnaea 32: 123 (1863) & in DC.: 1216 (1866), *p.p.*; J. J. Sm.: 615 (1910), *in obs.*

Sapium indicum Willd., Sp. Pl. 4: 572 (1805); Hook. f.: 471 (1888); Pax & Hoffm. v: 251 (1912); Merr.: 348 (1921) (excl. *Gibbs* 2659); Ridley: 317 (1924); Gagnep.: 396 (1926); Corner: 277 (1940); Airy Shaw in Kew Bull. 26: 330 (1972); Whitmore: 128, 129 (1973).
Stillingia indica (Willd.) Baill., Ét. Gén. Euphorb.: 513 (1858).

SR; SB; K1, 1a.—S. & E. India, SE. Asia, and throughout Malesia (exc. Philippines) to Solomon Is. Represented in Java by the only varietally distinct *E. virgata* Zoll. & Mor. ex Miq.; cf. J. J. Sm., *l.c. supra*, and Backer & Bakh. f.: 499 (1963).

Tree to 27 m. high, in primary *Nypa* forest in salt water, in seasonal swamps, on tidal river banks and seashores, on black soil or yellow clay alluvium, from sea-level up to 10 m. alt.

The large, smooth, almost globose fruits, up to 3 cm. in diameter and black when ripe, are very characteristic, distinguishing *E. indica* from all other Asiatic members of the tribe *Hippomaneae*.

The bole of the tree is thorny. The whole plant yields a copious white sticky latex. A black dye is obtained from the leaves by boiling in water. The husk of the unripe green fruits is used as a fish-poison, but the seeds are edible.

I am following J. J. Smith and Backer & Bakhuizen in treating this plant as an *Excoecaria*, since it appears on the whole to be closer to the type of that genus, *E. agallocha* L., than to the type of *Sapium*, *S. jamaicense* Sw., of the West Indies. It must be confessed, however, that generic limits in the tribe *Hippomaneae* are most unsatisfactory and in urgent need of a major overhaul.

Excoecaria philippinensis *Merr.* in Philipp. Journ. Sci. 1, Suppl.: 82 (1906).

Var. **euphlebia** *Merr.* in Philipp. Journ. Sci. 7, Bot.: 389 (1912); Pax & Hoffm. vii: 423 (1914); Merr.: 459 (1923); Meijer, 7: 44 (1967).

SB.—Philippines.

Tree of unknown stature, in primary forest on hillside, ridge-top, at 810 m. alt.

Leaves alternate, oblanceolate, entire, up to 24 cm. long; nerves spreading almost at right angles, prominent on both surfaces (in var. *philippinensis* they are less prominent beneath); flowers dioecious; female raceme terminal, up to 13 cm. long.

Known from a single gathering, *J. Singh* SAN 21032, from Bukit Ampuon in the Ranau district of Sabah.

Fahrenheitia *Reichb. f. & Zoll.*

Fahrenheitia pendula (*Hassk.*) *Airy Shaw* in Kew Bull. 20: 410 (1966) & 26: 270 (1972); Whitmore: 97 (1973).

Croton pendulus Hassk., Pl. Jav. Rar.: 266 (1848).
Fahrenheitia collina Reichb. f. & Zoll. in Linnaea 28: 600 (1856); Muell. Arg.: 1256 (1866).
Paracroton pendulus (Hassk.) Miq., Fl. Ind. Bat. 1(2): 382 (1859); Muell. Arg.: 1113 (1866); Boerl., Handl. Fl. Ned. Ind. 3(1): 233 (1900); J. J. Sm.: 586 (1910); Pax & Hoffm. iii: 12 (1911).
Tritaxis macrophylla Muell. Arg. in Flora 47: 482 (1864).
Trigonostemon macrophyllus (Muell. Arg.) Muell. Arg. in Linnaea 34: 213 (1865) & in DC.: 1106 (1866).
Ostodes crenato-serrata Merr. in Philipp. Journ. Sci. 4, Bot.: 283 (1909).
O. macrophyllus (Muell. Arg.) Benth. ex Pax & Hoffm. iii: 18 (1911); Merr. in Philipp. Journ. Sci. 11, C. Bot.: 73 (1916) & Enum.: 345 (1921) & 454 (1923) & Philipp. Journ. Sci. 29: 386 (1926); Ridley: 269 (1924); Stern in Am. Journ. Bot. 54: 670 (1967).
O. collinus (Reichb. f. & Zoll.) Pax & Hoffm. iii: 21 (1911).
O. pendula (Hassk.) A. Meeuse in Backer, Bekn. Fl. Java IVC, fam. 112: 109 (1943) & in Blumea 5: 508 (1945); Backer & Bakh. f.: 493 (1963); Meijer, 7: 28 (1967).

SR; B; SB; K1, 1a, 3.—Lower Burma (?), Lower Siam, and throughout W. Malesia.

Tree to 15 m. high, very common in primary riparian or Mixed Dipterocarp forest, or in seasonal swamps, on silty clay or shale-derived soil, white sandy soil, black sandstone soil, sandy loam, clay loam, or loam with limestone, up to 1500 m. alt.

Timber useful (Sabah). Sap irritating and painful. Plant medicinal (Sabah).

Glochidion *J. R. & G. Forst.*

1 Stems and/or lower leaf-surface (at least on nerves) pubescent or puberulous:
  2 Inflorescences pedunculate and often supra-axillary:
    3 Very robust and strongly tomentose; leaves large, up to 20 × 10 cm., usually cordate, venation very prominent below and distinctly impressed above; inflorescences manifestly branched; capsules mostly 3-locular, broadly lobed, thinly crustaceous, pubescent, pedicels slender, up to 12 mm. long . . . **G. superbum**
    3 Less robust; leaves smaller, rarely cordate, venation less prominent, not impressed above; inflorescences not or scarcely branched; capsules not or shallowly lobed, pedicels shorter:
      4 Leaves coriaceous; capsules mostly 4-locular, somewhat depressed **G. arborescens**
      4 Leaves thinly chartaceous; capsules 5–6-locular, very depressed, dividing into 10–12 narrow segments . . **G. angulatum**
  2 Inflorescences epedunculate, axillary or almost so:
    5 Strongly fulvous-pubescent; leaves mostly oblong, rarely exceeding 4–6 × 1·5–2 cm., truncate or subcordate at base, regularly and closely distichous . . . . **G. molle**
    5 Not fulvous; leaves not as above, not regularly and closely distichous:
      6 Style or styles elongate, 2–3 mm. long:

7 Style single, filiform, minutely 3-lobulate at apex **G. monostylum**

7 Styles 3–5, free or ± connate:

8 Styles 3, free, filiform; leaves grey-puberulous beneath, rounded or cordate at base . . . . . . **G. sericeum**

8′ Styles 3, connate for four-fifths; leaves variously puberulous **G. rubrum**

8″ Styles 5, connate for two-thirds; leaves pubescent but not grey-puberulous beneath, not cordate at base **G. tenuistylum**

6 Style or styles shorter, more robust, or obsolete:

9 Capsule subglobose, almost unlobed, about 1 cm. diam., hard, the stigmas forming a globose button at the apex **G. macrostigma**

9 Capsule distinctly lobed:

10 Capsule dehiscing into about 12 narrow segments; stylar column depressed-globose, deeply grooved **G. philippicum**

10 Capsule dehiscing into 3–6 broad segments; stylar column otherwise:

11 Capsule over 1 cm. in diameter; leaves up to 9 cm. broad:

12 Leaves ± oblong, up to 30 cm. long; plant coarsely hirsute; seeds bright red . . **G. calospermum**

12 Leaves broadly elliptic, 5–19 cm. long; plant shortly tomentellous; seeds brown . . . **G. elmeri**

11 Capsule less than 1 cm. in diameter; leaves less than 5 cm. broad:

13 Leaves with a dull, minutely granular surface, especially below (as in *G. lanceisepalum*); capsule borne on a very slender pedicel 4–6 mm. long **G. singaporense**

13 Leaves with a smooth or at least non-granular lower surface; pedicel of capsule less slender:

14 Styles 3, free, short, thickish, divaricate, obliquely truncate or attenuate at the apex **G. kunstlerianum**

14 Styles obsolete, represented by a broad, flat or slightly convex, glabrous stigmatic disk **G. glomerulatum**

1 Stems and lower leaf-surface glabrous or almost so:

15 Inflorescences distinctly peduncled and/or supra-axillary:

16 Female flower solitary in the inflorescence, long-pedicelled, with elongate fasciculate styles; stipules narrow, elongate, reflexed **G. andersonii**

16 Female flowers several or numerous; pedicels and styles shorter; stipules otherwise:

17 Capsule glabrous, 3–6 mm. in diameter:

18 Plant usually greenish when dry; leaves ± lanceolate, very oblique at the base; capsule very depressed, almost discoid . . . . **G. lanceilimbum**

18 Plant blackening on drying; leaves ± elliptic or ovate, less

oblique; secondary nerves often conspicuously transverse; capsule much less depressed . . . **G. borneënse**

17 Capsule pubescent, 6–10 mm. in diameter:

19 Capsule subglobose, not or scarcely lobed, with a corky or fibrous surface, sometimes almost ridged or shallowly lamellate; stylar column 1–2 mm. long **G. brunneum**

19 Capsule depressed, shallowly lobed, densely matted-pubescent, not corky or lamellate; styles very short, less than 1 mm . . **G. pubicapsa** var. **brunneiforme**

15 Inflorescences epedunculate, axillary:

20 Leaves broadly obovate, usually with a rounded apex; capsules rather large (1·5 cm. diam.), multilocular, dehiscing into 20–25 narrow segments . . . . **G. littorale**

21 Style less than 1 mm. long; mostly a coastal plant var. **littorale**

21 Style 3 mm. long; plant of mountain tops var. **culminicola**

20 Leaves ± narrowed or acute at apex:

22 Leaves with a dull, minutely granular surface, especially below (as in *G. singaporense*), thickish in texture, up to 30 cm. long; nerves distinctly impressed above; inflorescence occasionally slightly supra-axillary; capsule 5-locular **G. lanceisepalum**

22 Leaves not granular below:

23 Capsule multilocular, dehiscing into about 20 narrow segments . . . . . **G. littorale** var. **caudatum**

23 Capsule 3–12-locular:

24 Capsule subglobose, unlobed, 1–1·5 cm. diam., thick-walled, tardily dehiscent, often borne on long pedicels (up to 2·3 cm.); stylar column conspicuous, often ± funnel-shaped, persistent . . . **G. obscurum**

24 Capsule distinctly lobed, often ± depressed, with thinner walls, dehiscing normally; pedicels shorter; stylar column not funnel-shaped:

25 Styles united into a very slender column, 2–3 mm. long, entire or minutely trifid at the apex:

26 Leaves very glaucous beneath; ♀ tepals erect, unequal, irregularly connate; capsule papillose-puberulous . . . . **G. styliferum**

26 Leaves very slightly glaucous beneath; ♀ tepals spreading, ± equal, conspicuous, 4–5 mm. long, connate below into a cup, margins with a few minute teeth; capsule glabrous **G. trusanicum**

25 Styles not as above (if elongate then thicker or more deeply lobed at apex):

27 Plants confined to the *kerangas* (Heath forest) formation, with stiffly coriaceous leaves strongly glaucous beneath:

28 Capsule 5–6 mm. in diameter, scarcely lobed, almost black when dry; leaves drying greenish above, margins very strongly revolute . . . . **G. celastroïdes**

28 Capsule 1–2·3 cm. in diameter, distinctly lobed:

29 Capsule 2–2·3 cm. diam.; leaves up to 22 cm. long, obliquely oblong-oblanceolate, caudate-acuminate **G. auii**

29 Capsule 1–1·2 cm. diam.; leaves up to 14 cm. long, not caudate-acuminate:

30 Leaves mostly exceeding 8 × 4 cm., very thickly coriaceous **G. aluminescens**

30 Leaves mostly less than 8 × 4 cm., less thickly coriaceous **G. kerangae**

27 Plants not confined to the *kerangas* formation; leaves, if coriaceous or glaucous, less strongly so:

31 Leaves usually somewhat glaucescent beneath, up to 8 × 4 cm., firmly chartaceous; ♀ sepals distinctly keeled on the back; capsule 5–7 mm. diam., thinly crustaceous, shallowly 8–12-lobed, light chestnut in colour when dry **G. lutescens**

31 Leaves not glaucescent beneath; ♀ sepals not or less distinctly keeled on the back; capsule mostly more firmly crustaceous and more distinctly 3–8-lobed:

32 Plant of rocky coastal situations; leaves 5–11 cm. long, mostly oblong or lanceolate, rather gradually narrowed to the often obtuse apex; lateral nerves immersed or only slightly prominulous below; capsule 9–13 mm. diam., slightly inflated . **G. mindorense**

32 Plant of other situations; lateral nerves mostly more evident below:

33 Capsule only 4–5 mm. in diameter; leaves elliptic-oblong, 8–11 cm. long, very glossy, drying olivaceous-green . . . **G. xestophyllum**

33 Capsule larger; leaves usually not drying greenish:

34 Capsule ± inflated, 1–2 cm. diam.; leaves very glossy on both surfaces, ± chestnut-coloured when dry . . . **G. nitidum**

34 Capsule smaller, not or less inflated; leaves relatively dull, or at least less glossy, mostly dark brown when dry:

35 Capsules 4(–5)-lobed, often

borne in dense glomerules on bare portions of stem; leaves often rather large
G. pubicapsa var. **pubicapsa**

35 Capsules 3- or 6-lobed, not usually borne on bare stems; leaves mostly smaller:

36 Leaves not 'satiny' below, nerves not puckering; petiole about 2 mm. broad, bicarinate; style depressed **G. williamsii**

36 Leaves usually with a rather smooth, 'satiny' surface beneath, the nerves tending to pucker slightly on drying; petiole about 1 mm. thick; style sometimes stoutly columnar, 2 mm. long, with a trifid apex . . **G. rubrum**
(See also the closely related **G. azaleon**, **G. brooksii**, **G. cupreum**)

N.B. *G. korthalsii*, *G. mehipitense* and *G. punctatum* are not accounted for in the above key.

Glochidion aluminescens *Airy Shaw* in Kew Bull. 27: 16 (1972).

SR (SW.).—Endemic.

Shrub to 4 m. high, in *padang* or *kerangas* or Heath forest on humic podsols overlying plateau sandstone at 80–100 m. alt.

A striking species, subglabrous, with thick rigidly coriaceous leaves, 8–14 × 4–6·5 cm., purplish- or brownish-glaucous beneath, often drying a lurid yellowish green above; male flowers on long slender pedicels (up to 13 mm.); capsule depressed, tricoccous, deeply rounded-lobate, 10–12 mm. diam. Apparently extremely local, so far as is known confined to the Bako National Park near Kuching.

Glochidion andersonii *Airy Shaw* in Kew Bull. 29: 287 (1974).

SR (SW.).—Endemic.

Small tree (stature unknown), in secondary peat-swamp forest six years after felling, doubtless at low altitude.

Approaching forms of *G. perakense*, but differing profoundly in the narrow elongate reflexed stipules and solitary long-pedicelled female flowers with elongate fasciculate styles.

Glochidion angulatum *C. B. Rob.* in Philipp. Journ. Sci. 4, Bot.: 91 (1909); Merr.: 397 (1923); Airy Shaw in Kew Bull. 25: 485 (1971) & 27: 70 (1972).

SB.—Philippines, Celebes, Moluccas, New Guinea, Solomons.

Shrub or tree to 10 m. high, in fresh-water swamps or in poor sandy soil near the sea or in hill forests up to 150 m. alt.

Near *G. perakense*, but inflorescence even more markedly pedunculate and supra-axillary, and stems, midribs and inflorescences very shortly tomentellous. Leaves very thinly chartaceous or almost membranous; capsule very depressed, 5–6-locular, not or very shallowly lobed, dividing into 10–12 narrow segments, rusty or whitish-puberulous.

Only three Bornean specimens seen: Marotai, *Maidin* (N. Borneo For. Dept.) 2344; Supu, *Apostol* (N.B. For. Dept.) 3440; and Gomantong Caves Hill, Kinabatangan, *G.H.S. Wood* A4609. The plant is indicated by Maidin as 'poison'.

Glochidion arborescens *Bl.*, Bijdr.: 584 (1825); J. J. Sm.: 114 (1910); Ridley: 211 (1924); Backer & Bakh. f.: 462 (1963); Airy Shaw in Kew Bull. 26: 273 (1972); Whitmore: 100 (1973).

G. bancanum Miq., Fl. Ind. Bat., Suppl.: 449 (1861).
Phyllanthus arborescens (Bl.) Muell. Arg. in Flora 48: 370 (1865) & in DC.: 279 (1866).
P. silheticus Muell. Arg., *ll.cc.* 378 (1865) & 297 (1866).
P. teysmanni Muell. Arg.: 1270 (1866).
? *P. zeylanicus* sec. Muell. Arg.: 281 (1866), quoad loc. 'Borneo', *an G. zeylanicum* Juss.?
Glochidion sclerophyllum Hook. f.: 313 (1887).
? *G. zeylanicum* sec. Boerl.: 275 (1900), quoad loc. 'Borneo'; Merr.: 329 (1921); *an* Juss.?
G. silheticum (Muell. Arg.) Croiz. in Journ. Arn. Arb. 21: 492 (1940).

SR; SB; K1.—Assam, Lower Siam, Malaya, Banka, Java, ? Celebes.

Tree to 24 m. high, in primary riverine or Mixed Dipterocarp forest on basalt or granodiorite rocks, or black or brownish soil, up to 900 m. alt.

Mostly shortly fulvous-tomentellous or pubescent. Leaves coriaceous, often reddish-brown when dry, often rather smooth above. Inflorescences pedunculate and slightly to markedly supra-axillary. Capsule mostly 4-locular, 5–7 mm. diam., somewhat depressed, very shallowly lobed, minutely puberulous, the styles forming a very short beak splitting into four short apiculi at dehiscence.

Two collectors note the production of a sticky yellow exudate from the cut stems.

Glochidion auii *Airy Shaw* in Kew Bull. 27: 19 (1972).

SR (NE.).—Endemic.

Small tree of 4·5 m., in Heath forest at 30 m. alt.

Near *G. kerangae*, but leaves much larger, up to 22 cm. long, obliquely oblong-oblanceolate and caudate-acuminate, and capsule 2–2·3 cm. in diameter.

Glochidion azaleon *Airy Shaw* in Kew Bull. 27: 15 (1972).

K1.—Endemic.

Treelet of 3 m., in forest at 1000 m. alt.

Close to *G. rubrum*, but leaves rigidly coriaceous, often more or less convolute when dry, and capsules smaller, much less deeply lobed, in fact merely shallowly 6-sulcate, with a slender stylar column 2 mm. long, the stigmatic lobules almost obsolete.

Glochidion borneënse (*Muell. Arg.*) *Boerl.*: 276 (1900); J. J. Sm.: 164 (1910); Merr.: 328 (1921); Backer & Bakh. f.: 463 (1963); Meijer, 7: 46 (1967); Airy Shaw in Kew Bull. 29 : 289 (1974).

Phyllanthus borneënsis Muell. Arg. in Flora 48: 377 (1865) & in DC.: 296 (1866).
P. polycarpus Muell. Arg., *ll.cc.*: 387 (1865) & 309 (1866).
Glochidion microbotrys Hook. f.: 319 (1887); Ridley: 209 (1924); Whitmore: 99, 100 (1973).
G. trilobum Ridley in Bull. Misc. Inf., Kew, 1923: 364 (1923), *p.p.*, & Fl. Malay Penins. 3: 210 (1924), *p.p.*, excl. specim. Burkilliano.

SR (SW.); **SB; K1.**—Malaya, Banka, Java.
Tree to 30 m. high, in forest on sandstone up to 740 m. alt.

Almost glabrous, or minutely puberulous, often drying blackish-brown, or sometimes chestnut; leaves 5–13 cm. long, smooth and shining above, chartaceous, the secondary nerves rather close and conspicuously transverse (at right angles to midrib). Inflorescences shortly peduncled, axillary or supra-axillary, rather many-flowered; capsule small, 3–4·5 mm. diam., 2–3 mm. long, shallowly 3-lobed; styles very short.
Indicated by one collector (in Sabah) as medicinal.

Glochidion brooksii *Ridley* in Bull. Misc. Inf., Kew, 1925: 88 (1925).

G. laevigatum var. *cuspidatum* Ridley: 215 (1924).
G. fuscum sec. Airy Shaw in Kew Bull. 27: 12 (1972), Whitmore: 100 (1973), *p.p.*, *vix* (Muell. Arg.) Boerl.

SR.—Sumatra, Malaya.
Small tree to 6 m. high, in primary Mixed Dipterocarp forest, sometimes on basalt, up to 870 m. alt.

Closely related to the common *G. rubrum*, but leaves somewhat more coriaceous, blackish above when dry, smooth below, the nerves not puckering in drying, cuneate or at least not rounded at the base, often long-acuminate, and fruits larger, up to 13 mm. diam.
This is one of the critical entities surrounding *G. rubrum.* I now think it best to maintain it provisionally as distinct, rather than merge it with *G. fuscum*, which seems to be confined to Sumatra and is distinguished by its broad, unilaterally cordate leaves. *G. brooksii* makes some approach to the *G. williamsii* complex, but is usually recognizable by the long acumen and blackish colour of the leaf upper surface when dry.

Glochidion brunneum *Hook. f.*: 312 (1887); Ridley: 207 (1924); Whitmore: 100 (1973).

G. pedunculatum Merr. in Philipp. Journ. Sci. 11, C. Bot.: 67 (1916) & Enum.: 328 (1921), **synon. nov.**

G. pedunculatum Ridley (*pro sp. nov.*), in Bull. Misc. Inf., Kew, 1923: 364 (1923) & Fl. Malay Penins. 3: 207 (1924), **synon. nov.**

SR; SB; K1.—Malaya, Sumatra, Natoena Is.

Small tree to 6 m. high, in primary forest (sometimes marshy), on limestone, or loam with lime, up to 45 m. alt.

Almost completely glabrous; leaves coriaceous, smooth, usually drying fuscous or plumbeous above and reddish-brown below, 6–15(–20) cm. long; inflorescence shortly peduncled and very shortly supra-axillary, glabrous or occasionally shortly pubescent, rather dense-flowered; capsule globose or subglobose, 1 cm. diam., not or scarcely lobed, puberulous, with a corky or fibrous surface, sometimes bearing very shallow thickish ridges or lamellae; stylar column conspicuous, 1–2 mm. long, very briefly 3–4-fid at the apex.

Bark used locally for dyeing nets (Sabah).

In the absence of fruit this species is sometimes difficult to distinguish from *G. perakense* Hook. f., which, however, is not yet definitely known from Borneo.

Glochidion calospermum *Airy Shaw* in Kew Bull. 27: 13 (1972).

G. gigantifolium sec. Meijer, 7: 45 (1967), *non* (Vidal) Merr.

SR; SB; K1, 3.—Endemic.

Shrub or small tree to 4 m. high, in primary Mixed Dipterocarp forest or mossy forest, on basalt-derived soil, or black or sandy soil, or limestone, up to 1500 m. alt.

A large-leaved, coarsely hirsute plant; branches up to 6 mm. thick; leaves more or less oblong or oblong-elliptic, up to 30 cm. long and 9 cm. broad, firmly chartaceous, nerves densely ferrugineous-strigose beneath; stipules very narrowly subulate, up to 15 mm. long. Inflorescences dense-flowered, axillary, sessile. Capsule 12–15 mm. wide, 9–11 mm. long, 3–4-locular, densely hirsute, reddish when fresh; seeds 5 mm. long, bright orange-red, smooth and shining.

Near *G. gigantifolium* (Vid.) Merr. of the Philippines, but indumentum stronger, stipules longer, capsules dehiscing completely into only 6–8 segments instead of only partially into as many as 12, and seeds bright orange. Also related, but less closely, to *G. elmeri*, differing from that in its much larger leaves, stronger indumentum and brightly coloured seeds.

A specimen from SW. Sarawak (*Ghazalli* 13419) has broader, obovate leaves with a purplish-glaucous undersurface, and somewhat broader stipules; its taxonomic status is at present uncertain.

Glochidion celastroïdes (*Muell. Arg.*) *Pax* in Engl. & Prantl, Pflanzenf. iii. 5: 24 (1890); Winkler in Engl., Bot. Jahrb. 50, Suppl.: 203 (1914), *in obs.*; Airy Shaw in Kew Bull. 27: 19 (1972), *in obs.*

Phyllanthus celastroïdes Muell. Arg. in Flora 48: 390 (1865) & in DC.: 314 (1866).

K1a.—Banka, Billiton.

No ecological information available.

Differs from *G. kerangae* in the more glossy upper surface of the leaves and their greenish colour when dry, their more strongly revolute margins, the more prominent nerves beneath, and the much smaller capsules, which are only 5–6 mm. diam., scarcely lobed and almost black when dry.

Apparently a very scarce species.

Glochidion cupreum *Airy Shaw* in Kew Bull. 27: 15 (1972).

G. aff. *fuscum* sec. Airy Shaw, *ibid.*: 13 (1972), *in obs.*, quoad *Kostermans* 13859.

K1.—Endemic.

Shrub or small tree to 5 m. high, in forest on sandy loam at 100–300 m. alt.

Closely related to *G. rubrum* and *G. fuscum*; capsules large as in the latter species (10–11 mm. diam.), but covered with a fine, short, chocolate-coloured indumentum, as in certain forms of *G. rubrum*. To be compared also with *G. korthalsii* (Muell. Arg.) Boerl.

Glochidion elmeri *Merr.*, Pl. Elmer.: 139 (1929); Meijer, 7: 45 (1967).

SB.—Endemic.

Shrub or small tree to 10 m. high, in primary forest on basalt or reddish soil or black sandy soil, ascending to 1500 m. on Kinabalu.

Habit lax; branches and leaves beneath shortly tomentellous; leaves broadly elliptic, chartaceous, 5–19 cm. long, up to 9 cm. broad; female flowers sessile, tomentellous, with narrow elongate sepals, up to 5 mm. long; capsule 1–2 cm. diam., 3-locular, somewhat depressed, whitish-pubescent, white or reddish when ripe.

Differs from *G. calospermum* in its smaller more elliptic leaves, shorter indumentun and ochraceous-brown seeds.

Glochidion glomerulatum (*Miq.*) *Boerl.*: 276 (1900); J. J. Sm.: 146 (1910); Ridley: 209 (1924); Corner: 286 (1940); Adelbert & Meeuse apud Backer in Blumea 5: 507 (1945); Backer & Bakh. f.: 463 (1963); Airy Shaw in Kew Bull. 26: 275 (1972); Whitmore: 100 (1973).

Agyneia? glomerulata Miq., Fl. Ind. Bat., Suppl.: 447 (1861).
Phyllanthus glomerulatus (Miq.) Muell. Arg. in Flora 48: 375 (1865) & in DC.: 293 (1866).
Phyllanthus nanogynus Muell. Arg., *ll.cc.*: 376 (1865) & 293 (1866).
Glochidion nanogynum (Muell. Arg.) Hook. f.: 318 (1887); Ridley: 214 (1924); Beille: 619 (1927).
G. palustre Koord. in Bull. Jard. Bot. Buitenz. III, 1: 145 (1918); cf. Adelbert & Meeuse and Backer & Bakh. f., *ll.cc.*

SR; SB; K1.—Siam, Indochina, Malaya, Sumatra, Java.

Shrub or tree to 20 m. high, in primary or secondary hill or swamp forest on black or brown sandy soil up to 1500 m. alt.

Without strongly marked vegetative characters; branchlets slender, shortly tomentellous or glabrescent; leaves mostly ovate-oblong or elliptic, 4–10(–15) cm. long, 2–3(–6) cm. wide, chartaceous, shortly tomentellous or puberulous below, mostly drying brown. Male flowers on rather long pedicels, the perianth-segments usually recurving in a characteristic manner. Female flowers in dense sessile glomerules, the ovary usually densely whitish or rusty-puberulous, crowned by the broad flat glabrous stigmatic disk. Capsule 3-locular, shallowly or deeply 6-lobed, 4–5(–9) mm. diam., puberulous, retaining the flat or slightly convex glabrous stigmatic disk in the depressed apex.

Apparently scarce in Borneo, Sumatra and Java, but commoner in Malaya. Develops pneumatophores in swamp forest in Java (*G. palustre* Koord.).

The closely related *G. wallichianum* Muell. Arg., of Lower Siam and N. Malaya, hardly differs except in the very different style, which is shortly columnar, slightly clavate and very shortly 3-lobed at the apex. Certain specimens of *G. desmocarpum* Hook. f., however, from Penang, seem almost intermediate. I had not appreciated this stylar character when I wrote the note on *G. wallichianum* in Kew Bull. 26: 281 (1971). There is also a greater tendency in *G. wallichianum* for the inflorescences to extend or 'spread' along the branch from the leaf-axil, forming a kind of 'caterpillar', but this sometimes occurs also in *G. glomerulatum* (cf. *Kochummen* KEP 76200, from Kuala Lumpur). (Cf. also *G. insigne.*)

Glochidion kerangae *Airy Shaw* in Kew Bull. 27: 18 (1972).

SR (C.); **B; SB.**—Endemic.

Tree to 15 m. high, in *kerangas* (tropical Heath forest) or *Shorea albida* swamp or open savannah-like vegetation, on giant podsol or very poor white sands or chocolate-coloured sandy soil, up to 100 m. alt.

Quite glabrous; leaves mostly elliptic, 4–8(–12) × 2–4(–5) cm., mostly cuneate at the base and obtuse or obtusely cuspidate at the apex, margin much reflexed, coriaceous, conspicuously glaucous-pruinose beneath. Inflorescences axillary, few-flowered; male flowers borne on filiform pedicels 1–1·5 cm. long; female flowers sessile. Capsule depressed, 10–12 mm. diam., 3–4-locular, slightly pruinose, pedicel up to 2 cm. long. A sticky red sap is said to ooze from the inner bark when cut.

A characteristic species of the *kerangas* formation.

Glochidion korthalsii (*Muell. Arg.*) *Boerl.*: 276 (1900); Merr.: 328 (1921).

Phyllanthus korthalsii Muell. Arg. in Flora 48: 377 (1865) & in DC.: 297 (1866).

K1a.—Endemic?

No ecological information.

Stated by Mueller to differ from *G. borneënse* in the shape and much greater size (8 mm. diam., 3·5 mm. long) of the capsules, which are rufescent-puberulous, in the longer (1 mm.) stylar column, and in the entire style-lobes. The inflorescences are said to be 'almost as in *G. zeylanicum*'. It is impossible to guess at the identity of this plant without seeing the type. It could be a form of *G. rubrum*, or *G. cupreum*.

Glochidion kunstlerianum *Gage* in Rec. Bot. Surv. Ind. 9: 220 (1922); Ridley: 210 (1924); Airy Shaw in Kew Bull. 23: 7 (1969) & 27: 7 (1972); Whitmore: 102 (1973).

K1.—Malaya, Sumatra.

No ecological information for Borneo.

Branchlets slender, elongate, shortly rusty-pubescent. Leaves oblong or elliptic-oblong, 4–9 cm. long, shortly pubescent beneath. Flowers in axillary glomerules; capsules depressed, 3-lobed, 7 mm. diam., 4 mm. long, finely pubescent, shortly pedicellate; styles free, short, rather stout, divaricate, apically attenuate or conical or obliquely truncate, usually puberulous. The styles are very distinctive for this species.

Glochidion lanceilimbum *Merr.* in Philipp. Journ. Sci. 26, Bot.: 462 (1925); Airy Shaw in Kew Bull. 27: 72 (1972).

G. perakense sec. Airy Shaw in Kew Bull. 26: 279 (1971), *pro parte, vix* Hook. f.

SR; SB.—Anamba Is., S. Philippines (Palawan, Mindanao), ? Celebes, Moluccas, New Guinea, Bismarcks, Solomons, Queensland.

Shrub or small tree to 9 m. high, in primary forest or forest margins or secondary scrub or thickets, on black soil or in poor sandy soil near the sea, up to 900 m. alt. on Kinabalu.

Closely related to *G. perakense*, into which it sometimes seems to merge, and yet distinct and recognizable over much of its range. Usually glabrous, but occasionally puberulous; leaves more or less lanceolate-elliptic, up to 15 × 4·5 cm., very oblique at the base, smooth and glossy, usually remaining greenish when dry. Inflorescences usually slightly supra-axillary and shortly peduncled, but sometimes almost axillary and sessile. Capsule very depressed, sometimes almost disc-like, 5–6 mm. diam., 3 mm. thick, not or scarcely lobed, usually glabrous, but with a roughish surface; stylar column slender, very short, less than 1 mm. long.

The green colour of the leaves on drying, and the very depressed capsule, usually distinguish this plant from *G. perakense*, but it must be admitted that the *perakense–lanceilimbum* complex is one of the most difficult in the genus.

Glochidion lanceisepalum *Merr.* in Journ. As. Soc. Str. Br. 86: 319 (1922).

Glochidion sp., probably undescribed, Merr., Pl. Elmer.: 140 (1929) (*Elmer* 20097).

SR; B; SB; K1a(?).—Endemic.

Shrub or tree to 6 m. high, in primary Mixed Dipterocarp forest or pole forest, on sandstone or bleached basalt-derived clay or dry soil, up to 1260 m. alt.

A distinct and striking species, not closely related to any other, almost glabrous except sometimes for the female flowers. Branchlets elongate, smooth, somewhat angled above; leaves large, elongate, mostly oblong-lanceolate, sometimes ovate-elliptic, up to 30 cm. long and 9 cm. wide, base cuneate to rounded or even unilaterally subcordate, acuminate or shortly caudate at apex, thickly coriaceous, with a dull, minutely granular surface, especially beneath, brown when dry; primary nerves mostly steeply ascending, prominent below, distinctly impressed above; inflorescences axillary or slightly supra-axillary; ovary glabrous or whitish-puberulous; stylar column 1 mm. long, slender, glabrous or puberulous, mostly 5-toothed at the apex; capsule slightly depressed, 10–12 mm. diam., mostly 5-locular, intruded at base and apex, almost glabrous.

The minutely granular leaf-surface is similar to that of *G. singaporense*, but that species is manifestly pubescent. It is strange that Merrill in 1929 failed to recognize the collections *Elmer* 20097 and *Ramos* 1248, 1423, from Sabah, as *G. lanceisepalum*, since he had himself described the species, only a few years previously, from a Sarawak specimen with which the Sabah collections agree perfectly.

Glochidion littorale *Bl.*, Bijdr.: 585 (1825); Hook. f.: 308 (1887); J. J. Sm.: 109 (1910); Hutch. in Journ. Linn. Soc., Bot. 42: 134 (1914); Merr.: 328 (1921) & 399 (1923) & Philipp. Journ. Sci. 29: 380 (1926) & Pl. Elmer.: 139 (1929); Ridley: 207 (1924); Beille: 610 (1927); Corner: 287 (1940); Backer & Bakh. f.: 461 (1963); Meijer, 7: 46, tab. (1967); Airy Shaw in Kew Bull. 26: 278 (1972); Whitmore: 101 (1973).

Phyllanthus litoralis [*sic*] (Bl.) Muell. Arg. in Flora 48: 370 (1865) & in DC.: 280 (1866).

Var. **littorale**:

SR; B; SB; K1, 3.—India, Ceylon, SE. Asia, and throughout W. Malesia. Shrub to 6 m. high, on tidal river banks, in thickets near mangrove swamps, among rocks by the sea-beach, in dense shrubbery on sea-cliffs, or in scrub land, on black or poor soil, from sea-level up to 30 m. alt. Also once collected in a flat swamp near a river bank, in primary forest on brown soil, at 900 m. alt. (Sabah, Sandakan Distr., *Ampuria* SAN 32645).

The coriaceous, glabrous, broadly obovate leaves, up to 8·5 × 6·5 cm., usually with a rounded apex, and rather large multilocular capsules with a short (1–5 mm.) pedicel and a very short stylar column, are diagnostic for this well-known coastal species.

The fruit is said to be edible. A preparation of the plant is used for bathing at childbirth. The timber is used for firewood.

Var. **culminicola** *Airy Shaw* in Kew Bull. 29: 290 (1974).

SR (SW.).—Endemic.
Shrub to 3·5 m. high, in elfinwood or pygmy forest or low scrub on burnt ground on exposed mountain tops, at 240–1800 m. alt.

Differs from the type in the style up to 3 mm. long, the slightly smaller and more coriaceous leaves, and the mountain habitat.

Var. **caudatum** *Airy Shaw* in Kew Bull. 29: 290 (1974).

SR (SW.).—Endemic.
Tree of 4·5 m., in secondary *Shorea albida* forest or peat-swamp forest at very low altitude.

Differs from the type in the leaves up to 11 × 5 cm., conspicuously abruptly acutely caudate (cauda 5–15 mm. long), and in the pedicels of the female flowers up to 13 mm. long.

Glochidion lutescens *Bl.*, Bijdr.: 585 (1825); J. J. Sm.: 169 (1910); Backer & Bakh. f.: 463 (1963).

Anisonema hypoleucum Miq., Fl. Ind. Bat., Suppl.: 449 (1861), **synon. nov.**

Glochidion glaucifolium Muell. Arg. in Linnaea 32: 65 (1863); Hook. f.: 321 (1887); Beille in Lecomte: 618 (1927); **synon. nov.**
Phyllanthus lutescens (Bl.) Muell. Arg. in Flora 48: 377 (1865) & in DC.: 296 (1866).
P. hypoleucus (Miq.) Muell. Arg. in Flora 48: 374 (1865).
P. laevigatus Muell. Arg., *l.c.* (1865), **synon. nov.**
P. glaucifolius (Muell. Arg.) Muell. Arg., *l.c.*: 378 (1865).
P. kollmannianus Muell. Arg., *l.c.* (1865), **synon. nov.**
Glochidion laevigatum (Muell. Arg.) Hook. f.: 319 (1887); Ridley: 215 (1924) (excl. var. *cuspidatum* Ridl.).
G. hypoleucum (Miq.) Boerl.: 275 (1900); Pax & Hoffm. in Mitt. Inst. Bot. Hamburg 7: 226 (1931); Airy Shaw in Kew Bull. 23: 8 (1969) & 26: 276 (1972) & 27: 71 (1972); Whitmore: 101 (1973).
G. breynioïdes C. B. Rob. in Philipp. Journ. Sci. 4, Bot.: 95 (1909); Merr. in Philipp. Journ. Sci. 11, C. Bot.: 68, 281 (1916) & Enum.: 328 (1921) & 398 (1923) & Pl. Elmer.: 139 (1929); **synon. nov.**
G. kollmannianum (Muell. Arg.) J. J. Sm.: 166 (1910); Merr., *l.c. supra* (1916) & Enum.: 328 (1921) & Philipp. Journ. Sci. 29: 379 (1926); Backer & Bakh. f.: 463 (1963); Meijer, 7: 46 (1967).
G. hollandianum J. J. Sm. in Nova Guinea 12: 544, t. 228A (1917), **synon. nov.**

SR; B; SB; K1, 1a.—SE. Asia, S. China, and throughout Malesia to New Guinea.

Shrub or tree to 21 m. high, in rich primary hill Dipterocarp forest, or Heath woodland or peat-swamp forest or low thin forest or secondary scrub, sometimes on steep rocky mountain-sides, on igneous- (andesitic) derived soil or blackish soil or brown sandy soil or yellow sandy loam (sometimes periodically inundated), once noted on limestone, up to 1050 m. alt.

Glabrous, smooth; leaves firmly chartaceous, up to 8 × 4 cm., more or less acute, usually more or less glaucescent beneath; sepals of female flowers usually distinctly keeled on the back; ovary glabrous; capsule 5–7 mm. diam., depressed, shallowly 8–12-lobed, often slightly widened upwards, thinly crustaceous, light chestnut when dry, styles very abbreviated or almost obsolete.

The glaucescent undersurface of the leaves and especially the dorsally keeled female sepals are characteristic for this widespread species. Blume's original description is brief and inadequate, but Mueller saw Blume's type material and mentions these two points expressly in his revised, detailed description of *G. lutescens*. This now appears to be the earliest name for this much-described plant. Cf. also Pax & Hoffmann, *l.c. supra* (1931), sub *G. hypoleucum*.

Glochidion macrostigma *Hook. f.*: 313 (1887); Ridley: 214 (1924); Whitmore: 100 (1973); Airy Shaw in Kew Bull. 29: 289 (1974).

G. capitatum J. J. Sm.: 133, 135 (1910); Backer & Bakh. f.: 461 (1963).

SR; SB; K1.—Malaya, Java.

Large tree, 15 m. high or more, in primary forest on igneous- (andesitic) derived soils up to 840 m. alt.

Closely related to *G. obscurum*, but capsules usually smaller and more shortly pedicelled, and styles very short and truncated, forming a globose button on the top of the fruit. In the flowering stage the styles are considerably larger than the ovary.

Glochidion mehipitense *Pax & Hoffm.* in Mitt. Inst. Bot. Hamburg 7: 225 (1931).

K3.—Endemic (?).

Tree of unknown stature in primary forest at 900 m. alt.

No authentic material of this species has been seen, but the description suggests the possibility that it might be identical with my *G. calospermum* (p. 122, *supra*). The latter is in fact known from **K3**, as well as from several other parts of Borneo. Pax & Hoffmann compared *G. mehipitense* with *G. insigne* J. J. Sm., of Java, stating that it differed from this 'durch die Vielzahl der Staubblätter'—presumably a slip for 'Vierzahl', since the stamens are given as four. The stamens both of *G. calospermum* and of *G. insigne* are normally three in number, but without knowing whether the number four is constant in *G. mehipitense* I would not necessarily take this as a significant difference. Similarly with the number of the female sepals—five in *mehipitense* and six in *calospermum*—and the number of loculi in the ovary and capsule—four in *mehipitense* and three to four in *calospermum* and *insigne*. It does not, however, seem possible to establish the identity of Pax & Hoffmann's species without seeing type material.

Glochidion mindorense *C. B. Rob.* in Philipp. Journ. Sci. 4, Bot.: 98 (1909) & 6, Bot.: 208 (1911); Merr.: 401 (1923); Meijer, 7: 46 (1967); Airy Shaw in Kew Bull. 27: 21 (1972) & 29: 291 (1974).

Subsp. **mindorense**: Airy Shaw, *l.c.* (Other subspecies in New Guinea and Queensland.)

SB.—Philippines, Celebes, Moluccas, Lesser Sunda Is.

Shrub or tree to 12 m. high, in primary forest on coral rock or rocky soil, mostly near the seashore, from sea-level up to 15 m. alt.

A coastal plant especially characteristic of some of the off-shore islands of Sabah; noted by one collector as 'seasonal'. Leaves often somewhat oblong in outline, 5–11 cm. long, firmly chartaceous or subcoriaceous, almost glabrous, somewhat resembling the leaves of *G. lutescens* or *G. rubrum*, gradually narrowed to the apex, which may be acute or obtuse or apiculate; nerves slender, evident but not or only slightly prominulous below, almost immersed; capsule depressed, 9–13 mm. diam., 5–6-locular, with 5–6 rounded lobes, minutely puberulous, deeply intruded at base and apex; styles exceedingly short.

Glochidion molle *Bl.*, Bijdr.: 586 (1825); C. B. Rob. in Philipp. Journ. Sci. 4, Bot.: 95 (1909); J. J. Sm.: 136 (1910); Merr. in Philipp. Journ. Sci. 11, C. Bot.: 282 (1916) & Enum.: 401 (1923): Backer & Bakh. f.: 464 (1963).

Phyllanthus mollis (Bl.) Muell. Arg. in Flora 48: 387 (1865) & in DC.: 308 (1866).

SB.—Philippines, Java, Celebes, Moluccas, Tenimber Is.
Tree to 5 m. high on river bank at 15 m. alt.

An easily recognized species. often remarkably mimicking *Sauropus villosus* (Blanco) Merr. Strongly fulvous-pubescent; leaves regularly distichous, oblong or lanceolate-oblong, occasionally almost ovate, 4–6(–9) cm. long, 1·5–2(–3) cm. broad, truncate or subcordate at the base; capsule depressed, tricoccous, 6-lobed, 7–9 mm. diam., thinly pilose.

Only one collection seen: Tigaman, 16 May 1933, *Balajadia* (North Borneo For. Dept.) 3293. It is remarkable that *G. molle* has apparently never been collected again in Sabah by the energetic Forestry Department. The species is abundant in the Philippines and Java. The above collector (Balajadia) states that the sap is used for dyeing clothes. J. J. Smith (*l.c.*) mentions that in Java the plant was formerly regarded as providing an excellent remedy for snake-bite.

Glochidion monostylum *Airy Shaw* in Kew Bull. 29: 288 (1974).

SB (Kinabalu).—Endemic.
Shrub or tree to 12 m. high, in primary hill forest on brown soil up to 1680 m. alt.

Superficially similar to forms of *G. glomerulatum*, *G. tenuistylum*, etc., but very distinct in its slender elongate style (cf. *G. styliferum*), minutely 3-toothed at the apex.

Glochidion nitidum *Merr.* in Philipp. Journ. Sci. 9, Bot.: 483 (1914) & Enum.: 401 (1923).

SB.—Philippines.
Shrub of 1·5 m., in forest (?) at 300 m. alt.

Glabrous; leaves coriaceous, elliptic, 5–10 cm. long and 2·5–4 cm. wide, cuneate to almost rounded at the base, very shortly and acutely caudate at the apex, smooth and glossy above, nerves conspicuously raised below, drying brownish; inflorescences axillary; capsule somewhat inflated, 1–2 cm. diam., 5–6-locular, with 5–6 rounded lobes, intruded at base and apex.

Further Bornean material is needed in order to settle definitely the status of the solitary specimen seen (Tambatu, Tambunan, alt. 300 m., 28 Feb. 1934, *Puasa-Angian* (North Borneo For. Dept.) 3877). It is also possible that *G. nitidum* is not distinct from *G. subfalcatum* Elm. (1908).

Glochidion obscurum (*Roxb. ex Willd.*) *Bl.*, Bijdr.: 585 (1825); Hook. f.: 317 (1887); J. J. Sm.: 122 (1910) (incl. var. *macrocalyx* J. J. Sm.), multis cum synonymis javanicis, *q.v.*; Ridley: 208 (1924); Beille: 623 (1927); Hend. in Journ. Malayan Br. Roy. As. Soc. 17: 70 (1939); Corner: 287 (1940); Backer & Bakh. f.: 461 (1963); Meijer, 7: 46 (1967); Airy Shaw in Kew Bull. 26: 279 (1972); Whitmore: 99, 101 (1973).

Phyllanthus obscurus Roxb. ex Willd., Sp. Pl. 4: 581 (1804); Muell. Arg. in Flora 48: 373 (1865) & in DC.: 287 (1866).
Glochidion glaucum Bl., Bijdr.: 587 (1825); Merr.: 328 (1921).
G. roxburghianum Muell. Arg. in Linnaea 32: 61 (1863).

G. obscurum var. *glabrum* Pax & Hoffm. in Mitt. Inst. Bot. Hamburg 7: 226 (1931), *e descr.*

SR; SB; K1, 1a, 3.—Indochina, Lower Siam, W. Malesia (except Philippines), Celebes, Moluccas, W. New Guinea (var.).

Shrub or tree to 15 m. high, in primary or secondary forest, most frequently on river margins, but also in hill forest, on black or clay-rich soil, up to 150 m. alt.

Widespread and variable, but usually recognizable from its very oblique, smallish leaves, glaucous-grey beneath, and globose, unlobed, thick-walled, tardily dehiscent, often long-pedicelled capsules, often crowned with a large conspicuous funnel-shaped style.—Pax & Hoffmann's var. *glabrum* was distinguished by its glabrous male and female sepals, the latter erect instead of reflexed, but I doubt if these points have any taxonomic significance.

Wood used for construction purposes in Sabah.

Glochidion philippicum (*Cav.*) *C. B. Rob.* in Philipp. Journ. Sci. 4, Bot.: 103 (1909); J. J. Sm.: 139 (1910); Merr.: 401 (1923); Meijer, 7: 45 (1967); Airy Shaw in Kew Bull. 27: 58, 73 (1972).

Bradleia philippica Cav., Ic. 3: 48, t. 371 (1797).

B. philippensis Willd., Sp. Pl. 4: 592 (1805), *nom. illegit.*

Glochidion philippinense Benth., Fl. Hongk.: 314 (1861), *nom. illegit.*

G. compressicaule Kurz ex Teijsm. & Binnend. in Nat. Tijdschr. Ned. Ind. 27: 45 (1864).

Phyllanthus philippinensis (Benth.) Muell. Arg. in Flora 48: 376 (1865) & in DC.: 295 (1866) (incl. vars.).

P. compressicaulis (Kurz ex Teijsm. & Binnend.) Muell. Arg., *ll.cc.*

P. quercinus Muell. Arg. in DC.: 306 (1866).

P. kurzianus Muell. Arg.: 1272 (1866).

Glochidion quercinum (Muell. Arg.) Boerl.: 276 (1900).

SB (incl. Banggi I.).—Formosa, Philippines, Talaud Is., ? Sumatra (Enggano I.; very doubtful), Java and throughout E. Malesia to Solomon Is. and N. Queensland.

Tree of 12 m. in secondary forest on sandy soil by a river bank, doubtless at low alt.

This first record of *G. philippicum* for Borneo rests upon two collections only. One is a somewhat poor collection from Banggi (Banguey) Island: R. Pangkalan, April 1885, *Dr. M. Fraser* 261, preserved at Kew. It bears the determination 'cf. Glochidion philippicum *Robinson*', signed by A. T. Gage, and although only leaves and male flowers are present I believe this name is correct. It represents a rather weakly puberulous form of the species. Merrill does not list *G. philippicum* in his account of the flora of Banggi in *Philipp. Journ. Sci.* 29: 379–380 (1926). The other collection is a recent fruiting specimen from the Lamag road, Sandakan, Sept. 1971, *Imbugan & Patrick* SAN 74254. The almost complete absence of this otherwise widespread species from Sumatra, Malaya and Borneo is presumably due to climatic factors.

Glochidion pubicapsa *Airy Shaw*, nom. nov., cum diagn. lat.

G. pubicarpum Elm., Leafl. Philipp. Bot. 10: 3733 (1939), *anglice, nom. non rite publ.*

Glochidion sp., Merr., Pl. Elmer.: 140 (1929) (*Elmer* 20769).

G. arborescenti Bl. peraffine, sed capsulis exceptis glaberrimum, inflorescentiis brevius vel haud pedunculatis. Typus: Philippines: Luzon, Irosin, Nov. 1915, *Elmer* 15498 (K).

Var. **pubicapsa:**

SR (NE.); **SB; K1.**—Philippines; ? Sumatra; ? Java.

Tree to 14 m. high, in primary or secondary riparian or hill forest (sometimes swampy), on brown or grey sandy soil or stony clay loam, up to 450 m. alt.

Very lax in habit, with long smooth angled branchlets, almost completely glabrous except for the capsules. Leaves chartaceous to stiffly coriaceous, very variable, ovate, elliptic or oblong, 6–23 cm. long, 3–9·5 cm. wide, light or dark brown when dry, rounded to cuneate at the base, acute to caudate at the apex. Inflorescences rather dense-flowered, sessile or almost so. Capsules often borne in dense glomerules on leafless branchlets, depressed, shallowly lobed, 4(–5)-locular, 6–10 mm. diam., said to be chalky white when mature, papillose-puberulous, with very short styles; seeds light brown, very glossy.

Bark used for dyeing fish-nets (Sabah).

I have taken the opportunity afforded by the invalid publication of the name *G. pubicarpum* (without Latin description, after 1935) to replace Elmer's hybrid Latin–Greek epithet by an all-Latin one, validating it by a Latin diagnosis.

Var. **brunneiforme** *Airy Shaw*, var. nov., a var. *pubicapsa* inflorescentiis distincte (3–5 mm.) pedunculatis, capsulis submajoribus densiuscule implexo-albido-pubescentibus manifeste recedit.

SR.—Fourth Division: Bt. Mersing, Anap, basalt ridge, in Mixed Dipterocarp forest, alt. 800 m., 23 Sept. 1964, *Sibat ak. Luang* S. 22198 (K, type): —Tree 18 m. tall, 90 cm. girth; fruit yellow.

This plant is superficially very like *G. brunneum*, especially in its glabrescence, but the capsule lacks the conspicuous stylar column of that species, and except for its matted whitish pubescence is very similar to that of var. *pubicapsa*. Further material may, however, indicate that the plant deserves specific rank.

Glochidion punctatum *Pax & Hoffm.* in Mitt. Inst. Bot. Hamburg 7: 226 (1931).

K3.—Endemic?

Tree on river bank at 35 m. alt.

I have seen no material of this taxon. The description suggests a species in the affinity of *G. eriocarpum* Champ., *G. luzonense* Elm., *G. fulvirameum* Miq., *G. meijeri* Airy Shaw, etc. The plant was described without fruit. Its status is indeterminable in the absence of authentic specimens.

Glochidion rubrum *Bl.*, Bijdr.: 586 (1825); C. B. Rob. in Philipp. Journ. Sci. 4, Bot.: 101 (1909); J. J. Sm.: 149 (1910) (cum f. *longistyli* J. J. Sm.)

[*q.v.* for detailed synonymy]; Merr.: 402 (1923) & Philipp. Journ. Sci. 29: 380 (1926) ('var.?'); Beille: 621 (1927); Corner: 288 (1940); Backer & Bakh. f.: 464 (1963); Meijer, 7: 46 (1967); Airy Shaw in Kew Bull. 26: 279 (1972); v. Steenis, Mountain Fl. Java: t. 18, fig. 7 (1972); Whitmore: 101 (1973).

Phyllanthus diversifolius Miq., Fl. Ind. Bat., Suppl.: 448 (1861); Muell. Arg. in Flora 48: 378 (1865) & in DC.: 297 (1866).

P. penangensis Muell. Arg. in Flora 48: 388 (1865) & in DC.: 310 (1866).

Glochidion leiostylum Kurz, For. Fl. Brit. Burma 2: 345 (1877); Hook. f.: 324 (1887); Merr. in Philipp. Journ. Sci. 11, C. Bot.: 68 (1916) & Enum.: 328 (1921); Ridley: 212 (1924).

G. coronatum Hook. f.: 326 (1887); Ridley: 212 (1924); Corner: 286 (1940).

G. diversifolium (Miq.) Merr. in Philipp. Bur. For. Bull. 1: 29 (1903).

G. thorelii Beille: 622 (1927).

G. penangense (Muell. Arg.) Airy Shaw in Kew Bull. 23: 6 (1969) (excl. var. *tenuistylum*); Whitmore: 101 (1973).

SR (C. & NE.)**; B; SB; K1, 1a,** ?**2, 3.**—SE. Asia and throughout W. Malesia to the Moluccas and Lesser Sunda Is.

Shrub or tree to 15 m. high, in primary or secondary riparian or mixed Lowland Dipterocarp forest or peat forest (sometimes periodically inundated), or on river banks beyond tidal influence, on clay loam or silty clay, frequently on sandy soil (brown or blackish, stony or rocky), or on volcanic soil or tufa, ascending to 1500 m. on Kinabalu.

Noted by one collector as poisonous. Bark used for tanning nets.

An exceedingly variable species, generally recognizable but sometimes making a deceptive approach to species such as *G. lutescens* or *G. mindorense*, and doubtfully demarcated from 'satellite' taxa such as *G. brooksii*, *G. cupreum*, *G. fuscum*, etc. Style-length (e.g. *G. leiostylum*) and pubescence (*G. penangense*) seem quite unreliable for specific distinction, being apparently uncorrelated with other characters. The most constant features seem to be the rather satiny undersurface of the leaves, with a distinct tendency for the nerves to pucker on drying, and the rather small capsule with 3(–6) broad rounded lobes. The species is apparently scarce in Sarawak. It evidently favours poor sandy soils.

Glochidion sericeum (*Bl.*) *Zoll. & Mor.* in Nat. en Geneesk. Arch. Neêrl. Indic 2: 585 (1845); Hook. f.: 326 (1887); Bocrl.: 276 (1900); J. J. Sm.: 170 (1910); Merr.: 329 (1921); Ridley: 215 (1924); Pax & Hoffm. in Mitt. Inst. Bot. Hamburg 7: 226 (1931); Backer & Bakh. f.: 462 (1963); Meijer, 7: 45 (1967); Whitmore: 101 (1973).

Glochidionopsis sericea Bl., Bijdr.: 588 (1825).

Phyllanthus sericeus (Bl.) Muell. Arg. in Flora 48: 390 (1865) & in DC.: 314 (1866).

SR; SB; K1.—Malaya, Sumatra, Java.

Shrub or tree to 10 m. high, in secondary peat-swamp forest following exploitation (not recorded from primary peat-swamp forest), or in forest intermediate between Heath and Dipterocarp type, or in *Agathis* forest on waterlogged sandy acid soil, or on hill-ridges or slopes, once noted on a steep moist rock-wall, on red, black or yellow soil, or loam, up to 1050 m. alt.

Superficially similar to *G. philippicum*, with minutely densely grey-puberulous or ochraceous branchlets and leaf-undersurface, but leaves with an almost symmetrical rounded or cordate base and with more closely parallel primary nerves; inflorescences few-flowered; capsules 1–3 together, tricoccous, depressed, 5–7 mm. diam., with dense matted whitish pubescence; cocci bilobed; styles linear, almost free; pedicel slender, 5–8 mm. long.

Glochidion singaporense *Gage* in Rec. Bot. Surv. Ind. 9: 221 (1922); Ridley: 209 (1924); Airy Shaw in Kew Bull. 23: 7 (1969) & 27: 8 (1972); Whitmore: 101 (1973).

SB; K1.—Malaya, Sumatra.

Shrub or tree to 7·5 m. high, in primary forest on loam containing lime at 20 m. alt.

Stem and leaf-underside puberulous; leaves mostly lanceolate or oblong, sometimes elliptic, sometimes more or less asymmetric, thickish in texture, with a dull, minutely granular surface, especially below, as in *G. lanceisepalum*. Inflorescences axillary, many-flowered; capsules 3(–4)-locular, broadly 3(–4)-lobed, 4–5 mm. diam., minutely puberulous, thinly crustaceous, brown when dry, with a very short stylar column; pedicel slender, 4–6 mm. long.

The pubescence and smaller leaves and capsules distinguish this species from *G. lanceisepalum*. The dull, granular leaf-surface occurs in very few other species.

Glochidion styliferum *J. J. Sm.* in Bull. Jard. Bot. Buitenz. III, 1: 391, t. 40 (1920); Airy Shaw in Kew Bull. 27: 19 (1972).

SR; SB; K1.—Endemic.

Shrub or small tree to 6 m. high, in primary forest at unknown altitude.

Closely related to *G. sericeum*, but leaves larger, up to 12·5 cm. long, glaucous beneath but glabrous except for the minutely puberulous nerves; capsules up to 4 together, shortly papillose-puberulous; styles united in a very slender almost filiform column, 2 mm. long, shortly trifid at the apex.

A scarce species; only five gatherings so far known.

Glochidion superbum *Baill.*, Ét. Gén. Euphorb.: 638 (1858), *nomen subnudum, ex Muell. Arg.* in Linnaea 32: 64 (1863), *descr.*; Hook. f.: 323 (1887); Boerl.: 275 (1900); Merr.: 329 (1921); Ridley: 208 (1924); Corner: 289 (1940); Backer & Bakh. f.: 462 (1963); Meijer, 7: 45 (1967); Airy Shaw in Kew Bull. 26: 280 (1972); Whitmore: 99, 100 (1973).

G. dasyphyllum Miq., Fl. Ind. Bat., Suppl.: 451 (1861), *non* K. Koch (1853).
Phyllanthus superbus (Baill.) Muell. Arg. in Flora 48: 375 (1865) & in DC.: 292 (1866).

SR; B; SB; K1.—Lower Siam, Malaya, Sumatra, Banka.

Tree to 21 m. high, in primary or secondary hill forest, or along trails, on yellow sandy clay over Tertiary sandstone, or on yellow sandy loam, up to 325 m. alt.

Strongly tomentose, robust; leaves large, coriaceous, up to 20 × 10 cm. mostly cordate at base, venation very prominent below and distinctly impressed above. Inflorescence mostly long-peduncled (1–2 cm.), axillary, occasionally sessile or supra-axillary, distinctly branched, many-flowered. Capsules 5–7 mm. diam., 3(–4)-locular, broadly lobed, thinly crustaceous, strongly pubescent, borne on slender pubescent pedicels up to 12 mm. long.

Glochidion tenuistylum *Stapf* in Trans. Linn. Soc. II, 4: 223 (1894); Merr.: 329 (1921).

G. coronatum sec. Meijer, 7: 46 (1967), *non* Hook. f.
G. penangense var. *tenuistylum* (Stapf) Airy Shaw in Kew Bull. 23: 7 (1969).

SB (Kinabalu).—Endemic.
Small shrub at 900–1200 m.; no further information.

Shortly and thinly tomentellous; leaves thinly chartaceous, oblong to elliptic, 5–12 cm. long, 2–4·5 cm. broad, very shortly acuminate, acute, brown when dry; venation very slightly bullate; female flowers sessile, in few-flowered axillary glomerules; calyx tomentose, deeply divided into narrow segments; styles 5, linear, 2–3 mm. long, long-exserted, connate for two-thirds, free and somewhat divaricate above, acute, pilosulous; fruit unknown.

Only one further specimen seen, other than the type: Dallas, Kinabalu, 15 Dec. 1931, *J. & M. S. Clemens* 27553. The 5 long-exserted, subulate, fascicled, apically free styles are very distinctive. I was quite mistaken in reducing this species to *G. penangense*.

Glochidion trusanicum *Airy Shaw* in Kew Bull. 27: 20 (1972).

SR (NE.).—Endemic.
No ecological information; probably at low altitude.

Almost completely glabrous, with lax habit; leaves oblong to elliptic, attenuate at base and apex, up to 12 × 4 cm., membranous or thinly chartaceous, dull and olivaceous when dry, slightly glaucous beneath; male flowers unknown; female flowers subsessile, in few-flowered axillary fascicles, the tepals conspicuous, 4–5 mm. long, oblong or narrowly elliptic, connate below into a more or less campanulate cup, acute, dorsally keeled, sometimes with a few minute forward-pointing teeth on the margin; ovary 3-locular, glabrous; style slenderly columnar, 2 mm. long, scarcely thickened but minutely trifid at the apex; capsule perhaps 1·5 cm. diam., somewhat inflated.

The general appearance of this rare plant (only known from the rather poor type) somewhat recalls that of *Sauropus androgynus*, particularly the foliage. It appears to bear no close relationship to the few other long-styled Bornean species of *Glochidion*—*G. rubrum* forma, *G. styliferum*, *G. monostylum*—but may perhaps have some relation with *G. acutifolium* Alston, of Ceylon.

Glochidion cf. **williamsii** *C. B. Rob.* in Philipp. Journ. Sci. 3, Bot.: 199 (1908) (*excl. descr. capsulae*) & 4: 95 (1909) (*descr. caps.*); Merr.: 403 (1923); *vel sp. aff.*

G. lucidum sec. Meijer, 7: 46 (1967), *non* Bl.

SB (Kinabalu).—Philippines.
Small tree to 9 m. high, in rich primary forest on mountainside at 1200–2250 m. alt.

I refer here a variable complex of gatherings from Kinabalu, which superficially at least are very close to *G. williamsii*, but which unfortunately lack the very important item of female flowers, and the only available fruits are broken and fragmentary. A positive identification is therefore not possible. In another direction some of the specimens approach forms of *G. rubrum*, but differ in their broad bicarinate petioles. The plants are glabrous, rigid, with often more or less rhombic-elliptic leaves, 4–9·5 cm. long, 2–4 cm. wide, cuneate at the base and very shortly acuminate at the apex, stiffly coriaceous, mostly few-nerved, drying dark brown or chestnut. As noted above under *G. brooksii*, there seems to be a close approach between the Sarawak representatives of that species and the present population from Kinabalu, the only obvious difference being the long-acuminate leaves of the former. More complete material will be needed before this question can be resolved.

Glochidion xestophyllum *Airy Shaw* in Kew Bull. 23: 11 (1969); Whitmore: 99 (1973).

K1.—S. Malaya (?)

Tree of 15 m., on a low ridge on sandy soil at 20 m. alt.

Superficially resembling *G. nitidum* Merr. and *G. subfalcatum* Elm., of the Philippines, especially in the rather robust, smooth, angled stems and glossy more or less oblong coriaceous leaves, greenish when dry, but differing profoundly in the many-flowered inflorescences and especially in the totally different, small, slightly depressed capsules, which are only 5 mm. in diameter. I do not now think that this plant is related to *G. lanceilimbum*, as I first suggested (*l.c. supra*).

I have not yet seen the Johore collection referred to this species by Whitmore, *l.c.*

Homalanthus *Juss.*

1 Leaves membranous, venation finer and less conspicuous, glands at apex of petiole very small or absent; fruits usually borne on long slender pedicels; styles 2–4 mm. long . . . . **H. populneus**

1 Leaves more chartaceous, venation stronger and beautifully reticulate beneath, glands at apex of petiole prominent and conspicuous; fruits borne on shorter stouter pedicels; styles robust, 5–10 mm. long **H. caloneurus**

Homalanthus caloneurus *Airy Shaw* in Kew Bull. 21: 413 (1968); Meijer, 10: 231 (1968).

H. populneus var. *cordifolius* Heine in Fedde, Rep. Sp. Nov. 54: 235 (1951).

B; SB.—Endemic.

Tree of 6 m., in montane forest in lower montane zone at 1200–3000 m. alt.

Allied to *H. alpinus* Elm., of the Philippines, from which it differs in the conspicuous, beautifully reticulate pattern formed by the strong secondary and tertiary nerves on the lower leaf-surface. On the upper surface the nerves are incised, instead of being raised as in *H. alpinus*. The petiolar glands lie closely juxtaposed on the upper surface, instead of occupying a lateral or even inferior position. On some leaves a line of black punctiform glands is present on each side of the midrib. The last two features are reminiscent of certain *Passiflora* species: cf. de Wilde in *Blumea* 20: 232, fig. 2 (1972).

Homalanthus populneus (*Geisel.*) *Pax* in Engl. & Prantl, Pflanzenf. III. 5: 96, fig. 60 (1890); J. J. Sm.: 621 (1910); Pax & Hoffm. v: 46 (1912); Merr.: 347 (1921) & 460 (1923) & in Philipp. Journ. Sci. 29: 387 (1926); Holth. & Lam in Blumea 5: 202 (1942); Backer & Bakh. f.: 498 (1963); Meijer, 7: tab. (1967); Airy Shaw in Kew Bull. 21: 409 (1968) & 26: 281 (1972); Whitmore: 102 (1973).

Stillingia populnea Geisel., Croton. Monogr.: 80 (1807).
Omalanthus leschenaultianus Juss., Euph. Gen. Tent.: 50, t. 16, fig. 53 (1824).
Carumbium populifolium Reinw. in Bl., Cat. Gewassen Buitenz.: 105 (1823), *nomen*, & in Syll. Pl. Ratisb. 2: 6 (1825–6); Scheff. in Ann. Mus. Bot. Lugd.-Bat. 4: 127 (1868). [*Non Omalanthus populifolius* Grah. (1827).]
C. populneum (Geisel.) Muell. Arg.: 1144 (1866).
Homalanthus populifolius (Reinw.) Hook. f.: 469 (1888); Ridley: 313 (1924); *non Omal. populifolius* Grah. (1827).
H. giganteus sec. Boerl.: 295 (1900), *p.p.*, *non* Zoll.
Sapium sebiferum sec. Hutch. in Journ. Linn. Soc., Bot. 42: 136 (1914), *non* (L.) Roxb.
S. indicum sec. Merr.: 348 (1921), quoad ref. 'Gibbs in Journ. Linn. Soc.' et specim. *Gibbs* 2659, *non* Willd.

SR; B; SB; K1, 1a, 3.—Lower Siam (extreme S.), and throughout W. & E. Malesia, except New Guinea; Bismarck Archip.

Shrub or tree to 15 m. high, very common in primary and secondary forests, thickets, *kerangas*, swamps, etc., on white sand, clay-rich yellow soil, loam with lime, and limestone rocks and cliffs, up to 1750 m. alt. on Kinabalu.

The fruits are said to be used medicinally in Sabah, for treating wounds.

Boerlage's record of *H. giganteus* Zoll. for Borneo (as well as for Sumatra, Celebes and Amboina) was almost certainly erroneous.

Homonoia *Lour.*

Homonoia riparia *Lour.*, Fl. Cochinch.: 637 (1790); Muell. Arg.: 1023 (1866); Hook. f.: 455 (1887); J. J. Sm.: 547 (1910); Pax & Hoffm. xi: 114 (1917); Merr.: 448 (1923); Ridley: 309 (1924); Gagnep.: 330 (1925); Merr. in Trans. Amer. Philos. Soc. II, 24(2): 239 (1935); Corner: 258 (1940); Li, Woody Fl. Taiwan: 430, fig. 152 (1963); Backer & Bakh. f.: 492 (1963); Airy Shaw in Kew Bull. 26: 282 (1972); Whitmore: 103 (1973).

Croton salicifolius Geisel., Croton. Monogr.: 6 (1807).
Adelia neriifolia Heyne ex Roth, Nov. Pl. Spec.: 375 (1821); Wight, Ic. Pl. Ind. Or. 5: 20, t. 1868 (1852).
Lumanaja fluviatilis Blanco, Fl. Filip. ed. 1: 821 (1837), ed. 2: 568 (1845) & ed. 3, 3: 236, t. 338 (1879).
Ricinus salicinus Hassk., Cat. Hort. Bogor.: 237 (1844) & Pl. Jav. Rar.: 264 (1848).

Haematostemon salicinus (Hassk.) Baill., Ét. Gén. Euphorb.: 293 (1858).
Spathiostemon salicinus (Hassk.) Hassk., Retzia: 41 (1858).
S. salicinus var. *angustifolius* Miq., Fl. Ind. Bat., Suppl.: 452 (1861).

SB.—India, SE. Asia, S. China, Formosa, throughout W. Malesia to Moluccas and Lesser Sunda Is., and recently collected in E. New Guinea (Territory of New Guinea, Central Distr., Port Moresby subdistr., Pt. Moresby–Sogeri road, alt. 60 m., 11 May 1971, *Streimann & Kairo* LAE 51618).
Rheophytic shrub of 2 m., on river banks and in rocky stream beds, up to 200 m. alt. or more.

Apparently extremely local in Borneo (only four collections seen, all from Sabah), although stated to be locally common along rapid streams. The root is said to be used for making bolo handles.

Jatropha *L.*

1 Petals coherent to middle, greenish yellow; leaves shortly 3–5-lobed or unlobed, margin not ciliate . . . . J. curcas
1 Petals free or almost so, purple; leaves 3–5-partite, glandular-ciliate J. gossypiifolia

Jatropha curcas *L.*, Sp. Pl.: 1006 (1753); Muell. Arg.: 1080 (1866); Hook. f.: 383 (1887); J. J. Sm. 566 (1910); Pax in Engler, IV.147 (Heft 42): 77 (1910); Merr.: 344 (1921) & 449 (1923); Ridley: 251 (1924); Gagnep.: 324 (1925); Merr. in Philipp. Journ. Sci. 29: 384 (1926) & in Trans. Amer. Philos. Soc. II, 24(2): 239 (1935); Corner: 259 (1940); Backer & Bakh. f.: 494 (1963); Airy Shaw in Kew Bull. 26: 283 (1971).

SB.—Native of Tropical America.
Shrub or tree to 5 m. high, planted at low altitudes.

Jatropha gossypiifolia *L.*, Sp. Pl.: 1006 (1753); Muell. Arg.: 1086 (1866); Hook. f.: 383 (1887); J. J. Sm.: 562 (1910) (var. *elegans* Muell. Arg.); Pax in Engler, IV.147 (Heft 42): 26 (1910); Merr.: 344 (1921) & 449 (1923); Ridley: 254 (1924); Gagnep.: 326 (1925); Corner: 260 (1940); Backer & Bakh. f.: 494 (1963) (var. *elegans*); Airy Shaw in Kew Bull. 26: 283 (1971).

SB; K1a.—Native of Tropical America.
Shrub to 2 m. high, cultivated at low altitudes.

Koilodepas *Hassk.*

1 Stipules 9–12 mm. long, deeply and finely pectinate; fruiting calyx membranous, much enlarged, up to 2·5 cm. long . . **K. pectinatum**
1 Stipules 3–7 mm. long, entire or shortly and sparsely pectinate; fruiting calyx almost unchanged, not large and membranous:
2 Leaves very smooth, entire, green when dry, quite glabrous, up to 33 cm. long; stamens much exserted, filaments narrowly subulate, divaricate-recurved . . . . **K. laevigatum**

2 Leaves less smooth (nerves more prominent below), entire or dentate, usually brown or purplish when dry, often ± stellate-pubescent; stamens less or scarcely exserted:

3 Leaves distinctly elevated (convex) above and hollowed (concave) below at the base on each side of the midrib; ♀ calyx deeply lobed, lobes subulate (elongate and narrowly linear in fruit in var. *stenosepalum*) . . . . **K. brevipes**

3 Leaves without marked basal convexities; ♀ calyx shortly and obscurely lobed:

4 Leaves usually drying brownish, usually distinctly dentate at least in the distal half (Sarawak & Sabah) . . **K. longifolium**

4 Leaves drying purplish, entire or almost so (SE. Borneo) **K. frutescens**

Koilodepas brevipes *Merr.* in Philipp. Journ. Sci. 30: 80 (1926) & Pl. Elmer.: 156 (1929); Airy Shaw in Kew Bull. 14: 389 (1960); Meijer, 7: 47 (1967).

Var. **brevipes**:

K. stenosepalum Airy Shaw in Kew Bull. 14: 390 (1960), *p.p.*, *typo excl.*

SB; K1.—Endemic.
Shrub or tree to 4·5 m. high, in primary Dipterocarp forest on sandstone or sandy loam or ultrabasic soil or dark red-brown or yellowish-black soil, especially on dry ridges, from sea-level up to 300 m. alt.

Closely related to *K. longifolium*, but apparently differing in the more shortly petiolate (5 mm.) leaves, which are distinctly poculiform (hollowed below, convex above) at the extreme base, and in the deeply lobed female calyx, with subulate lobes.

Var. **stenosepalum** (*Airy Shaw*) *Airy Shaw*, comb. & stat. nov.

K. stenosepalum Airy Shaw in Kew Bull. 14: 390 (1960), *quoad typum*, *non* 16: 356 (1963); Meijer, 7: 47 (1967).

K1.—Endemic.
Shrub of 2 m., in forest under 100 m. alt.

Additional material received over the past ten years indicates that *K. stenosepalum* is not specifically separable from *K. brevipes*. The elongate, narrowly linear fruiting sepals developed in this form are, however, extraordinarily distinctive.

Koilodepas frutescens (*Bl.*) *Airy Shaw* in Kew Bull. 14: 385 (1960).

Calpigyne frutescens Bl., Mus. Bot. Lugd.-Bat. 2: 193 (1856): Miq., Fl. Ind. Bat. 1(2): 193 (1859); Muell. Arg.: 1255 (1866); Boerl.: 244 (1900); Pax & Hoffm. vii: 255 (1914) & xiv: 35 (1919); Merr.: 343 (1921).
Ptychopyxis frutescens Croiz. in Journ. Arn. Arb. 23: 49 (1942), *pro nom. nov.*

K1a.—Endemic.
Shrub of 3 m.; no further information.

This little-known species agrees with *K. bantamense* Hassk. and *K. wallichianum* Benth. in the purplish colour assumed by the leaves on drying, but differs from the former in the absence of black discoid glands on the female calyx at flowering time, and from the latter in the broadly cuneate or rounded base of the leaves, in the primary nerves less than 1·5 cm. apart, ascending steeply and only slightly curved, and in the fulvous rather than canescent inflorescences. Further collections from the type locality are needed in order to determine the status of this taxon.

Koilodepas laevigatum *Airy Shaw* [ex Meijer, 7: 47 (1967), *anglice, in clavi*, &] in Kew Bull. 23: 83 (1969), *latine.*

Aporosa brevipetiolata Merr. ex Keith, Prelim. List N. Borneo Pl. Names (N. Born. For. Rec. 2): 48, 79, 124 (1938); ed. 2: 116, 192, 292 (1952); *nomen nudum.*

Koilodepas longifolium var. *integrifolium* Airy Shaw in Kew Bull. 14: 388 (1960) & 16: 355 (1963).

SR; SB.—Endemic.

Tree to 12 m. high, in Mixed Dipterocarp forest on black-brown sandy soil in flat or undulating country up to 400 m. alt.

Distinguished from other entire-leaved species by the large size of the leaves (up to 33 × 11·5 cm.), which have an unusually smooth surface due to the less sharply prominent nerves and tend to remain green on drying, and by the long-exserted, narrowly subulate, divaricate-recurved stamens.

Koilodepas longifolium *Hook. f.*: 420 (1887); Pax & Hoffm. vii: 270 (1914); Ridley: 275 (1924); Airy Shaw in Kew Bull. 14: 388 (1960) (var. *longifolium*, excl. var. *integrifolium*) & 26: 284 (1972); Meijer, 7: 47 (1967); Whitmore: 104 (1973).

Coelodepas glanduligerum Pax & Hoffm., *l.c.*; Ridley, *l.c.*

C. subcordatus [sic] Gage in Rec. Bot. Surv. Ind. 9: 239 (1922), *p.p.*, quoad *Curtis* 1374.

Nephrostylus poilanei Gagnep. in Bull. Soc. Bot. France 72: 467 (Aug. 1925) & in Lecomte: 327, fig. 37 (Dec. 1925).

Koilodepas wallichianum sec. Airy Shaw in Kew Bull. 16: 354 (1963), *p.p., an* Benth.?

K. cf. *stenosepalum* sec. Airy Shaw in Kew Bull. 16: 356 (1963).

SR; SB.—Lower Siam, Malaya, Annam, Banka.

Shrub or small tree to 10 m. high, in Lowland or Mixed Dipterocarp or primary *kerangas* forest, on sandstone and diorite scree, or clay loam or blackish soil, up to 90 m. alt.

Nearest to *K. brevipes*, but differing in its leaves being flat at the base, with longer petioles (about 1 cm.), and in the female calyx scarcely or obscurely lobed at the flowering stage. The limits of this species are not yet clear.

Koilodepas pectinatum *Airy Shaw* in Kew Bull. 23: 82 (1969).

SB.—Endemic.

Tree to 9 m. high, in primary or secondary forest on stony clay loam or black or yellowish soil at 60–240 m. alt.

This differs from all other species in the elongate (9–12 mm.), deeply and finely pectinate stipules. (Small, shortly pectinate stipules occur in forms of *K. longifolium.*) The female calyx is membranous and broadly accrescent (up to 2·5 cm. long), as in *K. calycinum* Bedd. (S. India) and *K. hainanense* (Merr.) Croiz. (Hainan) (Sect. *Hyalodepas*).

Koilodepas sp. nov.? cf. Airy Shaw in Kew Bull. 23: 84 (1969), *in obs.*

SB.—Endemic.

Small tree to 7·5 m. high, in primary forest on black soil at 24 m. alt.

The material consists only of a twig with three large leaves, up to 40 × 15 cm., and a large fruit 4 cm. diam. and 2·7 cm. long. It probably represents a new species, but scarcely warrants description.

Macaranga *Thou.*

by T. C. Whitmore*

Forty-seven species of *Macaranga* are now known from Borneo, of which one has two subspecies, three have two varieties, and one has three forms. Of these 53 taxa, 16 have been described since 1965 by Airy Shaw and a further ten by the present author, i.e. nearly half in total. Several taxa are known from only one or a few collections; there is a dearth of material from Kalimantan (three species from there have not been seen and are not well accounted for in the present treatment), and in consequence there is every reason to expect that Borneo has more species yet awaiting discovery.

Five clear species groups can be recognized; a further nine species are individually distinctive. The biggest group is the *Pachystemon* circle of affinity with 21 species, three of them with two varieties. Sect. *Pachystemon* accounts for about 45 per cent of the genus in Borneo, and has as far as we know speciated more here than anywhere else.

Malaya has 27 species of *Macaranga*, just half as many as in Borneo. Of these 12 are of the *Pachystemon* group, about the same proportion as in Borneo.

A single key is given to all the species, but they are described by groups, as this makes it easier to comprehend the variation.

It will be appreciated that the present account can only be regarded as the first step towards a definitive account of *Macaranga* in Borneo. It will have served its purposc if it stimulates botanists in Indonesia and Malaysia to collect these highly accessible, gregarious small trees, some of which are apparently of very restricted distribution.

I am very grateful to the Bentham–Moxon Trustees for a grant towards expenses incurred in the preparation of this account.

Synopsis of the groups of *Macaranga* in Borneo

1. § **Pachystemon** (*Bl.*) *Muell. Arg.*, (a) *s. str.*

1. *M. aëtheadenia* Airy Shaw
2. *M. beccariana* Merr.

* I am greatly indebted to Dr. Whitmore for contributing this valuable account of *Macaranga*, a genus which he has studied specially in the field, both in Malaya and in Borneo, for a number of years. I have left Dr. Whitmore free to depart somewhat in his treatment from the style adopted in the remainder of this enumeration.—H.K.A.S.

3. *M. calcicola* Airy Shaw
 var. *calcicola*
 var. *calcifuga* Whitmore
4. *M. depressa* (Muell. Arg.) Muell. Arg.
 forma *depressa*
 forma *glabra* Whitmore
 forma *strigosa* Whitmore
5. *M. havilandii* Airy Shaw
6. *M. hullettii* King ex Hook. f. subsp. *borneënsis* Whitmore
7. *M. hypoleuca* (Reichb. f. & Zoll.) Muell. Arg.
8. *M. indistincta* Whitmore
9. *M. kingii* Hook. f.
 var. *kingii*
 var. *platyphylla* Airy Shaw
10. *M. lamellata* Whitmore
11. *M. motleyana* (Muell. Arg.) Muell. Arg.
12. *M. petanostyla* Airy Shaw
13. *M. quadricornis* Ridley
14. *M. recurvata* Gage
15. *M. trachyphylla* Airy Shaw
16. *M. triloba* (Bl.) Muell. Arg.

2. § **Pachystemon,** (b) the *M. caladiifolia* alliance
17. *M. caladiifolia* Becc.
18. *M. curtisii* Hook. f. var. *glabra* Whitmore
19. *M. rostrata* Heine
20. *M. sarcocarpa* Airy Shaw

3. The **M. pruinosa** group
21. *M. hosei* King ex Hook. f.
22. *M. pearsonii* Merr.
23. *M. pruinosa* (Miq.) Muell. Arg.
24. *M. puberula* Heine

4. § **Pseudo-Rottlera** (*Reichb. f. & Zoll.*) *Pax & Hoffm.*
25. *M. anceps* Airy Shaw subsp. *puncticulata* Whitmore
26. *M. baccaureifolia* Airy Shaw
27. *M. brevipetiolata* Airy Shaw
28. *M. fulva* Airy Shaw
29. *M. lowii* King ex Hook. f.
 var. *lowii*
 var. *kostermansii* Airy Shaw
30. *M. praestans* Airy Shaw
31. *M. rarispina* Whitmore
32. *M. repando-dentata* Airy Shaw
33. *M. strigosissima* Airy Shaw

5. § **Stachyella** (*Miq.*) *Pax & Hoffm.*
34. *M. costulata* Pax & Hoffm.
35. *M. kinabaluensis* Airy Shaw

6. Species which are individually distinctive
36. *M. amissa* Airy Shaw
37. *M. conifera* (Zoll.) Muell. Arg.

38. *M. endertii* Whitmore
39. *M. gigantea* (Reichb. f. & Zoll.) Muell. Arg.
40. *M. gigantifolia* Merr.
41. *M. tanarius* (L.) Muell. Arg.
42. *M. trichocarpa* (Reichb. f. & Zoll.) Muell. Arg.
43. *M. winkleri* Pax & Hoffm.
44. *M. winkleriella* Whitmore

7. Species incertae sedis (no material seen)
45. *M. brachythyrsa* Pax & Hoffm.
46. *M. eloba* Pax & Hoffm.
47. *M. gossypiifolia* Pax & Hoffm.

8. Specimens not placed because inadequate
Endert 3266 (K, L)—a species of the *Pachystemon* group
SAN 61678 (K, L)

Key to *Macaranga* species in Borneo

M. brachythyrsa, *M. eloba* and *M. gossypiifolia* have been left out. No material has been seen and the descriptions are inadequate.

1 Leaves penninerved, all or most with conspicuous and usually large superficial glands near insertion of petiole; base cordate to weakly (3 mm.) or rarely more strongly (to 1 cm.) peltate 2
Leaves palminerved; without such glands 15
2 Leaves with main nerves below strigose 3
Not so 5
3 Leaves harshly whitish-strigose above as well as below, 10–14 cm. long; ovary densely whitish tomentose . . . 33. **M. strigosissima**
Leaves glabrous above except for midrib and main nerves 4
4 Leaves 7–17 cm. long, thinly whitish strigose-pubescent below; peduncle of ♀ inflorescence 1·5–3 cm. long . . . 32. **M. repando-dentata**
Leaves 5–7 cm. long, softly fulvous-strigose on midrib below; peduncle of ♀ inflorescence less than 1 cm. long . . . . . 28. **M. fulva**
5 Stipules crowded, numerous, usually linear 6
Stipules seldom crowded and numerous, narrowly triangular or caducous 9
6 Leaves to 13 cm. long only 7
Leaves much larger, 15–36 cm. long 8
7 Leaves finely pubescent on nerves below, dull . . . 29. **M. lowii**
Leaves entirely glabrous, shiny . . 29. **M. lowii** var. **kostermansii**
8 Peduncle 8 cm.; fruits spiny . . . . 31. **M. rarispina**
Peduncle to 2 cm. only; flowers and fruits congested; fruits smooth 30. **M. praestans**
9 Leaves not puncticulate below 10
Leaves closely and finely puncticulate below 11
10 Leaves markedly truncate at base . . . . 19. **M. rostrata**
Leaves tapering to base . . . . 25. **M. anceps**
11 Leaves big, ovate, reaching 11(–14) × 14(–20) cm., base broadly cordate 26. **M. baccaureifolia**
Leaves smaller, elliptic, not longer than 9 cm. 12

12 Petiole short, only 1–2·5 cm. . . . . . 27. **M. brevipetiolata**
Petiole at least 4 cm. 13
13 Inflorescence a raceme; flowers distant and few; styles to 1 cm. long 25. **M. anceps** subsp. **puncticulata**
Inflorescence a panicle; flowers numerous and close; styles very short 14
14 Bracteoles of inflorescence to 10 mm. long, sometimes floccose. Primary forest, common on slopes of Mt. Kinabalu 35. **M. kinabaluensis**
Bracteoles of inflorescence to 5 mm. long, glabrous. Secondary forest (usually) . . . . 34. **M. costulata**
15 Leaves peltate, usually at least 2·5 cm. peltate, rarely 0·7 cm. only . 16
Leaves not peltate or rarely to 0·7 cm. peltate, sometimes cordate . 48
16 Leaves with 3 or more lobes which are usually prominent but occasionally only as lateral cusps 17
Leaves not lobed (sometimes a few leaves cusped) 35
17 Leaves with broadly truncate base bearing large volcano-like glands 1. **M. aëtheadenia**
Not so 18
18 Leaves beetroot red below . . . . 13. **M. quadricornis**
Not so 19
19 Internodes very short; twigs pale in colour 3. **M. calcicola** var. **calcicola**
Not so 20
20 Leaves with an intensely white paint-like substance on the lower surface 21
Not so 22
21 Leaves with 3 very deep and narrow lobes . . 2. **M. beccariana**
Leaves with 3 broad lobes, reaching about half-way 7. **M. hypoleuca**
22 Leaves 5-lobed 23
Not so 24
23 Stipules to 2 cm. long, stiff, recurved like a pair of horns. W. Sarawak only . . . . 9. **M. kingii**
Stipules to 1 cm. long, papery . . . . . 4. **M. depressa** (rarely)
24 Leaf-blade huge, to 1 m. across; stipules papery, persistent, 2 cm. across, up-pointing. Only known from Sandakan and environs 40. **M. gigantifolia**
Not so 25
25 Twigs solid 26
Twigs hollow, ant-inhabited 28
26 Stipules 1·5 cm. across × 4 cm. long; leaves shallowly lobed, 30 cm. or more across . . . . 39. **M. gigantea**
Stipules much smaller, 1 cm. or less across; leaves lobed to half-way or more, not exceeding 15 cm. across, usually less 27
27 Stipules pointed, as long as or longer than broad; fruits horned, in heads 4. **M. depressa** (usually)
Stipules rounded, broader than long; fruits unarmed, not clustered into heads . . . . 23. **M. pruinosa**
28 Twigs very pale in colour; leaf-surface below with yellow granular glands plus hairs on the nerves; lobes shallow . 8. **M. indistincta**
Not so 29
29 Leaves very rough above, like sandpaper . . 15. **M. trachyphylla**
Leaves above not or scarcely rough 30

30 Leaves lobed to more than half-way 31
Not so 32
31 Leaves velvety below; reticulations close, raised and conspicuous. Only known from Mts. Kinabalu and Trus Madi . . 24. **M. puberula**
Leaves glabrous below, reticulations distant, not raised and faint 5. **M. havilandii** (sometimes)
32 Bracteoles of inflorescence caudate 33
Bracteoles of inflorescence acuminate 34
33 Fruits often slightly glaucous, with no glands; inflorescence bracts early caducous . . . . 5. **M. havilandii** (usually)
Fruits not glaucous, the lobes each with a gland patch; a few inflorescence bracts long-persistent . . . . . . 11. **M. motleyana**
34 Leaves only cusped, not fully lobed . 3. **M. calcicola** var. **calcifuga**
Leaves fully 3-lobed . . . . 16. **M. triloba**
35 Stipules to 2 cm. long, stiff, recurved like a pair of horns 9. **M. kingii** var. **platyphylla**
Not so 36
36 Leaves bullate, velvety below; fruits fleshy . . . 20. **M. sarcocarpa**
Not so 37
37 Leaves velvety below 38
Leaves glabrous below, or occasionally with sparse hairs on the nerves 39
38 Inflorescence bracteoles broad, margins fimbriate; fruits to 1 cm diam., softly spiny . . . . 41. **M. tanarius**
Inflorescence bracteoles linear, entire, distorted by 1–several large raised glands; fruits to 3 mm. diam., smooth . 38. **M. endertii**
39 Twigs solid 40
Twigs hollow, ant-inhabited 42
40 Stipules persistent; leaves big, often longer than 20 cm., secondary nerves more or less straight, prominent . . . 14. **M. recurvata**
Stipules caducous; leaves much smaller 41
41 Leaves with yellow granular glands below, not puncticulate 18. **M. curtisii** var. **glabra**
Leaves without such glands but minutely puncticulate 17. **M. caladiifolia** (sometimes)
42 Fruits tiny, not greater than 2 mm. diam., bilobed 43
Fruits 5 mm. diam. or more 44
43 Stipules as broad as long, 8 mm. diam.; styles on fruit 1 mm. Limestone plant . . . . 44. **M. winkleriella**
Stipules much longer than broad, 20 mm. long; styles on fruit shorter than 1 mm. Not on limestone . . . . . 43. **M. winkleri**
44 Stipules caducous 45
Stipules persistent 46
45 Leaves 25–30 cm. long . . . . 5. **M. havilandii** (rarely)
Leaves to 10(–15) cm. long . . . 17. **M. caladiifolia** (sometimes)
46 Stipules broadly triangular; fruits horned 47
Stipules markedly longer than broad; fruits lamellate 10. **M. lamellata**
47 Inflorescence bracteoles caudate; styles fused in fruit 6. **M. hullettii** subsp. **borneënsis**
Inflorescence bracteoles acute; styles free in fruit. Mainly known from Mt. Kinabalu . . . . 12. **M. petanostyla**

48 Leaves 3-lobed 49
Leaves not lobed 50
49 Stipules persistent; inflorescence branches glabrous . 21. **M. hosei**
Stipules caducous; inflorescence branches with tawny, soft, spreading hairs . . . . 22. **M. pearsonii**
50 All or most leaves cordate at the base, sometimes only minutely so 36. **M. amissa**
Not so 51
51 Leaves ovate, with a long tip, margins toothed, velvety all over below; shrub with sarmentose branches . . . . 42. **M. trichocarpa**
Leaves elliptic, not long-tipped, margins entire, glabrous; small tree 37. **M. conifera**

1. Section **Pachystemon**: (a) *sens. str.*

This group is the heart of *Macaranga* in Borneo, and is fairly clearly defined (see Airy Shaw in Kew Bull. 25: 538–9 (1970)), though through group 2 (the *M. caladiifolia* alliance) it merges into Sect. *Semiglobosae* Pax & Hoffm. Seventeen species are known, of which five have been recently described by Airy Shaw, and three (plus a subspecies and two varieties) are newly described by myself. Several species have only been collected once or a few times, and it seems probable that many more await discovery.

1. **Macaranga aëtheadenia** *Airy Shaw* in Kew Bull. 29: 322 (1974).

SR (C.).—?? Java.
Tree to 20 m., in Mixed Dipterocarp forest on basalt-derived or clay-rich shale-derived soils at 150–300 m. alt.

Highly distinctive in the massive volcano-like glands on the leaf-margin, especially along the base. Known from around Kapit and on the basalt of Bt. Mersing.
Borssum Waalkes 178 from Java, Bogor Residency, Tjisarua Selatan, 1400 m., may be a variety of *M. aëtheadenia*. Full investigation must await study of the genus in Java.

2. **Macaranga beccariana** *Merr.* in Webbia 7: 315 (1950); Meijer, 7: 48 (1967).

M. hypoleuca sec. Muell. Arg.: 992 (1866) & Pax & Hoffm. vii: 311 (1914) *quoad specim.* 'Borneo (Lowe)' *tantum, non Mappa hypoleuca* Reichb. f. & Zoll.
M. hypoleuca var. *borneënsis* Hutch. ex Gibbs in Journ. Linn. Soc. 42: 136 (1914); Merr.: 342 (1921); Pax & Hoffm. xvii (Euph.–Addit. vii): 184 (1924).

SR; B; SB.—Endemic.
Tree to 15 m. high, in primary (?) or secondary Dipterocarp forest or secondary Heath forest up to 900 m. alt.

Extremely distinctive; otherwise only *M. hypoleuca* has the same curious paint-like covering to the leaves below.

3. **Macaranga calcicola** *Airy Shaw* in Kew Bull. 25: 536 (1971).

SR (SW.); **K1.**—Sumatra (east coast).
Small tree of 4–10 m. (once reaching 20 m.), in thin bushy pyrogenic vegetation, always on limestone, up to 150 m. alt.

To the collections from the limestone hills of SW. Sarawak cited by Airy Shaw may now be added *Kostermans* 6023 from Gunong Tepian Lobang, NE. of Sangkulirang, east Kutai, which differs only in its leaves drying dark brown not greenish; also, *Bartlett* 8075 from Pargambiran, Asahan, Sumatra —a remarkable disjunction in range.
Besides the distinctive features recorded by Airy Shaw note also the pale-coloured twigs with very short internodes, only 5–8 mm. long near the apex. The former character is shared with the superficially confusable *M. indistincta* Whitmore (*q.v.*).

Var. **calcifuga** *Whitmore* in Kew Bull. 29: 445 (1974).

SR.—Endemic.

The single puzzling collection, SAR 29457, from Lowland Dipterocarp forest in the Semengoh Arboretum near Kuching, is closest to *M. calcicola*, differing in being entirely glabrous, having larger leaves (12 × 16–20 × 26 cm.) and the twigs drying dark with 15 mm. internodes, a more usual length for *Macaranga*.

4. **Macaranga depressa** (*Muell. Arg.*) *Muell. Arg.*: 989 (1866); Pax & Hoffm. vii: 382 (1914); Merr.: 341 (1921); Whitmore in Kew Bull. 29: 445 (1974).

Pachystemon depressus Muell. Arg. in Flora 47: 465 (1864).
M. depressa var. *β genuina* Muell. Arg.: 989 (1866); Pax & Hoffm. vii: 383 (1914).

This is known in three forms:

forma **depressa**: Whitmore, *l.c.*

K1a.—Endemic.

Twigs, petioles and main leaf-nerves below pubescent.

forma **glabra** *Whitmore* in Kew Bull. 29: 446 (1974).

SB.—Endemic. All collections except one from around Sandakan.

Twigs and petioles glabrous.

forma **strigosa** *Whitmore l.c.* (1974).

M. divergens Muell. Arg.: 989 (1866); Pax & Hoffm. vii: 382 (1914); Merr.: 341 (1922); Meijer, 7: 48 (1967); Whitmore in Biol. Journ. Linn. Soc. 1: 228 (1969).

SR; K1 (?).—Endemic.

Strigose hairs on twigs (at least tips), petioles and main leaf-nerves below.

Macaranga quadricornis is close but is a mountain species which, in Borneo, is nearly always glabrous. It also differs in having the fruits less markedly clustered into heads.

Macaranga depressa and *M. quadricornis* differ subtly from each other in leaf shape. In *M. quadricornis* the body of the blade is substantial, the lobes penetrate less deeply and are very broad at their base. *M. depressa* leaves have a marked tendency to a truncate base, occasionally developed as a pair of basal lobes. The two lateral lobes penetrate deeply and are less massive at the base, so the blade appears less substantial. *M. depressa* forma *strigosa* is intermediate. If a quantitative measure could be devised to express these differences, and if far more collections were available, I have little doubt that, despite some overlap (e.g. 5-lobed *M. quadricornis* from Johore), the mean blade dimensions would be found to differ significantly.

5. **Macaranga havilandii** *Airy Shaw* in Kew Bull. 23: 112 (1969).

M. havilandii var. *resecta* Airy Shaw in Kew Bull. 25: 524 (1971), **synon. nov.**

SR.—Endemic.
Small tree to 3 m. in primary forest up to 270 m. alt.

6. **Macaranga hullettii** *King ex Hook. f.*: 452 (1887); Pax & Hoffm. vii: 383 (1914); Meijer, 7: 47 (1967) & 10: 232 (1968); Whitmore: 107 (1973).

? *M. bartlettii* Merr. in Papers Mich. Acad. Sci. 1933, 19: 161 (1934).

(Subsp. *hullettii* in Malaya, Sumatra (?).)

Subsp. **borneënsis** *Whitmore* in Kew Bull. 29: 446 (1974).

M. hullettii sec. Meijer, 7: 47 (1967) & 10: 232 (1968), *non* King ex Hook. f.
? *M. cornuta* sec. Meijer, 7: 48 (1967), *non* Muell. Arg.

SR; SB; K1, 3. —Endemic.
Treelet or small tree 1–8 m. tall in primary often riverine forest, lowlands to 450(–1100) m., and in Heath or moss [upper montane] forest at 800–1230 m. alt.

None of the specimens I have seen possess lobed leaves, in Malaya many have. The only exception is SAN 66862, which is atypical in *M. hullettii* in possessing glands, albeit sparsely, on the leaf-undersurface. This specimen is provisionally placed here.

7. **Macaranga hypoleuca** (*Reichb. f. & Zoll.*) *Muell. Arg.*: 992 (1866) (*excl. specim.* 'Borneo (Lowe)'); Hook. f.: 448 (1887); Pax & Hoffm. vii: 311 (1914) (*excl. specim.* 'Borneo (Lowe)'); Merr.: 342 (1921) (*excl. var.*); Ridley: 300 (1924); Merr., Pl. Elmer.: 159 (1929); Corner: 266 (1940); Whitmore in Malayan Nature Journ. 20: 93, 97 (1967); Meijer, 7: 48 (1967) & 10: tab. opp. p. 64 (1968); Airy Shaw in Kew Bull. 26: 289 (1972); Whitmore: 106 (1973).

Mappa (?) *hypoleuca* Reichb. f. & Zoll. in Verhand. Natuurk. Vereen. Ned. Ind. 1: 30 (1856) & in Linnaea 28: 309 (1856).

SR; SB; K1, 2.—Lower Siam, Malaya, Sumatra.

Tree to 24 m. high, in primary or secondary forest, rarely as high as 2400 m. alt.

8. **Macaranga indistincta** *Whitmore* in Kew Bull. 29: 447 (1974).

Pachystemon depressus Muell. Arg. var. *mollis* Muell. Arg. in Flora 47: 465 (1864).

M. depressa var. *mollis* (Muell. Arg.) Muell. Arg.: 989 (1866); Pax & Hoffm. vii: 383 (1914).

SR; SB.—Sumatra.

Small tree to 9 m. tall, in lowlands to 900 m. alt., probably in primary forest.

This is a very characteristic member of Sect. *Pachystemon* with no striking features. The very shallowly lobed leaves drying greenish-brown below and the pale twigs (with, however, long internodes) resemble those of *M. calcicola*.

9. **Macaranga kingii** *Hook. f.*: 451 (1887); Pax & Hoffm. vii: 383 (1914); Ridley: 300 (1924); Corner: 266 (1940); Whitmore in Malayan Nature Journ. 20: 98, t. XI (1967); Airy Shaw in Kew Bull. 25: 533 (1971); Whitmore: 106 (1973).

M. insignis Merr. in Philipp. Journ. Sci., C. Bot., 11: 69 (1916) & Enum.: 342 (1921).

SR.—Malaya.

Treelet of 25 m., lowlands.

Var. **platyphylla** *Airy Shaw* in Kew Bull. *l.c.* (1971).

SR.—Endemic.

Tree to 6 m.

10. **Macaranga lamellata** *Whitmore* in Kew Bull. 29: 448 (1974).

SB.—Endemic.

Primary forest (? lowland, ? lower montane).

The long-pedicelled fruits, strikingly lamellate and reminiscent of the slow-combustion stoves formerly (and again in future?) used to heat large buildings, especially ecclesiastical ones, are unmistakable. Collected only once, and as recently as 1971! Borneo has surprises yet in store.

11. **Macaranga motleyana** (*Muell. Arg.*) *Muell. Arg.*: 994 (1866); Pax & Hoffm. vii: 310 (1914); Merr.: 342 (1921) & Pl. Elmer.: 160 (1929).

Mappa motleyana Muell. Arg. in Flora 47: 467 (1864).

Siam (SE. & S. Penins.), Malaya, Sumatra, Borneo.

Subsp. **motleyana**

SB; K1, 1a.—? Sumatra.
Tree of secondary forest, often by rivers and reaching 12–20 m. in height, or, rarely, in primary forest and then taller, to 21–33 m.

The Bornean material is all of the typical subspecies. Its restriction to the east and south coast is noteworthy. A second subspecies, *M. motleyana* subsp. *griffithiana* (Muell. Arg.) Whitmore (in Kew Bull. 29: 448 (1975)) occurs in Malaya.
Beccari 514 from Sumatra, Padang, is very close to *M. motleyana* and differs solely in the stipules being longer than broad.

12. **Macaranga petanostyla** *Airy Shaw* in Kew Bull. 25: 537 (1971).

SB; K1(?)—Endemic.
Tree to 9 m., riversides and Dipterocarp (i.e. primary) forest, 1100–1200 m. alt. on Mt. Kinabalu.

To the two fruiting collections cited by Airy Shaw can be added *Clemens* 40318, in young male flower, which is quite clearly this species, especially from the leaf and the stout, flattened, tawny-pilose inflorescence rachis with opposite branches.
I interpret *Kostermans* 12691 from west Kutai as a variety of this species and possibly a depauperate specimen. It differs in its lowland station (100 m. alt.), more slender twigs and small inflorescence; and in the rather smaller leaf, 23 × 15 cm., not shiny and with the lateral lobes represented by mere cusps, or absent, on both the Leiden and Kew sheets.

13. **Macaranga quadricornis** *Ridley* in Bull. Misc. Inf. Kew 1923: 367 (1923) & Flora 3: 300 (1924); Whitmore in Biol. Journ. Linn. Soc. 1: 222–231 (1969) & Tree Flora 2: 107 (1973).

M. ? tenuifolia sec. Corner: 269 (1940), *an* Muell. Arg.?

SR; SB.—Malaya.
Small tree of 8 m. in mountain forest, 1000–1300 m. alt.

This replaces *M. depressa* in the mountains of Borneo. It has been collected from Mt. Poë in west Sarawak to Mt. Kinabalu. Like the northern population of *M. quadricornis* in Malaya it is a mountain species and like that the leaves are [beetroot] red below (possibly not always in Borneo). All the Borneo collections seen are completely glabrous, except *Clemens* 20377 from Mt. Poë, which has short spreading hairs on the main nerves of the lower leaf-surface. Most Malayan collections are hairy in some degree. Characteristically the leaf-blade has a large central unlobed part. This gives *M. quadricornis* a different *Gestalt* from *M. depressa*, which is, however, difficult to quantify.

14. **Macaranga recurvata** *Gage* in Rec. Bot. Surv. Ind. 9: 246 (1922); Ridley: 301 (1924).

B; SB; K1.—Malaya, Celebes?
Tree occasionally attaining 30 m. in height, lowlands to 500 m. alt., secondary seasonal swamp forest, secondary (dryland) forest, *Agathis* forest.

This species is rather isolated in Sect. *Pachystemon*, with no close relatives.

15. **Macaranga trachyphylla** *Airy Shaw* in Kew Bull. 25: 534 (1971).

SR; B.—Endemic.
Small tree to 12 m. tall, in primary forest up to 300 m. alt.

The leaves are indeed rough on the upper surface, but this is a difficult character because it is only one of degree.

16. **Macaranga triloba** (*Bl.*) *Muell. Arg.*: 989 (1866); Hook. f.: 452 (1887); Pax & Hoffm. vii: 380 (1914); Merr.: 343 (1921) & 443 (1923); Ridley: 298 (1924); Gagnep.: 439 (1926); Corner: 268 (1940); Backer & Bakh. f.: 488 (1963); Whitmore in Malayan Nature Journ. 20: 94, 99 (1967); Meijer, 7: 48 (1967); Airy Shaw in Kew Bull. 26: 292 (1972); Whitmore: 107 (1973).

Ricinus trilobus Reinw. ex Bl., Cat. Gewassen Buitenz.: 108 (1823), *nomen.*
Pachystemon trilobus Bl., Bijdr.: 626 (1825).
P. bancanus Miq., Fl. Ind. Bat., Suppl.: 462 (1861).
Macaranga cornuta Muell. Arg.: 988 (1866).
M. bancana ('*-us*') (Miq.) Muell. Arg.: 990 (1866).
M. motleyana auct. sec. Meijer, 7: 48 (1967), *in obs.*, *non* Muell. Arg.

SR; SB; K1.—Lower Burma, Lower Siam and throughout W. Malesia.

This is more variable in Borneo than in Malaya, and around Kuching there is a form that sometimes has unlobed leaves which can only be called aberrant (e.g. *Beccari* 2 & 463, *Haviland* 457 & 458, *Native Collector* 168). *M. triloba* is right at the heart of Sect. *Pachystemon*, but a full account of it must await evaluation of the highly variable material from Sumatra, Java and attendant smaller islands.

2. Section **Pachystemon**: (b) the *M. caladiifolia* alliance

This is a subgroup within the *Pachystemon* circle of affinity. *M. caladiifolia*, to which I now reduce several other species on the basis of careful investigation in the forest, is a typical 'Pachystemon' *Macaranga*. I have already remarked on its close relationship to *M. curtisii* (in Kew Bull. 25: 240 (1971)). The typical form of this latter (with leaves velvety and drying chocolate-brown below and with spreading hairs on the twigs) has not been found in Borneo, but the montane var. *glabra* is now known from Mt. Kinabalu. This differs from *M. caladiifolia* in the leaves having no puncticulations but instead possessing raised, yellow granular glands on the lower surface, and the twigs are always solid; it is only known from mountain forests. *M. curtisii* belongs to Sect. *Semiglobosae* Pax & Hoffm. The status of this section and its distinction against Sect. *Pachystemon* needs further study. *M. sarcocarpa*, a very distinct but rare species, completes this subgroup in Borneo.

17. **Macaranga caladiifolia** *Becc.* in Malesia 2: 46, t. 3 (1884); Pax & Hoffm. vii: 383 (1914); Merr.: 341 (1921); Meijer, 7: 47 (1967); Whitmore: 111 (1973).

M. caladiifolia var. *pilosula* Pax & Hoffm. vii: 384 (1914), **synon. nov.**
M. caladiifolia var. *truncata* Pax & Hoffm., *l.c.*, **synon. nov.**
M. tenuiramea Pax & Hoffm. vii: 384 (1914); Merr.: 342 (1921); **synon. nov.**

M. puncticulata Gage in Rec. Bot. Surv. Ind. 9: 246 (1922); Ridley: 302 (1924); **synon. nov.**

? *M. myrmecophila* Diels in Engl., Bot. Jahrb. 60: 309 (1926).

SR; B; SB; K2, 3.—Malaya.

Bushy-crowned treelet or small tree to 10 m.; peat-swamp forest, fresh-water swamp forest, Lowland Heath forest, ?? mesophytic rain forest.

Fruits bright pink, seeds red.

At first sight the union of the Malayan *M. puncticulata* with *M. caladiifolia* might appear mistaken. But careful study in the forest over a number of years both by the author and by Dr. J. A. R. Anderson has led to the conclusion that but one variable species is involved. The extreme condition of leathery, slightly bullate, ovate-acute leaf, brown below, is united by a continuous series of intermediates to the thin, pruinose, rather truncate leaf of typical *M. caladiifolia*, and great variation can be found on a single tree (compare, for example, the Singapore and Kepong sheets of SFN 36636 from Pontian, Johore). The twigs are usually thick, hollow and ant-inhabited, but are sometimes slender and solid, in both Malaya and Sarawak. The twigs, petioles and main nerves below vary from almost glabrous to pubescent. *M. caladiifolia* is rare in primary peat-swamp forest and gregarious in clearings. It has not proved possible to correlate morphological variation with the considerable range of habitats.

The varieties of *M. caladiifolia* described by Pax & Hoffmann fall well within the variability of this species as now interpreted, and it seems highly probable from the description that *M. myrmecophila* Diels (based on *Hackenberg* 104 from Sampit) does also. The type of *M. tenuiramea* is very similar indeed to *M. caladiifolia* and that species must also be reduced. The Kew sheet has the apical part of the twig solid and slender and the basal part hollow and markedly swollen.

Macaranga caladiifolia presents a thought-provoking contrast to the *M. pruinosa* group in Borneo. In the former we have now shown, partly from careful study in the forest, that a wide range of morphological forms, which have led to several species being described, are in fact completely interlinked, and the correct interpretation appears to be that we are dealing with one polymorphic species. In the latter group, as described below, there do seem to be qualitative differences, albeit small ones, in Borneo, and although these are of the same order of magnitude as variants of *M. caladiifolia* disjunctions exist and the species which have been described can stand.

18. **Macaranga curtisii** *Hook. f.* var. **glabra** *Whitmore* in Kew Bull. 25: 240 (1971) & Tree Flora 2: 111 (1973).

SB.—Malaya.

Small tree 10 m. tall, in primary rain forest at *c.* 1590 m. on Mt. Kinabalu.

Close to *M. caladiifolia* (*q.v.*) but distinctive at a glance from its more rotund leaves, drying chocolate-brown below. *M. brooksii* Ridley of Sumatra is closely allied, but has persistent stipules and the young twigs and leaves below are somewhat glaucous.

Creagh s.n., 'Gaya 5/95', from Sabah, received at Kew August 1895, is a sheet I cannot place. It falls somewhere in the *M. curtisii* circle of affinity. The leaves are greyish below from a dense fine indumentum, in which it approaches *M. bicolor* Muell. Arg. of the Philippines.

19. **Macaranga rostrata** *Heine* in Fedde, Repert. Sp. Nov. 54: 233 (1951); Meijer, 7: 49 (1967).

SR; SB.—Endemic. Much collected from the slopes and foothills of Mt. Kinabalu, twice from Gunong Alab further south on the Crocker Range, and twice from Sarawak (Lawas and Bukit Mersing).

Tree to 20 m. high, in lowland Mixed Dipterocarp forest or primary lower montane forest or rarely in secondary forest, from 800 to 1920 m. alt.

A highly distinctive species of Sect. *Pachystemon*, with very prominent, 10 mm. curved horns on the fruits. The Sarawak gatherings have hollow, ant-inhabited twigs. All those from Sabah have solid twigs, except one, *Poore & Ho* s.n., 8 Feb. 1964, which is possibly a sapling.

20. **Macaranga sarcocarpa** *Airy Shaw* in Kew Bull. 23: 88 (1969).

SR (SW.).—Endemic.

No ecological information available; probably at low altitude.

Highly distinctive in its deeply reticulate, velvety lower leaf-surface and roundish fleshy fruits. It is remarkable that no collection has been made since 1894.

3. The **Macaranga pruinosa** group

So far four species of this group have been recorded from Borneo. They are quite distinct, and although only differing in a few characters no intermediates are known. The distinctions are clearly emphasized by the following key:

1 Leaves deeply peltate at base; twigs, petioles and panicle branches hairy. Swampy (? always) lowlands or mountains 2
 Leaves not peltate, or only very slightly so, base cordate; twigs and petioles glabrous, pruinose, twigs hollow. Lowlands, nearly always dry land 3
2 Twigs solid (in Malaya occasionally hollow); stipules very soon falling or persistent. Lowlands; swampy land (? always in Borneo) **M. pruinosa**
 Twigs hollow, ant-inhabited; stipules very soon falling. Mountains (Mts. Kinabalu & Trus Madi) . . . . **M. puberula**
3 Panicle branches glabrous; stipules persistent . . . . **M. hosei**
 Panicle branches with tawny, soft, spreading hairs; stipules caducous **M. pearsonii**

21. **Macaranga hosei** *King ex Hook. f.*: 449 (1887); Pax & Hoffm. vii: 309 (1914); Merr.: 342 (1921); Ridley: 298 (1924); Whitmore in Malayan Nature Journ. 20: 95, 97 (1967); Meijer, 7: 48 (1967); Airy Shaw in Kew Bull. 26: 289 (1972); Whitmore: 111 (1973).

M. pseudo-pruinosa Pax & Hoffm. vii: 308 (1914).

SR; B; SB; K3.—S. Penins. Siam, Malaya, Sumatra.

Small tree to 8 m., in regenerating forest (? always), dry lowlands.

Very abundantly represented at Kew, in contrast to the closely related *M. pruinosa*.

22. **Macaranga pearsonii** *Merr.* in Philipp. Journ. Sci. 29: 383 (1926) & Pl. Elmer.: 160 (1929).

M. hosei sec. Airy Shaw in Kew Bull. 26: 289 (1971), *non* King ex Hook. f.

SB; K1, 2.—Endemic.

Tree to 22 m. tall, 0·9 m. girth, in primary or (more usually) secondary forest, lowlands, and once at 1200 m. alt., nearly always on dry land.

This was reduced to *M. hosei* by Airy Shaw in his account of the *Euphorbiaceae* of Siam, but it is in fact distinct as shown in the key above.

23. **Macaranga pruinosa** (*Miq.*) *Muell. Arg.*: 992 (1866); Pax & Hoffm. vii: 309 (1914); Airy Shaw in Kew Bull. 26: 289 (1972); Whitmore: 106 (1973).

Mappa pruinosa Miq., Fl. Ind. Bat., Suppl.: 457 (1861).
M. maingayi Hook. f.: 449 (1887); Meijer, 7: 48 (1967); Airy Shaw, *l.c.* (1972).
M. formicarum Pax & Hoffm. vii: 308 (1914); Merr. (sphalm. '*formicarium*'): 341 (1921); **synon. nov.**

SR; K3.—Malaya.

Tree to 0·6 m. girth, sometimes gregarious, in secondary peat-swamp forest.

I suspect this species is much more common than the five specimens at Kew suggest. Only one of the collections gives habitat notes and was from swampy ground, another from Pontianak probably was. I guess that, just as in Malaya, this is the gregarious *Macaranga* of secondary forest on wet ground. The Bornean sheets examined differ from Malayan specimens in the slightly puberulous axes, which are not pruinose. I am reluctant to formalize these differences by creating a new taxon, though they are of the same order of magnitude as the differences used to distinguish the published species of this alliance in Borneo. I was wrong in stating (in Biol. Journ. Linn. Soc. 1: 225 (1969)) that in Borneo this taxon with deeply peltate leaves grows on dry land and its cordate-leafed ally (*M. hosei*) in swamps.

24. **Macaranga puberula** *Heine* in Fedde, Repert. Sp. Nov. 54: 232 (1951).

SB.—Endemic.

Tree to 6 m. in primary forest. Restricted to the Crocker Range at Mts. Kinabalu and Trus Madi, 900–1800 m. alt.

This is close to *M. pruinosa*, but I retain it distinct on its hollow twigs, stipules always rapidly caducous, and montane habitat. There are only three collections at Kew.

4. Section **Pseudo-Rottlera**

This section, in many ways intermediate to *Mallotus*, has been thoroughly dealt with by Airy Shaw (in Kew Bull. 19: 315–328 (1965)). At that time he described all the Bornean species except *M. lowii* (which, as in Malaya, is by far the commonest one), and excepting also the new species and subspecies which I have recently added.

Members of Sect. *Pseudo-Rottlera* have a distinctive *Gestalt*, with their essentially penninerved, usually elliptic leaves, and their often stiff setaceous stipules. They also stand apart from most other *Macaranga* in being small, hard-wooded trees of primary forest, not pioneers.

25. **Macaranga anceps** *Airy Shaw* in Kew Bull. 19: 316 (1965).

subsp. **puncticulata** *Whitmore* in Kew Bull. 29: 450 (1974).

SR.—Endemic.
Small tree in ridge forest at low altitude.

The type variety is only known from Simalur Island, to the west of Sumatra.

26. **Macaranga baccaureifolia** *Airy Shaw* in Kew Bull. 19: 316 (1965) & 21: 404 (1968); Meijer, 7: 50, tab. (1967); Whitmore: 112 (1973).

SB (Lahad Datu, Tawau area).—Malaya.
Shrub or tree of 9–15 m., in primary or (once) secondary forest at 90–180 m. alt.

A highly distinctive member of Sect. *Pseudo-Rottlera* from its long-petioled, ovate-oblong, shallowly cordate leaves, reminiscent of some species of *Baccaurea* and *Elaeocarpus*.

27. **Macaranga brevipetiolata** *Airy Shaw* in Kew Bull. 19: 326 (1965); Meijer, 7: 50 (1967).

SR; SB; K1.—Endemic.
Shrub or tree to 14 m. high, in primary or secondary [?] Mixed Dipterocarp forest on dry sandstone or sandy clay or sandy loam up to 300 m. alt.

28. **Macaranga fulva** *Airy Shaw* in Kew Bull. 19: 322 (1965).

K1.—Endemic.
Treelet of 4 m., on low sandy ridge.

Still only known from the type collection. A distinctive member of Sect. *Pseudo-Rottlera* from the fulvous tomentum on the twig tips and on the midrib below of the rather small, coriaceous, slightly bullate, elliptic leaves.

29. **Macaranga lowii** *King ex Hook. f.*: 453 (1887); Pax & Hoffm. vii: 364 (1914); Ridley: 304 (1924); Airy Shaw in Kew Bull. 19: 323 (1965); Whitmore in Malayan Nature Journ. 20: 95, 98 (1967); Meijer, 7: 50 (1967); Airy Shaw in Kew Bull. 23: 107 (1969) & 26: 290 (1972); Whitmore: 112 (1973).

Mallotus auriculatus Merr. in Philipp. Journ. Sci., Bot. 7: 396 (1912); Pax & Hoffm. vii: 194 (1914); Ridley, Flora 5: 333 (1925).

Mallotus affinis Merr. in Philipp. Journ. Sci., Bot. 13: 82 (1918); Merr.: 338 (1921); Pax & Hoffm. xvii (Euph.–Addit. vii): 183 (1924); Merr., Pl. Elmer.: 157 (1929).

Macaranga poilanei Gagnep. in Bull. Soc. Bot. France 69: 703 (1923) & in Lecomte: 448 (1926); Croiz. in Journ. Arn. Arb. 23: 51 (1942).

Mallotus tsiangii Merr. & Chun in Sunyatsenia 1: 63 (1930).
Macaranga auriculata (Merr.) Airy Shaw in Kew Bull. 19: 325 (1965) & 26: 287 (1972); Whitmore in Gard. Bull. Sing. 26: 62 (1972).

SR; SB; K1.—Hainan, Indochina, SE. & Penins. Siam, Malaya, Philippines.
Shrub or tree to 21 m. high, in primary Lowland Dipterocarp forest or disturbed Mixed Dipterocarp forest (sometimes on dikes between rice-fields). on clay-rich shale-derived soil, dark brown sandy soil, sandy loam or clay loam, up to 1350 m. alt.

In Malaya there is a complete intergradation between *M. auriculata* and *M. lowii* (see Whitmore, *l.c. supra* (1972)), though the Borneo collections at Kew do not show it.

Var. **kostermansii** *Airy Shaw* in Kew Bull. 23: 107 (1969).

M. glaberrima sec. Airy Shaw in Kew Bull. 19: 322 (1965), *p.p.*, quoad specimina borneënsia; Meijer, 7: 50 (1967); nec *Rottlera glaberrima* Hassk.

K1.—Endemic.
Small tree to 15 m. high, in dryland forest on sandy loam up to 70 m. alt.

Living wood dry, used as firewood.

30. **Macaranga praestans** *Airy Shaw* in Kew Bull. 19: 318 (1965) & in Hook. Ic. Pl. 38, t. 3718 (1974).

SR; B.—Endemic.
Treelet or small tree to 6 m., locally semi-gregarious, in deep shade in primary lowland rain forest, up to 120 m. alt.

31. **Macaranga rarispina** *Whitmore* in Kew Bull. 29: 450 (1974).

'Aff. *M. anceps* Airy Shaw' sec. Airy Shaw in Kew Bull. 19: 317 (1965).

SR.—Endemic.
This species links *M. anceps* with *M. praestans*, having characters of both.

32. **Macaranga repando-dentata** *Airy Shaw* in Kew Bull. 19: 321 (1965); Meijer, 7: 50 (1967).

SR; B; SB; K1.—Endemic.
Small tree of 6–7 m., locally semi-gregarious in primary lowland rain forest (in Sabah over ultrabasic rocks) up to 300 m. alt.

33. **Macaranga strigosissima** *Airy Shaw* in Kew Bull. 19: 320 (1965).

SR (SW. & C.).—Endemic. Known from two collections.
Tree 9–12 m. high, locally common in primary lowland rain forest up to 150 m. alt.

The harshly strigose pubescence of all parts including the upper surface of the leaf makes this a highly distinctive species of Sect. *Pseudo-Rottlera*.

5. Section **Stachyella**

M. costulata and *M. kinabaluensis* represent Sect. *Stachyella* in Borneo. These two species are closely related to *M. laciniata* and *M. heynei* of Malaya and to *M. javanica*. The shape and ornamentation of the bracteoles of the inflorescence is a vital feature in distinguishing between these species.

34. **Macaranga costulata** *Pax & Hoffm.* vii: 338 (1914); Merr.: 341 (1921).

M. javanica sec. Meijer, 7: 49 (1967), *non* (Bl.) Muell. Arg.

SR; SB; K1.—Endemic.
Shrub or tree to 25 m. high, in primary or more often secondary forest or scrub, clearings, *ladang* in shifting cultivation, etc., on clay or dark brown soil, once noted on an exposed ridge on coral limestone, up to 1800 m. alt.

In *M. costulata* the bracteoles (caducous in females) are narrowly oblong, glabrous, *c.* 5 mm. long, and bear 1–several large dorsal glands, which distort the outline; often there is only one, apically, and then the bracteole is spoon-shaped. The leaf is usually truncate, sometimes broadly so, but a few collections have been made with narrowly elliptic leaves, strongly reminiscent of much *M. heynei*. The base is narrowly peltate, sometimes only minutely so, when it is also subcordate; there are two large glands near the petiole insertion; innovations are rusty-tomentose, mature leaves glabrous and slightly glaucous below. The species is characteristically found in secondary forest, in dense stands, at low to medium elevations, but a few collections have been made in high primary rain forest.

35. **Macaranga kinabaluensis** *Airy Shaw* in Kew Bull. 23: 91 (1969); Meijer, 10: 232 (1968), *nomen, in obs.*

M. javanica auctt., sec. Meijer, 7: 49 (1967), *non* (Bl.) Muell. Arg.

SB.—Endemic.
Tree to 24 m. high, in primary *Tristania* forest at 1300–1600 m., once recorded at 180 m. alt.

Quite distinct from *M. costulata*. The bracteoles are similarly narrowly oblong, with conspicuous dorsal glands, but are bigger (sometimes reaching 10 mm.) and more conspicuous, more coriaceous, and often floccose. The lamina is stiffly coriaceous with the margin reflexed and sometimes without glands at the petiole insertion. *M. kinabaluensis* is unique amongst Malayan and Bornean species of Sect. *Stachyella* in being, so far as is known, a primary forest species. The very broad leaf of SAN 46634, commented on by Airy Shaw (*l.c.*) (the only collection not from the vicinity of Mt. Kinabalu, and, from 180 m., the only collection from below 1300 m.), is linked to the more typical form by SAN 60533, and does not warrant taxonomic recognition.

6. Species which are individually distinctive

36. **Macaranga amissa** *Airy Shaw* [ex Meijer, 7: 49 (1967), *nomen, in clavi,* &] in Kew Bull. 21: 403 (1968); Schaeffer in Blumea 19: 192 (1971); Whitmore & Airy Shaw in Kew Bull. 25: 237 (1971); Whitmore, Tree Flora 2: 112 (1973).

Endospermum perakense King ex Hook. f.: 458 (1887); Pax & Hoffm. iv: 35 (1912); Ridley: 306 (1924); *non Macaranga perakensis* Hook. f. (1887).
Macaranga sp., Corner in Gard. Bull. Str. Settlem. 10: 299 (1939).
'*Macaranga* aff. *populifolia*', Corner, Ways. Trees: 269 (1940).
Macaranga sp., Airy Shaw in Kew Bull. 14: 396 (1960).

SB.—Malaya.
Tree of 12 m., in secondary forest on swampy ground at very low altitude. Apparently rare.

37. **Macaranga conifera** (*Zoll.*) *Muell. Arg.*: 1005 (1866); Pax & Hoffm. vii: 392 (1914); Meijer, 7: 49 (1967); Whitmore: 112 (1973).

Mappa conifera Zoll. in Linnaea 29: 466 (1857).
Pachystemon populifolius Miq., Fl. Ind. Bat., Suppl.: 462 (1861).
Mappa populifolia (Miq.) Muell. Arg.: 1006 (1866); Hook. f.: 450 (1887); Pax & Hoffm. vii: 322 (1914); Merr.: 342 (1921); Ridley: 304 (1924).

SR (SW.); **B; SB; K1.**—Sumatra, Malaya.
Shrub or tree to 30 m. high, in primary or semi-secondary Lowland Dipterocarp forest (sometimes marshy or periodically inundated) on black or greyish or dark brown sandy soil or sandstone or tufa or loam with coral limestone, up to 750 m. alt.

38. **Macaranga endertii** *Whitmore* in Kew Bull. 29: 449 (1974).

SR; K1.—Endemic.
Small tree to 15 m., on slopes of lowland limestone hills, in forest.

Highly distinctive, with a slight resemblance to *M. dipterocarpifolia* Merr. and its allies, of the Philippines.

39. **Macaranga gigantea** (*Reichb. f. & Zoll.*) *Muell. Arg.*: 995 (1866); Pax & Hoffm. vii: 307 (1914); Merr.: 341 (1921); Corner: 265 (1940); Whitmore in Malayan Nature Journ. 20: 94, 96 (1967); Meijer, 7: 49 (1967); Whitmore & Airy Shaw in Kew Bull. 25: 241 (1971); Airy Shaw in Kew Bull. 26: 288 (1972); Whitmore: 107 (1973).

Mappa gigantea Reichb. f. & Zoll. in Linnaea 29: 465 (1857).
M. megalophylla Muell. Arg. in Flora 47: 467 (1864).
M. rugosa Muell. Arg. in Linnaea 34: 197 (1865).
Macaranga megalophylla (Muell. Arg.) Muell. Arg.: 995 (1866); Hook. f.: 449 (1887); Ridley: 298 (1924).
M. incisa Gage in Rec. Bot. Surv. Ind. 9: 245 (1922); Ridley: 301 (1924).

SR; B; SB; K1, 2?, 3.—Siam, Malaya, Sumatra, Celebes.
Tree to 21 m. high, in primary or secondary lowland or riverine or hill forest, on sandy clay or leached yellow sandy soil up to 600 m. alt.

40. **Macaranga gigantifolia** *Merr.* in Philipp. Journ. Sci. 7, Bot.: 391 (1912); Pax & Hoffm. vii: 318 (1914); Merr.: 441 (1923).

SB; K1.—Philippines, Sangië & Talaud Is.?, Celebes?, Moluccas?
Tree of 7·5 m., on hillside near stream at 300 m. alt. Very common around Sandakan (W. Meijer, *pers. comm.*) and near Tanahgrogot.

41. **Macaranga tanarius** (*L.*) *Muell. Arg.*: 997 (1866); Hook. f.: 447 (1887); Pax & Hoffm. vii: 352 (1914) (excl. synon. *M. clavata* Warb.); Merr.: 342 (1921) & 443 (1923); Ridley: 302 (1924); Gagnep.: 442 (1926); Hend. in Journ. Malayan Br. Roy. As. Soc. 17: 71 (1939); Corner: 268 (1940); Backer & Bakh. f.: 488 (1963); Whitmore in Malayan Nature Journ. 20: 94, 99 (1967); Meijer, 7: 49, tab. (1967); Airy Shaw in Kew Bull. 23: 99 (1969), *in clavi*, & 26: 291 (1972); Whitmore: 111 (1973).

Ricinus tanarius L. in Stickm., Herb. Amboin.: 14 (1754) & in Amoen. Acad. 4: 125 (1759).
Mappa tanarius (L.) Bl., Bijdr.: 624 (1825).
M. tomentosa Bl., *l.c.* (1825).
Macaranga molliuscula Kurz in Journ. As. Soc. Bengal 42: 245 (1873).

SR; B; SB; K1, 2, 3.—Andamans, Nicobars, SE. & Lower Siam, Cochinchina, S. China, Formosa & Ryu-Kyu Is., throughout Malesia to N. Australia & Melanesia.

Tree to 12 m. high, in primary or more often disturbed secondary forest (sometimes swampy), or in dry forest in a teak area, or on tidal river banks, once noted as abundant on a marine beach behind the littoral *Casuarina* belt, on black or yellow-brown sandy soil or on a sandstone ridge, or on rocky soil or basalt, up to 2100 m. alt.

Branchlets sticky. Timber used for house construction, bark for cordage.

42. **Macaranga trichocarpa** (*Reichb. f. & Zoll.*) *Muell. Arg.*: 1003 (1866); Hook. f.: 450 (1887); Pax & Hoffm. vii: 358 (1914); Ridley: 303 (1924); Gagnep.: 441 (1926) (incl. var. *trilobulata* Gagnep.); Corner: 268 (1940); Whitmore in Malayan Nature Journ. 20: 95, 99 (1967); Meijer, 7: 49 (1967); Airy Shaw in Kew Bull. 23: 97 (1969) & 26: 291 (1972); Whitmore: 112 (1973).

Mappa trichocarpa Reichb. f. & Zoll. ex Zoll. in Verhand. Natuurk. Vereen. Ned. Ind. 1: 8 (1856); Zoll. in Linnaea 28: 307 (1856).
Mappa zollingeriana Miq., Fl. Ind. Bat., Suppl.: 457 (1861).
Macaranga minutiflora Muell. Arg. in Flora 47: 466 (1864); Pax & Hoffm., *l.c.*: 358 (1914).
Mappa borneënsis Muell. Arg. in Flora 47: 468 (1864).
Macaranga borneënsis (Muell. Arg.) Muell. Arg.: 1003 (1866).
Macaranga helferi Muell. Arg.: 1004 (1866).

SR; SB; K1–3.—SE. Siam, Indochina, Lower Burma, Malaya, Sumatra.

Shrub or tree to 7·5 m. high, in primary or secondary Mixed Dipterocarp forest on yellow loam, sandy loam, sandy clay, black soil or a tufa plateau, up to 480 m. alt.

43. **Macaranga winkleri** *Pax & Hoffm.* vii: 355 (1914); Merr.: 343 (1921); Pax & Hoffm. in Mitt. Inst. Bot. Hamburg 7: 228 (1931, ante 15 Mart.) & in Engl. & Harms, Pflanzenf. ed. 2, 19c: 133 (1931); Meijer, 7: 47 (1967); Airy Shaw in Kew Bull. 23: 96 (1969) & 27: 90 (1972).

SR; SB; K1, 3.—Endemic.

Tree to 18 m. high, in primary or secondary forest, up to 1800 m. alt. Apparently very common.

See also *M. winkleriella*.

44. **Macaranga winkleriella** *Whitmore* in Kew Bull. 29: 449 (1974).

SR.—Endemic.
Treelet of 1·5 m. on limestone.
Known from a single collection in fruit (CWL 457) from the Melinau Gorge in Baram District.

Close to *M. winkleri* but clearly distinct, notably in its smaller stipules and longer styles.

7. Species incertae sedis

These are three species described from collections by the two H. Winklers, uncle and nephew. None is recognizable from the descriptions. Until and unless material is found (and at both Berlin and Hamburg the *Euphorbiaceae* were destroyed in World War II) these will remain *species incertae sedis.*

45. **Macaranga brachythyrsa** *Pax & Hoffm.* vii: 320 (1914); Merr.: 341 (1921).

K1a.—Endemic.
Small tree in Heath forest.

Based on *Winkler* (*avunc.*) 3319 from Heath forest near Djihi in SE. Borneo. Placed by Pax & Hoffmann in Sect. *Semiglobosae.*
The leaves are huge, 40–50 × 20–23 cm., narrowly triangular, ovate, cuspidate-acuminate with a broadly peltate base, glabrous; the rachis is compressed and the inflorescence branches opposite; and the stipules are very soon caducous. In these respects *M. brachythyrsa* is highly reminiscent of *M. petanostyla* of Mt. Kinabalu. But one cannot be sure without seeing material, especially in view of its totally different habitat.

46. **Macaranga eloba** *Pax & Hoffm.* in Mitt. Inst. Bot. Hamburg 7: 228 (1931).

K3.—Endemic.

Based on *Winkler* (*nep.*) 968a, 20 Dec. 1924, Bt. Raja, 1400 m., in West Kalimantan. Placed by Pax & Hoffmann in Sect. *Pachystemon.* This could be yet another species of Sect. *Pachystemon* or conceivably yet another manifestation of *M. caladiifolia.*

47. **Macaranga gossypiifolia** *Pax & Hoffm.* vii: 311 (1914); Merr.: 342 (1921).

K1a.—Endemic.

South-east Borneo, between Simpokak and Semurung, *Winkler* (*avunc.*) 2996. From its description this could be anything, or a complete novelty. Pax & Hoffmann place it after *M. hypoleuca* in Sect. *Pruinosae* and suggest an affinity with *M. hosei.*

Mallotus *Lour.*

Key to Sections

1 Leaves penninerved (rarely trinerved in § *Axenfeldia*):
  2 Leaves glandular-punctate above, opposite or alternate; fruit frequently indehiscent and winged . . . . . . § **Polyadenii**
  2 Leaves not glandular-punctate above, opposite (but in § *Hancea* rarely glandular-punctate above and sometimes *apparently* alternate); fruit an unwinged dehiscent capsule:
    3 Leaves truly opposite, but exceedingly unequal, one member of each pair of normal size and shape, rarely granular-glandular beneath, the reduced member either suborbicular or even stipuliform; stems terete; nodes often swollen . . . . § **Hancea**
    3 Leaves opposite, equal or somewhat unequal, but similar in shape, granular-glandular beneath; stems and nodes somewhat flattened § **Axenfeldia**
1 Leaves palmately nerved or trinerved, mostly granular-glandular beneath:
  4 Leaves all opposite, epeltate: ♀ calyx not spathaceous; plant not smelling of fenugreek . . . . § **Rottleropsis**
  4′ Leaves opposite and alternate on the same branch; ♀ calyx spathaceous, deciduous; dried plant smelling of fenugreek . . § **Stylanthus**
  4″ Leaves all alternate:
    5 Capsule smooth; ♀ calyx not spathaceous; plant not smelling of fenugreek . . . . § **Rottlera**
    5 Capsule echinate:
      6 Stamens 45–100; ♀ calyx not or scarcely spathaceous, persistent; plant not smelling of fenugreek . . . . . . § **Mallotus**
      6 Stamens 15–45; ♀ calyx spathaceous, deciduous; plant smelling of fenugreek . . . . § **Stylanthus**

§ Axenfeldia

1 Leaves shortly cuspidate-acuminate . . . . **M. resinosus**
1 Leaves gradually very long-acuminate . . . . . . **M. laevigatus**

Mallotus laevigatus (*Muell. Arg.*) *Airy Shaw* in Kew Bull. 19: 328 (1965); Meijer, 7: 52 (1967).

Rottlera glaberrima sec. Reichb. f. & Zoll. ex Zoll., Over de soorten van Rottlera, in Verhand. Natuurk. Vereen. Ned. Ind. 1: 29 (1856) & in Linnaea 28: 313 (1856); Miq., Fl. Ind. Bat. 1(2): 393 (1859); *non* Hassk.
Mallotus glaberrimus sec. Muell. Arg. in Linnaea 34: 192 (1865); J.J. Sm.: 435 (1910); Pax & Hoffm. vii: 192 (1914); Merr.: 338 (1921); Backer & Bakh. f.: 484 (1963); *non Rottlera glaberrima* Hassk.
M. moritzianus β laevigatus Muell. Arg. in Linnaea 34: 191 (1865).

SR (Merrill); **K**– (J. J. Smith).—Sumatra, Java.
No ecological information available for Borneo.

Closely related to *M. resinosus*, differing chiefly in the long-drawn-out acumen of the leaves.

I have seen neither of the specimens upon which J. J. Smith's and Merrill's records (for Indonesian Borneo and Sarawak respectively) were based, nor have I seen any other Bornean material that I could refer to this species. Pending confirmation, the occurrence of *M. laevigatus* in Borneo must be regarded as suspect.

Mallotus resinosus (*Blanco*) *Merr.*, Sp. Blanco.: 222 (1918), & Enum.: 436 (1923); Meijer, 7: 53 (1967); Airy Shaw in Kew Bull. 26: 294 (1972).

Adelia resinosa Blanco, Fl. Filip. ed. 2: 562 (1845); Muell. Arg.: 731 (1866).
Claoxylon muricatum Wight, Ic. Pl. Ind. Or. 5(2): 24, t. 1886 (1852).
Axenfeldia intermedia Baill., Ét. Gén. Euphorb.: 419 (1858).
Rottlera muricata (Wight) Thw., Enum. Pl. Zeyl.: 273 (1864).
Mallotus dispar var. *psiloneurus* Muell. Arg. in Linnaea 34: 191 (1865).
M. muricatus (Wight) Muell. Arg., *l.c.* (1865) & in DC.: 972 (1866), *pro parte*; Hook. f.: 436 (1887); J. J. Sm.: 428 (1910); Pax & Hoffm. vii: 190 (1914); Backer & Bakh. f.: 484 (1963).
Coelodiscus muricatus (Wight) Gagnep.: 369 (1925), *p.p.*

SB.—S. India & Ceylon, Andamans & Nicobars; N. Malaya, Indochina, Philippines, Celebes, Java, Lesser Sunda Is., New Guinea and N. Queensland.
Tree of 5–10 m., on limestone hills at 15 m. alt.; said to be locally 'very common' in Sabah, but only two collections seen.

The flattened branchlets, opposite, unequal, short-petioled, penninerved, glabrous, cuneate-obovate leaves, usually somewhat repand-dentate distally and with a few conspicuous macular glands towards the base, are characteristic of this limestone-loving species.

§ Hancea

1 Reduced leaf of each opposite pair orbicular or ovate:
  2 Glabrous or thinly puberulous (rarely subtomentellous); leaves sinuate; ♂ inflorescence short; ♂ buds oblong, acute . **M. miquelianus**
  2 Densely tomentellous; leaves entire; ♂ inflorescence elongate; ♂ buds subglobose, obtuse . . . . **M. havilandii**
1 Reduced leaf of each opposite pair subulate, stipuliform:
  3 Leaves with numerous minute glandular granules or pits on the upper surface and a group of macular glands at the base:
    4 Plant shortly fulvous-tomentellous throughout (except upper leaf-surface); ♂ inflorescence up to 5 cm. long; stamens about 60 **M. beccarii**
    4 Plant almost glabrous (except inflorescences); ♂ inflorescence about 1 cm. long; stamens about 30 . . . **M. brachythyrsus**
  3 Leaves without granules or pits on the upper surface:
    5 Inflorescences shortly and densely viscid-glandular; aculei of capsule capitate-glandular and setulose . . . . **M. griffithianus**
    5 Inflorescences eglandular; aculei of capsule not capitate-glandular:
      6 Glabrous (rarely sparsely pilose on the midrib beneath); leaves chartaceous to coriaceous; stipules narrow, acute **M. penangensis**

6 Softly pubescent or tomentellous:
  7 Stipules and floral bracts narrower, subulate, acute, fulvous-pubescent, soon caducous . . . . . **M. eucaustus**
  7 Stipules and floral bracts broader, oblong, obtuse or subobtuse, blackish when dry, ± persistent . . . . **M. stipularis**

Mallotus beccarii *Airy Shaw* in Kew Bull. 27: 86 (1972).

SR.—Endemic.

Treelet of 1·2 m., on hillside on yellow clayey soil at 225 m. alt.

Leaves elliptic or oblanceolate, with 13–14 pairs of nerves, a group of conspicuous macular glands at the base, and numerous glandular granules (or at least pits) on the upper and sometimes also on the lower surface. The margin is shallowly and bluntly repand-dentate, each tooth bearing a conspicuous hydathodal gland, each of which is supplied by a short vein radiating from the 'loop' or submarginal anastomosis between two main lateral nerves. The branchlets, petioles, midribs beneath, stipules and inflorescences are very shortly fulvous-velutinous.

Closely related to *M. brachythyrsus*, but differing in the short fulvous-tomentellous indumentum of all parts except the upper leaf-surface. The stipules, reduced leaves and male inflorescences are also appreciably longer than in that species, and the stamens are twice as numerous.

Mallotus brachythyrsus *Merr.* in Sarawak Mus. Journ. 3: 526 (1928).

SR (SW.).—Endemic.

Shrub or small tree to 2·5 m., in forest up to 150 m. alt.; no data as to substratum.

Differs from the closely related *M. beccarii* in the almost complete glabrescence of all parts except the inflorescences. The inflorescences, stipules and reduced leaves are considerably shorter than those of *M. beccarii*, and there are only 30 instead of 60 stamens.

I am indebted to Dr. R. Ornduff, of the University of California, Berkeley, Calif., U.S.A., for the loan of the type of this very local species. Only one other specimen has been seen, likewise from Mt. Poi (Gunong Pueh), the type locality: *Purseglove* P. 4665, collected 22 Sept. 1955. This shows that there may be as many as 13 lateral nerves, as against the 8 of the original description. The female plant has yet to be obtained.

Merrill was in error in referring this species to § *Axenfeldia*. It is unquestionably a member of § *Hancea*.

Mallotus eucaustus *Airy Shaw* in Kew Bull. 23: 80 (1969).

SR; SB; K1.—Endemic.

Tree to 24 m. high, in primary Mixed or Lowland Dipterocarp forest, on shale or sandy ridges, or on a tufa plateau, up to 285 m. alt.

Shortly and softly (sometimes sparsely) pubescent throughout (except upper leaf-surface); leaves entire (rarely shallowly dentate towards apex), 12–14-nerved, often with distinctly reflexed margin, very shortly petioled; stipules acute; male inflorescence 1–2 cm. long, with long narrow bracts; female sepals very narrowly linear-subulate; capsule covered with very numerous short aculeoli and also setulose-pubescent.

Mallotus griffithianus (*Muell. Arg.*) *Hook. f.*: 433 (1887); Pax & Hoffm. vii: 178 (1914); Ridley: 291 (1924); Meijer, 7: 52, tab. (1967); Whitmore: 116 (1973).

Diplochlamys griffithianus Muell. Arg. in Flora 47: 539 (1864) & in DC.: 1024 (1866).
Buettneria uncinata Masters in Hook. f., Fl. Brit. Ind. 1: 377 (1874).
Mallotus impar Pax & Hoffm. vii: 396 (1914); Merr.: 339 (1921) & Pl. Elmer.: 157 (1929); **synon. nov.**
M. woodii Merr. in Philipp. Journ. Sci. 13, C. Bot.: 81 (1918) & Enum.: 340 (1921) & Pl. Elmer.: 158 (1929), *q.v.*

SR; B; SB; K1.—Malaya.
Tree to 12 m. high, in primary and secondary Lowland Dipterocarp forest, dry Mixed Dipterocarp forest, or Heath forest, on dark soil, sandstone, yellow sandy clay, sandy loam with small gritty stones, once noted on a tufa plateau, up to 360 m. alt.

Notable for the short viscid glandular covering usually developed on the axes of the inflorescences, and for the long glandular-capitate aculei (which are themselves setulose) on the capsules. The short viscid glands sometimes extend to the stems and petioles, and may leave a yellow stain on the mounting paper. Occasionally the glandular covering is almost absent.
Mallotus griffithianus is related to *M. kingii* Hook. f. (Lower Siam & Malaya); it has nothing to do with *M. subpeltatus* (Bl.) Muell. Arg., despite Pax & Hoffmann's confident assertion (*l.c.*).

Mallotus havilandii *Airy Shaw* in Kew Bull. 20: 39 (1966) & 23: 82 (1969).

SR (SW.).—Endemic.
Small tree to 7·5 m. high, in an open place on a ridge between two vertical limestone walls, or on a scree slope of a limestone hill, with limestone boulders and rock, on chert-derived soil, at 75–150 m. alt.

Differs from the related *M. miquelianus* in the short, dense tomentellous indumentum of all parts except the upper leaf-surface, in the elongate male inflorescence (up to 17 cm. long), in the elongate floral bracts (male and female), and in the subglobose buds of the male flowers. It is now known from four gatherings, all from near Kuching and all from limestone.

Mallotus miquelianus (*Scheff.*) *Boerl.*: 290 (1900); Pax & Hoffm. vii: 200, fig. 29 (1914); Merr.: 339 (1921) & 434 (1923) & in Philipp. Journ. Sci. 29: 382 (1926) & Pl. Elmer.: 158 (1929); Meijer, 7: 53 (1967); Airy Shaw in Kew Bull. 26: 295 (1972); Whitmore: 114 (1973).

Rottlera miqueliana Scheff. in Ann. Mus. Bot. Lugd.-Bat. 4: 124 (1868).
Mallotus anisophyllus Hook. f.: 436 (1887); Boerl.: 290 (1900); Ridley: 293 (1924).

SR; SB; K1, 1a.—Lower Siam, and throughout W. Malesia (except Java).
Shrub or tree to 7·5 m. high, common in primary Mixed Dipterocarp forest (dry, swampy or periodically inundated), on brown or sandy soil, coral limestone or granodiorite, up to 600 m. alt.

A very characteristic species, differing from *M. havilandii* in its thinly puberulous (rarely subtomentellous) indumentum or more often complete glabrescence, and especially in the narrowly ovoid-oblong, acute buds of the male flowers. The leaf-outline is very similar to that of *M. beccarii.*

Wood very tough and durable, used for making walking-sticks. Native collectors in Sabah have twice noted the plant as poisonous. The flowers are said to be fragrant.

Mallotus penangensis *Muell. Arg.* in Linnaea 34: 186 (1865) & in DC.: 961 (1866); Hook. f.: 440 (1887); Pax & Hoffm. vii: 201 (1914); Ridley: 293 (1924) (incl. var. *grandifolia* Ridl.); Meijer, 7: 52 (1967); Whitmore: 116 (1973).

M. echinatus Elm. in Leafl. Philipp. Bot. 3: 925 (1910); Pax & Hoffm. vii: 204 (1914); Merr.: 433 (1923); **synon. nov.**
M. leptophyllus Pax & Hoffm. vii: 203 (1914); Merr.: 339 (1921); **synon. nov.**
M. pseudopenangensis Pax & Hoffm. vii: 203 (1914), **synon. nov.**
M. sarawakensis Pax & Hoffm. vii: 201 (1914); Merr.: 340 (1921); **synon. nov.**
M. xylacanthus Pax & Hoffm. vii: 203, 397 (1914), **synon. nov.**

SR; B; SB; K1.—Malaya, Sumatra, Philippines, Moluccas.

Shrub or tree to 20 m. high, very common in primary Lowland or Mixed Dipterocarp forest (occasionally swampy), or in primary riparian or montane forest or *kerangas*, on sandstone or yellow sandy clay, or on a silty clay river bank (beyond tidal influence), or on dacite or black or brown soil, up to 950 m. alt., ascending to 1200 m. on Kinabalu.

I find it impossible to maintain the above five synonyms as distinct. *M. leptophyllus* may represent a shade form or younger growth; it appears to grade insensibly into the typical form. The remainder seem to represent small individual variations.

Mallotus stipularis *Airy Shaw* [ex Meijer, 7: 53 (1967), *anglice, in clavi,* &] in Kew Bull. 21: 398 (1968), *latine,* & 26: 296 (1972); Meijer, 10: 231 (1968).

SR; SB.—Lower Siam, Sumatra; to be expected in Malaya.

Tree to 11 m. high, in primary Mixed Dipterocarp or submontane forest on clay-rich or basalt-derived soil up to 950 m. alt.

Differs from the common *M. penangensis* in the relatively broad, subobtuse, oblong or elliptic or lanceolate (not acutely subulate) stipules, and broader floral bracts. The leaves are laxly pubescent on the nerves beneath.

§ MALLOTUS (*Echinus*)

1 Aculei of the capsule relatively robust, short, conic-subulate, rigid, very shortly whitish-tomentellous; leaves frequently rhombic, acuminate, sometimes tricuspidate or trilobed, not peltate; indumentum very short and smooth . . . . **M. paniculatus**
1 Aculei of the capsule relatively slender, less rigid, more or less pubescent; leaves rarely rhombic or trilobed:
  2 Aculei shorter, very dense, forming a continuous woolly layer; leaves narrowly peltate; indumentum not floccose **M. macrostachyus**

2 Aculei longer, less dense, individually visible, forming a loose shaggy covering; leaves rarely peltate; indumentum often subfloccose
M. mollissimus

Mallotus macrostachyus (*Miq.*) *Muell. Arg.*: 963 (1866); Hook. f.: 429 (1887); Boerl.: 289 (1900); Pax & Hoffm. vii: 163 (1914); Merr.: 339 (1921); Ridley: 288 (1924); Gagnep.: 357 (1925); Corner: 271 (1940); Meijer, 7: 51 (1967); Airy Shaw in Kew Bull. 26: 297 (1972); Whitmore: 114 (1973).

Rottlera macrostachya Miq., Fl. Ind. Bat., Suppl.: 454 (1861); Scheff. in Ann. Mus. Bot. Lugd.-Bat. 4: 123 (1868).
Mallotus albus sec. Muell. Arg. in Linnaea 34: 188 (1865) & in DC.: 965 (1866); Hook. f.: 429 (1887); Boerl., *l.c.* (1900); Pax & Hoffm. vii: 168 (1914); Merr.: 338 (1921); *ubique excl. syn. Rottlera tetracocca* Roxb.; cf. Croiz. in Journ. Arn. Arb. 21: 503 (1940); *non Rottlera alba* Roxb. ex Jack = *Mallotus paniculatus* (Lam.) Muell. Arg.

SR; SB; Kɪa, 3.—Lower Siam, Malaya, Banka.
Shrub or small tree to 12 m. high, in primary Mixed Dipterocarp or riverine forest or secondary undergrowth, often on river banks, on black or brown or yellow sandy soil, once noted on basalt, from low levels to 900 m. alt.

Branches robust, shortly rufous-tomentellous or almost chocolate-coloured; leaves large, triangular-ovate, narrowly peltate, very long-petioled; inflorescences often leaf-opposed, the male much branched, the female simple; capsule globose, with a very dense covering of tomentose processes forming a uniform layer, ochraceous when dry.

Mallotus mollissimus (*Geisel.*) *Airy Shaw* in Kew Bull. 26: 297 (1972).

Croton mollissimus Geisel., Croton. Monogr.: 73 (March 1807); cf. Croiz. in Journ. Arn. Arb. 19: 141 (1938) & 21: 502 (1940), *in obs.*
C. ricinoïdes Pers., Syn. 2: 586 (Sept. 1807).
Rottlera ricinoïdes (Pers.) A. Juss., Euphorb. Gen. Tent.: 33 (1824).
Adisca zippelii Bl., Bijdr.: 611 (1825).
Chrozophora mollissima (Geisel.) Spreng., Syst. 3: 851 (1826).
Mallotus ricinoïdes (Pers.) Muell. Arg. in Linnaea 34: 187 (1865) & in DC.: 963 (1866); Hook. f.: 430 (1887); Boerl.: 289 (1900); J. J. Sm.: 409 (1910); Hutch. in Journ. Linn. Soc., Bot. 42: 136 (1914); Pax & Hoffm. vii: 170 (1914); Merr.: 340 (1921) & 436 (1923); Gagnep.: 356 (1925); Merr. in Philipp. Journ. Sci. 29: 383 (1926); Holth. & Lam in Blumea 5: 202 (1942); Backer & Bakh. f.: 482 (1963); Meijer, 7: 51 (1967).
Echinus mollissimus (Geisel.) Baill., Adansonia 6: 316 (1866).

SB; Kɪa?.—NW. Siam? Lower Burma? Indochina, and throughout Malesia (except Malaya) to New Guinea, Australia and Melanesia.
Tree to 26 m. high, in primary or secondary forest (sometimes swampy) or in brushwood, on brownish or yellowish soil, up to 1500 m. alt. on Kinabalu.

The processes on the fruit are long, coarse and shaggy, not forming a neat ball as in *M. macrostachyus*. The indumentum is soft and often subfloccose, not close and smooth as in *M. paniculatus*. The species is evidently extremely common in Sabah, but apparently almost absent from the rest of Borneo.

Timber used for firewood and construction; fibrous inner bark said to be useful as a strap for carrying. Leaves used medicinally: the water in which the leaves have been boiled is drunk as a cure for spleen; the bark is also used as an external cure for the same complaint.

Mallotus paniculatus (*Lam.*) *Muell. Arg.* in Linnaea 34: 189 (1865) & in DC.: 965 (1866); Merr.: 339 (1921) & 434 (1923) & in Trans. Amer. Philos. Soc. II, 24(2): 237 (1935); Corner: 272 (1940); Backer & Bakh. f.: 483 (1963); Meijer, 7: 51 (1967); Airy Shaw in Kew Bull. 26: 298 (1972); Whitmore: 114 (1973).

Croton paniculatus Lam., Encycl. Méth., Bot. 2: 207 (1786).
Echinus trisulcus Lour., Fl. Cochinch. 2: 633 (1790).
Mallotus cochinchinensis Lour., *l.c.*: 635 (1790); Hook. f.: 430 (1887); Boerl.: 289 (1900); J. J. Sm.: 413 (1910); Hallier in Meded. Rijks Herb. 1910: 7 (1911); Pax & Hoffm. vii: 166 (1914); Ridley: 289 (1924); Gagnep.: 355 (1925); Pax & Hoffm. in Mitt. Inst. Bot. Hamburg 7: 227 (1931).
Trewia triscuspidata Willd., Sp. Pl. 4: 835 (1805).
Rottlera alba Roxb. ex Jack, Mal. Misc. 1: 26 (1820); Roxb., Fl. Ind. 3: 829 (1832); cf. Airy Shaw, *l.c. supra* (1972), *in adnot.*
R. paniculata (Lam.) Juss., Euphorb. Gen. Tent.: 33 (1824).
Mappa cochinchinensis (Lour.) Spreng., Syst. 3: 878 (1826).
Mallotus albus (Roxb. ex Jack) Muell. Arg. in Linnaea 34: 188 (1865) & in DC.: 965 (1866), *tantum quoad synon.*
M. chinensis 'Lour.' ex Muell. Arg.: 965 (1866), *in synon., sphalm.*
Croton appendiculatus Elm., Leafl. Philipp. Bot. 1: 312 (1908).

SR; SB; K1, 1a, 3.—SE. Asia, S. China, Formosa, and throughout Malesia to New Guinea and Queensland.

Tree to 12 m. high, in primary or secondary Lowland Dipterocarp forest on sandy clay from low levels up to 1500 m. on Kinabalu.

Stems, petioles and inflorescences very shortly rufous-stellate-furfuraceous; leaves rhombic-ovate, often tricuspidate or trilobed, acuminate, rufous or whitish-tomentellous beneath, sometimes narrowly peltate, with two large macular glands at the base; capsule very shortly whitish-tomentellous, bearing numerous coarse relatively short subulate processes.

Leaves used in native medicine (unspecified) in Sabah.

§ Polyadenii (*Coccoceras*)

1 Capsules borne on long pedicels (up to 2·5 cm. long), with smooth, rounded, tardily dehiscent cocci; petioles relatively slender **M. leucodermis**
1 Capsules borne on short pedicels, indehiscent, the loculi angled, winged or appendaged; petioles relatively more robust:
 2 Capsule larger, trigonous, the angles narrowly winged **M. muticus**
 2 Capsule smaller, the loculi produced into three long, narrowly subulate, acute, widely spreading appendages or wings . **M. sumatranus**

The Bornean species of § *Polyadenii* cannot be identified with absolute certainty in the absence of female flower or fruit, although certain features of the leaves of *M. leucodermis* can give a fairly strong presumption in its favour.

Mallotus leucodermis *Hook.f.*: 441 (1887); Pax & Hoffm. vii: 180 (1914); Ridley: 291 (1924); Airy Shaw in Kew Bull. 16: 350, *in obs.*, 352 (1963) & 20: 39 (1966) & 21: 397 (1968); Meijer, 7: 53 (1967); Whitmore: 116, fig. 10 (1973), *excl. descr. fruct.*

Coccoceras muticum var.? *pedicellata* Hook. f.: 424 (1887); Pax & Hoffm. vii: 210 (1914); Ridley: 291 (1924).

? **SR** (C.); **SB.**—Malaya, Sumatra (Simalur), Solomon Is. (S. Ysabel).
Tree to 27 m. high, in primary or secondary forest on brown (or basalt-derived?) soil, up to 45 m. alt.

Leaves on the average narrower than the other species, often almost cuneate-obovate, entire or almost so, borne on very slender petioles (1 mm. diam.); capsules borne on very long pedicels (up to 2·5 cm.), woody, deeply tricoccous, cocci smooth, rounded, unwinged (not 3-winged and beaked, as in Whitmore, *l.c.*), tardily dehiscent.

The male plant of *M. leucodermis* has not yet been certainly collected, unless the specimen forming the basis of the above doubtful record for Sarawak (*Sibat ak. Luang* S. 21923, from Bt. Mersing, Anap, 4th Division) proves to be correctly named, and even on this in the Kew duplicate all but the extreme base of the only inflorescence has been broken off.

Mallotus muticus (*Muell. Arg.*) *Airy Shaw* in Kew Bull. 16: 351 (1963); Meijer, 7: 53 (1967); Whitmore: 116, 117 (1973), *excl. fruct. descr. et fig.*

Coccoceras muticum Muell. Arg. in Flora 47: 470 (1864); Pax & Hoffm. vii: 210 (1914); Ridley: 294 (1924).
Mallotus borneënsis Muell. Arg.: 980 (1866); Boerl.: 290 (1900); Pax & Hoffm. vii: 198 (1914); Merr.: 338 (1921).
Rottlera borneënsis (Muell. Arg.) Scheff. in Ann. Mus. Bot. Lugd.-Bat. 4: 125 (1868).
? *Coccoceras* sp., Boerl.: 288 (1900).
Coccoceras borneënse J. J. Sm. (*pro sp. nov.*) in Bull. Jard. Bot. Buitenz. III, 6: 94 (1924).

SR; SB; K1, ?**3.**—Malaya, ? Sumatra.
Tree to 35 m. high, common in primary and secondary Lowland Dipterocarp forest on river banks, flat riverine sites, swampy or periodically inundated ground, etc., on clay-rich or sandy alluvial soil (blackish or purplish-brown), from sea-level up to 300 m. alt. Very characteristic for swamp forest in Sabah.

Leaves mostly broad, with a broadly rounded or even subcordate base, usually shallowly and distantly indentate-crenate; petiole robust (about 2 mm. diam.); capsule indehiscent, conspicuously 3-angled and narrowly winged (not shortly spiny and unwinged, as in Whitmore, *l.c.* and fig. 10).

Mallotus sumatranus (*Miq.*) *Airy Shaw* in Kew Bull. 16: 351 (1963).

Coccoceras sumatranum Miq., Fl. Ind. Bat., Suppl.: 456 (1861); Pax & Hoffm. vii: 209 (1914).

K1, 1a.—Sumatra.

Tree to 31 m. high, locally very common in swamp forest on river banks, on loam with coral limestone, up to 50 m. alt.

Leaves shallowly indentate-crenate, indistinguishable from those of *M. muticus*; petioles mostly robust (about 2 mm. diam.); capsule shortly pedicellate, indehiscent, with three long, narrowly subulate, acute, widely spreading wings (up to 2·5 cm. long).

§ Rottlera (*Philippinenses*)

Only species in Borneo . . . . **M. philippensis**

Mallotus philippensis (*Lam.*) *Muell. Arg.* (corr. Merr.) in Linnaea 34: 196 (1865) & in DC.: 980 (1866); Hook. f.: 442 (1887); J. J. Sm.: 450 (1910); Pax & Hoffm. vii: 184 (1914); Merr.: 340 (1921) & 435 (1923); Ridley: 291 (1924); Gagnep.: 362 (1025); Hend. in Journ. Malayan Br. Roy. As. Soc. 17: 71 (1939); Corner: 272 (1940); Merr. & v. Steenis in Webbia 8: 405 (1952), *in obs.*; Backer & Bakh. f.: 483 (1963); Meijer, 7: 53, tab. (1967); Airy Shaw in Kew Bull. 21: 392 (1968) & 26: 300 (1972); Whitmore: 115 (1973).

Croton philippense (!) Lam., Encycl. Méth., Bot. 2: 206 (1786).
C. punctatus Retz., Obs. Bot. 5: 30 (1789).
C. coccineus Vahl, Symb. Bot. 2: 97 (1791).
Rottlera tinctoria Roxb., Pl. Corom. 2: 36 (1802).
Croton montanus Willd., Sp. Pl. 4: 547 (1805).
Rottlera aurantiaca Hook. & Arn., Bot. Beech. Voy.: 270 (1841).
R. affinis Hassk. in Flora 25(2), Beibl. 2: 41 (1842).
Mappa stricta Reichb. f. & Zoll. in Verhand. Natuurk. Vereen. Ned. Ind. 1: 31 (1856) & in Linnaea 28: 310 (1856).
Aconceveibum trinerve Miq., Fl. Ind. Bat. 1(1): 389 (1859); cf. Merr. & v. Steenis, *l.c. supra.*
Macaranga stricta (Reichb. f. & Zoll.) Muell. Arg.: 1004 (1866).
Echinus philippinensis [Lam.] Baill. (orth. mut.) in Adansonia 6: 314 (1866).
Rottlera philippinensis (Baill.) Scheff. in Ann. Mus. Bot. Lugd.-Bat. 4: 124 (1868–9).
Mallotus reticulatus Dunn in Journ. Linn. Soc. London 38: 365 (1908).

SB.—W. Himalaya & Ceylon to Formosa, and throughout Malesia to Australia and Melanesia.

Tree to 21 m. high, in primary or secondary (*padang*) forest on brownish or rocky soil from low levels up to 270 m. alt.

Unmistakable when in fruit from the layer of crimson-scarlet granules (sometimes used as a dye) with which the capsules are covered. The restricted occurrence in Borneo of this otherwise widespread species is curious.

Timber sometimes used for house-posts of medium construction.

It is surprising that the widespread scandent species, *M. repandus* (Willd.) Muell. Arg., of this section, has not yet been reported from Borneo.

§ ROTTLEROPSIS (*Echinocroton* & *Plagianthera*)

1 Coastal or littoral plant; densely minutely greyish-ochraceous-tomentellous; leaves triangular or rhombic-ovate; ultimate nerves on lower surface forming a dense grey-papillose areolation . **M. tiliifolius**
1 Not coastal; ultimate nerves not densely papillose-areolate:
  2 Basal pair of nerves strong, elongate, reaching half-way or more up the lamina:
    3 Second and succeeding pair of nerves arising in the distal half of the lamina; margin usually repand-dentate, at least towards the apex; indumentum harsher and sparser, sometimes almost absent **M. korthalsii**
    3 Second and succeeding pair of nerves arising in the basal half of the lamina; margin entire or almost so; indumentum softer and denser . . . . **M. moritzianus**
  2 Basal pair of nerves not stronger or more elongate than the remainder:
    4 Leaves or each opposite pair somewhat unequal, subentire or slightly dentate towards apex; petiole of smaller leaf more than 1 cm. long; lower surface more obscurely granular-glandular . **M. wrayi**
    4 Leaves of each pair very unequal, often repand-dentate towards apex; petiole of smaller leaf less than 1 cm. long; lower surface more manifestly granular-glandular . . . . **M. dispar**

Mallotus dispar (*Bl.*) *Muell. Arg.* in Linnaea 34: 191 (1865) (excl. var. *psiloneuro* Muell. Arg.) & in DC.: 971 (1866); Boerl.: 289 (1900); J. J. Sm.: 432 (1910); Pax & Hoffm. vii: 152 (1914); Merr.: 338 (1921); Ridley: 286 (1924); Hend. in Journ. Malayan Br. Roy. As. Soc. 17: 30, 71 (1939); Backer & Bakh. f.: 484 (1963); Meijer, 7: 52 (1967); Airy Shaw in Kew Bull. 21: 380 (1968) & 26: 303 (1972); Whitmore: 115 (1973).

Rottlera dispar Bl., Bijdr.: 608 (1825).
Mallotus leucocalyx Muell. Arg.: 970 (1866); Pax & Hoffm. vii: 150 (1914) (excl. synon.); Merr.: 434 (1923); Ridley: 286 (1924); Merr. in Philipp. Journ. Sci. 29: 382 (1926); Hend., *l.c.* (1939).
Claoxylon? stipulosum Reichb. f. & Zoll. in Verhand. Natuurk. Vereen. Ned. Ind. 1: 20 (1856) & in Linnaea 28: 323 (1856).
Mallotus stipulosus (Reichb. f. & Zoll.) Muell. Arg. in Linnaea 34: 192 (1865) & in DC.: 970 (1866).
Rottlera stipulosa Scheff. in Ann. Mus. Bot. Lugd.-Bat. 4: 125 (1868–9).
Coelodiscus dispar (Bl.) Kurz in Journ. As. Soc. Beng. 42. ii: 244 (1873).

SR; SB; K1.—Siam, Indochina, and throughout W. Malesia; Celebes? Small tree to 11 m. high, in rocky primary Mixed Dipterocarp forest, almost always on limestone (sometimes with sand), occasionally on basalt, up to 750 m. alt.

Softly and shortly tomentellous, occasionally glabrescent; leaves of each pair very unequal (smaller one very short-petioled, sometimes even absent), often repand-dentate towards the apex; stipules often pallid; young capsules densely covered with fine puberulous processes, but almost devoid of them when ripe. Female inflorescences said to be held vertically on the horizontal branches.

Roots used medicinally in fever (Sabah).

Mallotus korthalsii *Muell. Arg.*: 976 (1866); Boerl.: 290 (1900); Pax & Hoffm. vii: 158 (1914); Merr.: 339 (1921) & Pl. Elmer.: 158 (1929); Meijer, 7: 51 (1967); Whitmore: 115 (1973).

Rottlera korthalsii (Muell. Arg.) Scheff. in Ann. Mus. Bot. Lugd.-Bat. 4: 124 (1868–9).

Mallotus smilaciformis Gage in Rec. Bot. Surv. Ind. 9: 242 (1922), *e descr.*; Ridley: 288 (1924); Whitmore: 115 (1973); **synon. nov.**

SR; SB; K1 (**1a**?).—Malaya, Philippines, Celebes.

Shrub (occasionally scrambling) or tree to 15 m. high, in primary or secondary Lowland or Mixed Dipterocarp forest (sometimes swampy) on limestone or basalt or sandy clay or sandstone, on black, brown or grey soil, mostly at low altitudes, ascending to 2000 m. on Mt. Poi.

Almost always recognizable by the exceptionally strongly trinerved leaf-base, the basal pair of nerves reaching beyond (often well beyond) the half-way mark, but the next pair (apart from the strong transverse connecting nerves) arising in the distal half of the lamina. The leaves are often repand-dentate in the upper half, but may also be subentire; nervation very prominent on the lower surface; texture chartaceous to thinly coriaceous; indumentum very short and minutely setulose, not dense, sometimes almost absent. Young capsules covered with a dense, rough, greyish, subfarinaceous covering, but when they are mature this is almost lost and there is a development of numerous short, fine, rather soft, curved prickles.

Authentic material of *M. smilaciformis* (described from Perak) has not been seen, but from description it appears to represent a glabrescent form of *M. korthalsii* with subcoriaceous glossy leaves, such as occurs in the First Division of Sarawak, etc.

Mallotus moritzianus *Muell. Arg.* in Linnaea 34: 190 (1865) (excl. var. *laevigato*) & in DC.: 971 (1866); J. J. Sm.: 439 (1910); Pax & Hoffm. vii: 152 (1914); Merr., Pl. Elmer.: 158 (1929); Whitmore: 115 (1973).

M. moritzianus var. *scaber* Muell. Arg., *l.c.* (1865).

Rottlera dispar sec. Zoll. & Mor., Syst. Verz.: 17 (1855); Reichb. f. & Zoll. in Verhand. Natuurk. Vereen. Ned. Ind. 1: 16 (1856) & in Linnaea 28: 318 (1856); *non* Bl.

Mallotus korthalsii sec. Meijer, 7: 51 (1967), *non* Muell. Arg.

SR; SB; K1.—Sumatra, Malaya, Sangië & Talaud Is., Java, Lesser Sunda Is.

Shrub or tree to 9 m. high, in relatively open primary Lowland Dipterocarp forest on limestone or clayey loam or sandstone or brownish soil (sometimes periodically inundated) up to 1000 m. alt.

Usually distinguishable from *M. korthalsii* by the duller, almost entire or obscurely sinuate leaves (sometimes suborbicular in outline), with rather shorter basal nerves, by the second pair of lateral nerves arising in the lower half of the lamina, by the denser, softer rufous indumentum, and by the oblong buds of the male flowers.

Mallotus tiliifolius (*Bl.*) *Muell. Arg.* in Linnaea 34: 190 (1865) & in DC.: 969 (1866); J. J. Sm.: 443 (1910); Pax & Hoffm. vii: 148 (1914); Merr.: 340

(1921) & 436 (1923); Ridley: 285 (1924); Merr. in Philipp. Journ. Sci. 29: 383 (1926); Corner: 273 (1940); Holth. & Lam in Blumea 5: 202 (1942); Backer & Bakh. f.: 484 (1963); Airy Shaw in Kew Bull. 26: 305 (1972); Whitmore: 114 (1973).

Croton tiliifolius Lam. var. *aromaticus* Lam., Encycl. Méth., Bot. 2: 206 (1786), *quoad synon. Rumph.*
Rottlera acuminata A. Juss., Euphorb. Gen. Tent.: 33 (1824) [*non Mallotus acuminatus* (Bl.) Muell. Arg. = *M. peltatus* (Geisel.) Muell. Arg.]
R. tiliifolia Bl., Bijdr.: 607 (1825).
R. blumei Decne in Nouv. Arch. Mus. [Paris] 3: 486 (1835).
Adelia papillaris Blanco, Fl. Filip., ed. 2: 562 (1845).
Mallotus playfairii Hemsl. in Journ. Linn. Soc. 26: 441 (1894).
M. papillaris (Blanco) Merr. in Philipp. Journ. Sci. 7, Bot.: 238 (1912) & Enum.: 434 (1923).
M. '*tiliaeformis* (Miq.) Muell. Arg.' (*sphalm.*), Meijer, 7: 51 (1967); cf. *op. cit.* 10: 232 (1968).

SR; B; SB.—Formosa, Hainan, Lower Siam, and throughout Malesia to N. Australia and Fiji.

Shrub or small tree to 7·5 m. high, in littoral forest beneath *Casuarina equisetifolia* on sandy soil or sandstone rocks on or near the seashore.

Densely minutely greyish-ochraceous-tomentellous; leaves broadly triangular-ovate or rhombic-ovate, base rounded to cordate, trinerved, the ultimate nerves forming a characteristic dense grey-papillose areolation under the lens; this last feature is absolutely diagnostic. Capsule softly echinate. Flowers strongly scented.

Mallotus wrayi *King ex Hook. f.*: 433 (1887); Ridley: 287 (1924); Airy Shaw in Kew Bull. 14: 361 (1960) & 16: 349 (1963); Meijer, 7: 52 (1967); Whitmore: 115 (1973).

? *M. lanceifolius* Hook. f.: 434 (1887); Ridley: 287 (1924).
M. caudatus Merr. in Philipp. Journ. Sci. 13, C. Bot.: 83 (1918) & Enum.: 338 (1921) & Pl. Elmer.: 157 (1929).
Kunstlerodendron cuspidata [!] Ridley: 287 (1924); cf. Kew Bull. 14: 361 (1960).

SR; B; SB; K1.—Sumatra, Malaya.

Shrub or small tree to 12 m. high, in primary Mixed Lowland Dipterocarp forest, on dry or seasonally swampy soils or edge of mangrove swamps, on clay loam, yellow sandy clay, clay-rich shale-derived soil, sedimentary rocks, sandstone, scree slope beneath igneous (andesitic) rock-face, or basalt, mostly at low altitudes, ascending to 1200 m. (and scarce) on Kinabalu.

Mallotus wrayi can be recognized by the basal nerves being no stronger than the remainder, and by the very slender (short or long), terete petioles, abruptly and conspicuously thickened and geniculate at the apex. The lamina is mostly entire, but occasionally dentate towards the apex, and is often very slenderly caudate. In the degree of development of indumentum the plant is very variable; it is sometimes almost glabrous. The species is almost absent from the First Division of Sarawak, but from the Rejang north-eastwards it becomes increasingly frequent, and in Sabah it is one of the commonest members of the genus. Flowers fragrant, much visited by bees.

I am not yet entirely convinced that *M. lanceifolius* Hook. f., from Penang, is identical with *M. wrayi*, as Whitmore (*l.c.*) believes. Further complete material from the type locality is needed.

§ STYLANTHUS

1 Manifestly shortly grey-tomentellous throughout (except upper leaf-surface); leaves peltate, sparsely granular-glandular above as well as below . . . . **M. lackeyi**
1 Glabrous or thinly pilosulous; leaves not granular-glandular above:
 2 Leaves clearly peltate, usually glaucous beneath . **M. floribundus**
 2 Leaves not peltate, not glaucous beneath . . . **M. oblongifolius**

Mallotus floribundus (*Bl.*) *Muell. Arg.* in Linnaea 34: 187 (1865) & in DC.: 962 (1866); Hook. f.: 432 (1887); Boerl.: 288 (1900); J. J. Sm.: 417 (1910); Pax & Hoffm. vii. 173 (1914); Merr.: 338 (1921) & 433 (1923); Ridley: 290 (1924); Corner: 271 (1940); Backer & Bakh. f.: 483 (1963); Meijer, 7: 51, tab. (1967); Airy Shaw in Kew Bull. 26: 306 (1972); Whitmore: 113 (1973).

Adisca floribunda Bl., Bijdr.: 610 (1825).
Rottlera floribunda (Bl.) Hassk., Cat. Pl. Hort. Bogor. Alter: 238 (1844).
Mappa floribunda (Bl.) Zoll. & Mor., Syst. Verz.: 17 (1855).
Mallotus anamiticus Kuntze, Rev. Gen. Pl. 2: 608 (1891); Pax & Hoffm. vii: 204 (1914); Merr. in Trans. Amer. Philos. Soc. II, 24(2): 236 (1935).
Coelodiscus anamiticus (Kuntze) Gagnep.: 375 (1926).

SR; SB; K1, 3.—SE. Asia and throughout Malesia to Melanesia.

Tree to 15 m. high, locally common in primary or secondary forest on rocky sandstone or brownish or black sandy soil, or on alluvial soil at foot of limestone hill, up to 150 m. alt.

Distinguished by its ovate or orbicular peltate leaves, usually glaucous beneath, with tawny bearding in the lower nerve-axils; inflorescences often precocious or borne with the young leaves. Smells strongly of fenugreek when dry.

Mallotus lackeyi *Elm.* in Leafl. Philipp. Bot. 4: 1298 (1911); Pax & Hoffm. vii: 176 (1914); Merr.: 433 (1923) & in Philipp. Journ. Sci. 29: 382 (1926) & Pl. Elmer.: 158 (1929); Meijer, 7: 51, tab. (1967); Whitmore: 114 (1973).

M. sanchezii Merr. in Philipp. Journ. Sci. 7, Bot.: 402 (1912).

SR (NE.); **SB; K1.**—N. Malaya, S. Philippines.

Shrub or tree to 12 m. high, in primary or secondary Mixed Dipterocarp forest, or on a rocky brook-bank, on limestone or black or brown sandstone or clay with sand, or on dacite scree, with clay-rich pale pink-grey soil between dacite boulders, from low levels up to 1200 (?2400) m. alt.

Readily recognizable in the section by the very short, but rather dense, whitish tomentellum on almost all parts except the upper leaf-surface, and by the presence of scattered glandular granules on the upper as well as the lower surface. The leaves are broadly peltate and often finely caudate. Capsule covered with dense tomentellous processes when young, deeply tricoccous and up to 1·5 cm. diam. when mature, the processes then often reduced to rather distant warts.

Mallotus oblongifolius (*Miq.*) *Muell. Arg.* in Linnaea 34: 192 (1865) & in DC.: 973 (1866); J. J. Sm.: 425 (1910); Pax & Hoffm. vii: 193 (1914); Merr.: 339 (1921); Ridley: 293 (1924); Backer & Bakh. f.: 484 (1963); Meijer, 7: 52 (1967); Airy Shaw in Kew Bull. 26: 306 (1972); Whitmore: 116 (1973).

Rottlera oblongifolia Miq., Fl. Ind. Bat. 1(2): 396 (1859).
Mallotus porterianus Muell. Arg. in Linnaea 34: 185 (1865) & in DC.: 960 (1866); Hook. f.: 432 (1887); Ridley: 292 (1924); Corner: 273 (1940).
M. furetianus Muell. Arg., *ll.cc.*: 190 (1865) & 968 (1866); Gagnep.: 352 (1925).
M. lambertianus Muell. Arg., *ll.cc.*: 190 (1865) & 968 (1866), *e descr.*, **synon. nov.**
M. helferi Muell. Arg., *ll.cc.*: 190 (1865) & 968 (1866).
M. stylaris Muell. Arg.: 973 (1866); Boerl.: 290 (1900).
Rottlera stylaris (Muell. Arg.) Scheff. in Ann. Mus. Bot. Lugd.-Bat. 4: 125 (1868–9).
M. puberulus Hook. f.: 435 (1887); Pax & Hoffm. vii: 172 (1914); Ridley: 290 (1924).
M. columnaris Warb. in Engl., Bot. Jahrb. 13: 349 (1891); Pax & Hoffm. vii: 176 (1914); Merr. in Philipp. Journ. Sci. 11, C. Bot.: 283 (1916); Airy Shaw in Kew Bull. 20: 38 (1966).
M. odoratus Elm., Leafl. Philipp. Bot. 4: 1299 (1911); Pax & Hoffm., *l.c.* (1914); Merr.: 434 (1923).
M. alternifolius Merr. in Philipp. Journ. Sci. 7, Bot.: 395 (1912).
M. camiguinensis Merr. in *op. cit.*: 397 (1912).
M. oblongifolius var. *helferi* (Muell. Arg.) Pax & Hoffm. vii: 194 (1914).
M. oblongifolius var. *siamensis* Pax & Hoffm., *l.c.* (1914).
M. maclurei Merr. in Philipp. Journ. Sci. 21: 347 (1922) & in Lingnaam Agric. Rev. 2: 29 (1924) & in Lingnan Sci. Journn. 5: 110 (1927).

SR; SB; K1.—Andamans, SE. Asia, and throughout Malesia to New Guinea.

Tree to 12 m. high, in primary Mixed Dipterocarp forest (sometimes seasonally swampy), or rocky river banks, mostly on limestone, sometimes on sandstone or basalt, or brown or yellowish soil, up to 250 m. alt.

Leaves ovate or elliptic or sometimes obovate, occasionally almost oblong, not peltate, not glaucous, glabrous or sometimes pilose on the nerves beneath; capsule tricoccous, glabrous, shortly and rigidly echinate, often with curved processes. Dried material mostly with a strong odour of fenugreek.

Mallotus oblongifolius was mistakenly included by Pax & Hoffmann (*l.c.*) in § *Axenfeldia*; cf. Kew Bull. 21: 388 (1968), *in obs.*

Mallotus (§ **Stylanthus**) prob. sp. nov.

SR.—Third Division: Bkt. Iju, Ulu Arip, Balingian, foot of hill at low alt.,

20 July 1965, *Sibat ak. Luang* S. 23224:—Tree 4·2 m. tall, 12·5 cm. girth; fruit cream.

Leaves alternate, cuneate-obovate, shortly caudate, 9–19 cm. long, 3·5–7 cm. wide, glabrous, with scattered glandular granules below, but without basal macular glands; primary nerves 4–6 pairs; petiole short, only 5–15 mm. long; flowers unknown; infructescences up to 14 cm. long; capsules 1–1·5 cm. diam., glabrous or minutely puberulous, ochraceous when dry, covered with long fine rigid dark brown aculei 5–7 mm. long; seeds globose, 6–7 mm. diam., smooth, shining, grey-brown with some obscure marbling. Plant smelling strongly of fenugreek.

I cannot refer this specimen to any known species, but the material is scarcely sufficient for description.

MANIHOT *Mill.*

1 Leaves narrowly (2–4 mm.) peltate, not glaucous beneath; lobes broader M. GLAZIOVII

1 Leaves epeltate, distinctly glaucous beneath; lobes narrower M. ESCULENTA

MANIHOT ESCULENTA *Crantz,* Inst. 1: 167 (1766); Merr. in Trans. Amer. Philos. Soc. II, 24(2): 240 (1935); Corner: 273 (1940); Backer & Bakh. f.: 496 (1963).

Jatropha manihot L., Sp. Pl.: 1007 (1753).
Janipha manihot (L.) Kunth in Humb., Bonpl. & Kunth, Nov. Gen. & Spec. 2: 85 (1817).
Manihot utilissima Pohl, Pl. Bras. Ic. & Descr.: 32, t. 24 (1827); Muell. Arg.: 1064 (1866); Hook. f.: 239 (1887); J. J. Sm.: 33 (1910); Pax & Hoffm. ii: 67 (1910); Merr.: 344 (1921) & 450 (1923); Gagnep.: 434 (1926).
Mandioca utilissima (Pohl) Link, Handb. 2: 436 (1831).
Jatropha stipulata Vell., Fl. Flum. 10: t. 82 (1835).
Manihot edule A. Rich. in Sagra, Fl. Cuban. ed. hisp. 3: 208 (1853).
Mandioca dulcis Parodi in An. Soc. Cienc. Argent. 4: 127 (1877).
Manihot manihot (L.) Cockerell in Bull. Torr. Bot. Cl. 19: 95 (1892).
M. aipi Rusby in Mem. Torr. Bot. Cl. 6: 120 (1896).

SR; SB.— Nativc of tropical Brazil.
Cultivated shrub of 2–3 m. at very low altitudes.

Root yields tapioca.

MANIHOT GLAZIOVII *Muell. Arg.* in Mart., Fl. Bras. 11(2): 446 (1874); J. J. Sm.: 32 (1910); Pax & Hoffm. ii: 89, fig. 31 (1910); Merr.: 450 (1923).

SR.—Native of N. Brazil.
Cultivated shrub of 4–5 m. on alluvium by river bank at very low altitude.

The Ceará rubber tree.

Margaritaria *L. f.*

Margaritaria indica (*Dalz.*) *Airy Shaw* in Kew Bull. 20: 387 (1966) & 25: 492 (1971) & 26: 308 (1972); Whitmore: 117 (1973).

Prosorus indicus Dalz. in Hook. Journ. Bot. & Kew Garden Misc. 4: 346 (1852); Airy Shaw in Kew Bull. 16: 342 (1963).
Phyllanthus indicus (Dalz.) Muell. Arg. in Linnaea 32: 52 (1863) & in DC.: 417 (1866); Hook. f.: 305 (1887); J. J. Sm.: 84 (1910); Merr.: 392 (1923); Beille: 596 (1927); Backer & Bakh. f.: 468 (1963).
Calococcus sundaicus Kurz apud Teijsm. & Binnend. in Nat. Tijdschr. Ned. Ind. 27: 48 (1864).
Phyllanthus sundaicus (Kurz) Muell. Arg.: 1272 (1866).
Glochidion longipedicellatum Yamam. in Journ. Soc. Trop. Agric. 5: 178 (1933).

SR; SB; K1.—India & Ceylon to Formosa; SE. Asia; scattered throughout Malesia (exc. Malaya) to New Guinea and Queensland.
Tree to 28 m. high, in primary forest on black sandy soil or loam containing lime, or on limestone rocks, up to 210 m. alt.

Deciduous dioecious tree, with entire membranous leaves, the flowers appearing with the young flushes. Distinguished from *Securinega* by the four sepals and four stamens and by the large, dry, globose, thinly crustaceous, long-pedicelled, indehiscent fruit, containing shining blue seeds.
A scarce species in Borneo; only one collection seen from Sarawak, two from Sabah and two from E. Borneo.

Megistostigma *Hook. f.*

Megistostigma cordatum *Merr.* in Philipp. Journ. Sci. 16, C. Bot.: 563 (1920) & Enum.: 446 (1923); Croiz. in Journ. Arn. Arb. 22: 426 (1941); Airy Shaw in Kew Bull. 23: 120 (1969) [*q.v.* for key to genus].

Sphaerostylis cordata (Merr.) Pax & Hoffm. in Engl. & Harms, Pflanzenf. ed. 2, 19c: 148 (1931).

SB.—Philippines.
Climber on roadside fence of cocoa plantation, at unknown altitude.

Known from a single collection each from Sabah (Ranau Distr.) and the Philippines (Samar). Distinguished from the remaining species of the genus by the ovate, entire, cordate, non-peltate leaves and bisexual inflorescence.

Melanolepis *Reichb. f. & Zoll.*

Melanolepis multiglandulosa (*Reinw. ex Bl.*) *Reichb. f. & Zoll.* in Verhand. Natuurk. Vereen. Ned. Ind. 1: 22 (1856) & in Linnaea 28: 324 (1856); Merr., Interpr. Rumph. Herb. Amboin.: 318 (1917) & Enum.: 340 (1921) & 431 (1923); Ridley: 284 (1924); Merr. in Philipp. Journ. Sci. 29: 382 (1926); Corner: 274 (1940); Holth. & Lam in Blumea 5: 203 (1942); Backer & Bakh. f.: 481 (1963); Meijer, 7: 51 (1967); Airy Shaw in Kew Bull. 26: 309 (1972); Whitmore: 118 (1973).

Croton multiglandulosus Reinw. ex Bl., Cat. Gewassen Buitenz.: 105 (1823).
Rottlera multiglandulosa (Reinw. ex Bl.) Bl., Bijdr.: 609 (1825).
Ricinus dioicus Wall. ex Roxb., Fl. Ind. 3: 690 (1832).
Adelia monoica Blanco, Fl. Filip. ed. 2: 561 (1845).
Melanolepis calcosa Miq., Fl. Ind. Bat. 1(2): 399 (1859).
M. angulata Miq., Fl. Ind. Bat., Suppl.: 455 (1861).
Mallotus moluccanus sec. Muell. Arg. in Linnaea 34: 185 (1865) & in DC.: 958 (1866); J. J. Sm.: 400 (1910); Hutch. in Journ. Linn. Soc., Bot. 42: 135 (1914); *non Croton moluccanus* L. [= *Aleurites moluccanus* (L.) Willd., *quoad synon. Burm. et loc.*].
M. calcosus (Miq.) Muell. Arg.: 958 (1866).
M. angulatus (Miq.) Muell. Arg., *l.c.* (1866).
Rottlera moluccana sec. Scheff. in Ann. Mus. Bot. Lugd.-Bat. 4: 122 (1868–9), *non Croton moluccanus* L.
R. angulata (Miq.) Scheff., *l.c.*: 124 (1868–9).
R. calcosa (Miq.) Scheff., *l.c.*: 125 (1868–9).
Melanolepis moluccana sec. Pax & Hoffm. vii. 142 (1914); Gagnep.: 347 (1925), *in clavi*; *non Croton moluccanus* L.
Mallotus multiglandulosus (Reinw. ex Bl.) Hurus. in Journ. Fac. Sci. Univ. Tokyo, Sect. 3 Bot., 6: 308 (1954).

SB; K1, 1a.— Lower Siam (extreme S.), Formosa, Ryu-Kyu Is., Guam, and throughout Malesia to the Bismarck Archip.

Tree to 12 m. high, mostly in secondary forest, on river banks, roadsides, etc., on clay loam, up to 300 m. alt.

Whole plant loosely floccose-stellate-tomentose; leaves alternate, acutely 3–5-lobed, not granular-glandular; flowers normally monoecious; capsule mostly dicoccous; leaves, sepals and capsules often purplish-tinged. Closely related to *Mallotus* § *Rottlera* (*Philippenses*).

Bark used as a cough remedy in Sabah.

The variety *pendula* (Merr.) Pax & Hoffm. (*l.c.*: 144), described from the Philippines, does not yet appear to have been observed in Borneo.

Moultonianthus *Merr.*

Moultonianthus leembruggianus (*Boerl. & Koord.*) *v. Steenis* in Bull. Bot. Gard. Buitenz. III, 17: 405 (1948); Airy Shaw in Kew Bull. 14: 394 (1960), *in obs.*; Meijer, 7: 24 (1967).

Erismanthus leembruggianus Boerl. & Koord. in Koord.–Schum., Syst. Verz. Abt. 2, Lief. 2: 30 (1910) & in Fedde, Repert. Sp. Nov. 10: 31 (1912); Pax & Hoffm. vi: 126 (1912) & in Engl. & Harms, Pflanzenf. ed. 2, 19c: 158 (1931).
Moultonianthus borneensis Merr. in Philipp. Journ. Sci. 11, C. Bot.: 70 (1916); Pax & Hoffm. xiv (Euph.–Addit. vi): 41 (1919); Merr.: 345 (1921); Pax & Hoffm. in Engl. & Harms, *l.c.*: 170 (1931).

SR; SB; K1.—Sumatra.

Tree to 20 m. high, in primary or secondary Mixed Lowland Dipterocarp forest, on black or yellowish soil or sandy loam or clay loam, once noted on a tufa plateau, up to 100 m. alt.

Male flowers scented.

Related to *Erismanthus* and *Syndyophyllum*. An unmistakable plant from its opposite, shallowly dentate, glabrous, coriaceous leaves, accompanied by large ovate stipuliform structures, its rather crowded male inflorescences of fascicled flowers with delicate oblong petals, and its very lax, 'leggy', few-flowered female inflorescences with greatly elongate pedicels, thickened upwards in fruit. Bisexual inflorescences (male below, female above) occur, but only very rarely. I strongly suspect that the branching of the stems is sympodial, not monopodial, and that the apparent 'stipules' are not such, but may be a pair of reduced decussate leaves, as suggested in Kew Bull., *l.c. supra* (1960).* This point could probably be determined by careful examination of living plants in the field.

Neoscortechinia *Pax & Hoffm.*

1 Leaves softly pubescent or minutely puberulous beneath:
  2 Leaves rather thin in texture, usually drying bright green or yellowish, more or less glandular-dentate . . . . . **N. sumatrensis** var. **sumatrensis**
  2 Leaves coriaceous, usually drying brownish, strongly crenate **N. nicobarica**
1 Leaves quite glabrous beneath:
  3 Leaves thin in texture, drying bright green, often narrow and elongate (up to 31 cm. long) . . . . **N. sumatrensis** var. **angustifolia**
  3 Leaves thicker in texture, usually drying brown or purplish (rarely green), rarely more than 15 cm. long:
    4 Petiole with an evident pair of raised glands at apex; leaves larger and thinner, not drying dark purplish-brown; inflorescence larger, denser-flowered, pedicels very slender . . . . **N. forbesii**
    4 Petiole without or with very small glands at apex; leaves smaller and thicker, dark purplish-brown when dry; inflorescence smaller, laxer-flowered, pedicels less slender . . . . . **N. kingii**
      5 Inflorescence denser, pubescent, pedicels 1 mm. long, thinner var. **kingii**
      5 Inflorescence laxer, less pubescent, pedicels 2–3 mm. long, thicker var. **pedicellata**

Neoscortechinia forbesii (*Hook. f.*) *Pax ex S. Moore* in Journ. Bot. Brit. & For. 62: 54 (1924), *in obs.*; C. T. White in Journ. Arn. Arb. 31: 93 (1950); Airy Shaw in Kew Bull. 16: 369 (1963) & 26: 310 (1972); Whitmore, Guide For. Brit. Sol. Is.: 70 (1966) & Tree Fl. Mal. 2: 120 (1973).

Scortechinia forbesii Hook. f. in Hook. Ic. Pl. 18: sub t. 1706 (Nov. 1887), *in obs.*; Merr. in Philipp. Journ. Sci. 11, C. Bot.: 76 (1916), *pro 'nomen'*; Pax & Hoffm. xiv (Euph.–Addit. vi): 53 (1919), *pro 'nom. nud.'*
Alcinaeanthus philippinensis Merr. in Philipp. Journ. Sci. 7, Bot.: 380 (1912).
A. parvifolius Merr. in Philipp. Journ. Sci. 9, C: 461 (1914).

* Cf. also Melville, On the nature of the bud scales in the *Cunoniaceae*, in *Kew Bull.* 26: 477–485 (1972).

Scortechinia parvifolia (Merr.) Merr. in Philipp. Journ. Sci. 11, C. Bot.: 76 (1916).
Neoscortechinia arborea var. *parvifolia* (Merr.) Pax & Hoffm. xiv (Euph.–Addit. vi): 52 (1919).
N. parvifolia (Merr.) Merr.: 456 (1923).
N. coriacea Merr., Pl. Elmer.: 164 (1929).

SR (C.); **SB; K1.**—Malaya, Sumatra, Philippines, New Guinea, Solomon Is.

Tree to 35 m. high, in primary Dipterocarp forest, sometimes in marshy ground on fringe of mangrove swamps, on black or yellow-brown soil, occasionally on sandstone and lime, once noted on a tufa plateau, from sea-level up to 1350 m. alt. on Kinabalu.

The thinner, less crenate or even subentire leaves, more or less cuneate at the base, and frequently reddish beneath when dry, and the smaller thinly puberulous or glabrescent inflorescence, distinguish this species from *N. nicobarica.*

Only two collections have been seen from Sarawak.

Neoscortechinia kingii (*Hook. f.*) *Pax & Hoffm.* xiv (Euph.–Addit. vi): 52 (1919); S. Moore in Journ. Bot. Brit. & For. 63, Suppl.: 100 (1925); Airy Shaw in Kew Bull. 16: 371 (1963); Whitmore: 119, 120 (1973).

Scortechinia kingii Hook. f. in Hook. Ic. Pl. 18: t. 1706 (Nov. 1887) & in Hook. f., Fl. Brit. Ind. 5: 367 (Dec. 1887); Merr. in Philipp. Journ. Sci. 11, C. Bot.: 76 (1916) & Enum.: 346 (1921); Ridley: 251 (1924).

The obsolete or minute glands at the apex of the petiole, the rather small obovate leaves, drying a dark purplish-brown, and the smaller, laxer, thinly puberulous inflorescences, distinguish *N. kingii* from the remaining species. The peat-swamp habitat is also distinctive.

Var. **kingii**: flower pedicels 1 mm. long, puberulous.

SR; B(?)**; K1.**—Malaya, Sumatra.

Tree to 25 m. high, abundant in primary peat-swamp forest or primary Lowland Dipterocarp forest, on low sandy ridges or sandy loam, up to 210 m. alt.

The doubtful record for Brunei is based upon a single collection, *Symington* KEP 35710, in which the leaves are exceptionally coriaceous. Further collections from this locality (Seriah oilfields) are desirable.

Var. **pedicellata** *Airy Shaw* in Kew Bull. 16: 371 (1963): pedicels longer and thicker, 2–3 mm. long; pubescence sparser.

SR.—Endemic.

No field data available.

Neoscortechinia nicobarica (*Hook. f.*) *Pax & Hoffm.* xiv (Euph.–Addit. vi): 53 (1919); Merr.: 456 (1923); Chatterjee in Kew Bull. 4: 564 (1950); Airy Shaw in Kew Bull. 16: 369 (1963) & 20: 413 (1966); Whitmore: 119, 120 (1973).

Scortechinia nicobarica Hook. f. in Hook. Ic. Pl. 18: sub t. 1706 (Nov. 1887) & in Hook. f., Fl. Brit. Ind. 5: 367 (Dec. 1887); Boerl.: 222 (1900); Merr. in Philipp. Journ. Sci. 11, C. Bot.: 76 (1916).
Alchornea arborea Elm., Leafl. Philipp. Bot. 4: 1274 (1911).
Alcinaeanthus arboreus (Elm.) Pax & Hoffm. vii: 415 (1914).
Scortechinia arborea (Elm.) Merr. in Philipp. Journ. Sci. 11, C. Bot.: 75 (1916).
Neoscortechinia arborea (Elm.) Pax & Hoffm. xiv (Euph.–Addit. vi): 52 (1919); Merr.: 456 (1923).
Scortechinia paniculata Ridley, Fl. Malay Penins. 5: 332 (1925).
Neoscortechinia sp., Merr., Pl. Elmer.: 165 (1929).

SR; SB; K1.—Lower Burma, Nicobars, Sumatra, Malaya, Philippines.
Tree to 35 m. high, in primary or secondary forest, sometimes marshy or periodically inundated, on clayey sand or sandstone or sandy loam or loam containing limestone, or brown soil, from low altitudes up to 210 m.

Differs from *N. forbesii* in the more coriaceous, strongly crenate leaves, rounded or even cordate at the base, and in the larger, fulvous- or cinereous-puberulous inflorescence.
Only one collection has been seen from Sarawak: Third Division, near Belaga airfield, *Ashton* S. 18268.
The wood is said to smell of fresh sugar-cane.

Neoscortechinia sumatrensis *S. Moore* in Journ. Bot. Brit. & For. 63, Suppl.: 99 (1925); Airy Shaw in Kew Bull. 16: 368 (1963); Whitmore: 119 (1973).

Var. **sumatrensis:** leaves pubescent beneath, shorter and slightly broader in outline.

SR; B; SB.—Sumatra, Malaya (S.).
Tree to 30 m. high, in Low Dipterocarp forest or in peat-swamp forest, on yellow sandy clayey soil or yellow sandy loam or black sandy soil, up to 300 m. alt.

Var. **angustifolia** *Airy Shaw* in Kew Bull. 16: 368 (1963): leaves glabrous beneath, longer and narrower in outline.

SB; K1.—Endemic.
Tree to 12 m. high, in primary forest on black soil up to 45 m. alt.

The relatively thin leaves and their green colour on drying are usually sufficient to distinguish *N. sumatrensis*. The softly pubescent undersurface of var. *sumatrensis* and the elongate and often narrow outline of var. *angustifolia* provide additional characters for recognition.

Omphalea *L.*

1 Erect shrub or tree; leaves elongate, up to 50 cm. long, and often relatively narrow; petiole 4–13 cm. long; floral bracts elongate, narrowly oblong **O. malayana**

1 Woody climber; leaves less than 20 cm. long (often only 10 cm. and broad in proportion); petiole 0·5–5 cm. long; floral bracts broader **O. bracteata**

2 Pedicel of female flowers not exceeding 1 cm. in length var. **bracteata**

2 Pedicel of female flowers slender, about 4 cm. long var. **pedicellaris**

Omphalea bracteata (*Blanco*) *Merr.*, Sp. Blanco.: 230 (1918) & Enum.: 457 (1923); J. E. Vidal in Not. Syst. 15: 449 (1959); Airy Shaw in Kew Bull. 20: 414 (1966) & 26: 310 (1972); Whitmore: 121 (1973).

Tragia bracteata Blanco, Fl. Filip., ed. 2: 480 (1845).
Omphalea philippinensis Merr. in Philipp. Journ. Sci. 3, C. Bot.: 236 (1908); Pax & Hoffm. v: 17 (1912).
O. sargentii Merr. in Philipp. Journ. Sci. 16: 574 (1920) & Enum.: 458 (1923) & in Philipp. Journ. Sci. 29: 386 (1926); cf. Airy Shaw in Kew Bull. 26: 311 (1972), *in obs.*; **synon. nov.**

Var. **bracteata**: pedicel of female flowers not exceeding 1 cm in length.

SB; K1.—Lower Burma, Indochina, Malaya, Philippines, Celebes.

Woody climber to 21 m., in primary hillside forest with *Dipterocarpus grandiflorus*, on red earth, or ultrabasic soil, or in the coastal alluvium zone, or in dry open forest on level ground, or in secondary forest (*belukar*) on brown soil, from sea-level up to 110 m. alt.

The climbing habit, shorter (less than 20 cm.) and relatively broader leaves, shorter (0·5–5 cm.) petioles and broader floral bracts distinguish this species from *O. malayana*. It is curious that (except for var. *pedicellaris*, below) it has not yet been found in Sarawak. The flowers are said to be fragrant. The stems exude a sticky brown latex.

There appears to be no taxonomic significance in the '*sargentii*' form as distinct from typical *O. bracteata*; they both evidently grow together in many localities in Sabah.

Var. **pedicellaris** *Airy Shaw*, var. nov., floribus femineis longe (circiter 4 cm.) et tenuiter pedicellatis distincta.

SR. Third Division: Kapit District; Belaga subdistr., hills on left bank of Rajang river 10 km. below Belaga, Segaham range, near Belaga airfield, approx. 2° 40′ N., 113° 50′ E., primary forest on sandstone substratum, alt. below 500 m., 26 Aug. 1958, *Jacobs* 5306 (K, type):—Big liana 15 m. or higher; no latex; flowers sordidly cream-coloured all over except the glossy golden-brown hairs outside.

It is possible that this solitary collection from central Sarawak may prove to be specifically distinct when further material becomes available. The slender elongate pedicels of the female flowers, which are already beginning to swell into young fruits, give the specimen a very distinct appearance.

Omphalea malayana *Merr.* in Philipp. Journ. Sci. 11, C. Bot.: 71 (1916) & Enum.: 346 (1921) & 458 (1923); Airy Shaw in Kew Bull. 20: 414 (1966); Whitmore: 121 (1973).

SR.—Malaya (P. Tioman), Philippines (Luzon).

Shrub or small tree to 4·5 m. high, in primary Mixed Lowland Dipterocarp forest, on sandy clay soil, up to 360 m. alt.

The erect habit, large, oblong to broadly oblanceolate leaves (up to 50 cm. long), elongate petioles (4–13 cm.), and elongate, narrowly oblong floral bracts distinguish *O. malayana* sharply from *O. bracteata*.

Phyllanthus *L.*

1 Herbs (sometimes ± woody at the base):
  2 Fruiting pedicels slender, 5–8 mm. long, often sharply deflexed; tepals small, narrow, not conspicuously white-edged; nerves of leaf not sharply raised; capsule smooth (§ *Macraea*) . . . . **P. virgatus**
  2 Fruiting pedicels much shorter, not or less sharply deflexed; tepals broader, conspicuously white-edged:
    3 Nerves of leaf not sharply raised; capsule not scaly (§ *Phyllanthus*) P. AMARUS
    3 Nerves of leaf finely and sharply raised; capsule usually scaly (§ *Urinaria*):
      4 Leaves thinner, more obtuse; stem slenderer, less strongly grooved, not purplish; plant not rheophytic . . . . . **P. urinaria**
      4 Leaves stiffer, more acute; stem stouter, more strongly grooved, purplish; plant rheophytic . . . . . **P. chamaepeuce**
1 Shrubs or trees:
  5 Fruit ± fleshy, drupaceous or baccate:
    6 Fruit a small berry, 3–5 mm. in diameter; leaves thin, membranous to chartaceous, to 5 × 1·5 cm.; flowers axillary (§ *Anisonema*) **P. reticulatus**
    6 Fruit a drupe, 1–3 cm. in diameter:
      7 Leaves linear-oblong, 12–20 × 2–5 mm., closely distichous; flowers mostly axillary; tepals 6; stamens 3, connate; no staminodes in the ♀ flower (§ *Emblica*) . . . . . **P. emblica**
      7 Leaves ovate, 5–9 × 2·5–4·5 cm., borne on slender deciduous branchlets resembling pinnate leaves; flowers mostly borne on short leafless branchlets on the thick older stems; tepals 4; stamens 4, free; ♀ flowers sometimes with staminodes (§ *Cicca*) P. ACIDUS
  5 Fruit a dry dehiscent capsule:
    8 Tepals of ♂ and ♀ flowers 5–6; branchlets sharply quadrangular; leaves stiffly coriaceous (§ *Paraphyllanthus*) . **P. pachyphyllus**
    8 Tepals of ♂ flower (and often of ♀) 4; branchlets not conspicuously quadrangular; leaves thinner:
      9 Leaves smooth, shining, chartaceous; tepals entire; ovary and capsule 5–8-locular; seeds smooth (§ *Scepasma*) **P. buxifolius**
      9 Leaves dull, membranous; tepals laciniate; ovary and capsule 3–4-locular; seeds often covered with transverse adpressed fibrils (§ *Eriococcus*):
        10 Capsule somewhat inflated, 1 cm. or more in diam.; seeds glabrous . . . . **P. kinabaluicus**

10 Capsule not inflated, less than 1 cm. in diam.; seeds with transverse adpressed fibrils:
  11 Leaves 3–12 cm. long; stems shortly rufo-puberulous **P. gracilipes**
  11 Leaves 0·5–3·5 cm. long:
    12 Ovary glabrous; leaves relatively broader; plant not rheophytic . . . . **P. pulcher**
    12 Ovary fibrillose-lanuginose; leaves relatively narrower; plant often semi-rheophytic . . . . **P. hullettii**

PHYLLANTHUS ACIDUS (*L.*) *Skeels* in U.S. Dept. Agric. Bur. Pl. Ind. Bull. 148: 17 (1909); Webster in Journ. Arn. Arb. 38: 66 (1957), *q.v.* for full synonymy; Backer & Bakh. f.: 467 (1963); Airy Shaw in Kew Bull. 26: 315 (1972); Whitmore: 122 (1973).

Averrhoa acida L., Sp. Pl.: 428 (1753).
Cicca disticha L., Mant.: 124 (1767); Ridley: 216 (1924).
C. acidissima Blanco, Fl. Filip.: 700 (1837).
Phyllanthus acidissimus (Blanco) Muell. Arg. in Linnaea 32: 50 (1863) & in DC.: 417 (1866).
P. distichus (L.) Muell. Arg.: 413 (1866); Hook. f.: 304 (1887); Beille: 594 (1927); Pax & Hoffm. in Engl. & Harms, Pflanzenf. ed. 2, 19c: 63 (1931).
Cicca acida (L.) Merr., Interpr. Rumph. Herb. Amboin.: 314 (1917) & Enum.: 328 (1921) & 396 (1923); Corner: 282 (1940).

B; SB.—Introduced; probably native of the coastal region of NE. Brazil (cf. Webster, *l.c. supra*: 70–72).

Tree to 6 m. high, in old secondary forest, on seashores, near villages, etc., on sandy loam, etc.; always as an escape from or remnant of cultivation.

Fruit edible, with apple-like flavour.

PHYLLANTHUS AMARUS *Schumacher & Thonn.* in Kongl. Danske Vidensk. Selsk. Skr. 4: 195 (1829); Webster in Journ. Arn. Arb. 37: 6 (1956) & 38: 313 (1957) [*q.v.* for full synonymy]; Webster & Airy Shaw in Kew Bull. 26: 92 (1971); Airy Shaw in Kew Bull. 26: 317 (1972); Whitmore: 122 (1973).

P. nanus Hook. f.: 298 (1887).
P. niruri sec. Boerl.: 273 (1900); Merr.: 327 (1921) & 394 (1923); Ridley: 199 (1924); Backer & Bakh. f.: 468 (1963); *saltem pro parte, non* L.

K1, 1a.—Native of America; now pantropical.
Herb 30–40 cm. high, on 'bankwall' at 15 m. alt.

Only two specimens seen from Borneo.

Phyllanthus buxifolius (*Bl.*) *Muell. Arg.* in Linnaea 32: 50 (1863) & in DC.: 426 (1866); C. B. Rob. in Philipp. Journ. Sci. 4, Bot.: 85 (1909); J. J. Sm.: 90 (1910); Merr.: 391 (1923); Pax & Hoffm. in Engl. & Harms, Pflanzenf. ed. 2, 19c: 65 (1931); Backer & Bakh. f.: 469 (1963).

Scepasma buxifolia [sic!] Bl., Bijdr.: 583 (1825); Baill., Ét. Gén. Euphorb.: 649, t. 25, figs. 10–15 (1858).

SB; K1a.—Philippines, Java.
Tree of 6 m., at 6 m. alt.; no details of habitat.

Leaves glabrous, glossy, distichous, chartaceous, very asymmetrical, 1–4 cm. long; flowers axillary, sessile.

Phyllanthus chamaepeuce *Ridley* in Trans. Linn. Soc. II, 3: 270, 345 (1893) & Fl. Malay Penins. 3: 200 (1924); Airy Shaw in Kew Bull. 23: 33 (1969) & 26: 317 (1972); Whitmore: 122 (1973).

P. quangtriensis Beille: 584 (1927).
P. urinaria sec. Hutch. in Journ. Linn. Soc., Bot. 42: 133 (1914); Webster in Journ. Arn. Arb. 38: 196 (1957), *vix* L.

SR; B; SB; K1, 1a.—Siam, Indochina, Malaya.
Woody herb to 20 cm., on rocks at river margin, on sandbanks or gravel banks, in river bed below regular high-water level; abundant on a slaty bank, preferably in light places; below 500 m. alt.

A rheophyte; closely related to *P. urinaria*, but differing in the slightly stiffer and more acute leaves, in the usually stouter, strongly grooved, purple stem, and especially in the totally different habitat.

Phyllanthus emblica *L.*, Sp. Pl.: 982 (1753); Muell. Arg.: 352 (1866); Hook. f.: 289 (1887); J. J. Sm.: 70 (1910); Hubert Winkler in Engl., Bot. Jahrb. 50, Suppl.: 208 (1914); Merr.: 327 (1921); Beille: 580 (1927); Pax & Hoffm. in Engl. & Harms, Pflanzenf. ed. 2, 19c: 64 (1931); Meijer, 7: 27 (1967); Airy Shaw in Kew Bull. 26: 319 (1972); Whitmore: 123, fig. 11 (1973).

Emblica officinalis Gaertn., Fruct. 2: 122 (1791); Corner: 282 (1940).
Dichelactina nodicaulis Hance in Walp., Ann. 3: 376 (1852).
Cicca emblica (L.) Kurz, For. Fl. Brit. Burma 2: 352 (1877).

SR; SB; K1a.—India, SE. Asia, S. China, Sumatra, Java & Lesser Sunda Is.
Tree to 48 m. high, in primary hill forest up to 1200 m. alt.

Leaves closely distichous, linear-oblong, up to 20 × 5 mm.; tepals six, stamens three, connate; fruit a drupe or fleshy indehiscent capsule.
Fruit edible, exceedingly acid; said to impart a sweet taste to water drunk afterwards; made into a sweetmeat with sugar or eaten raw as a condiment. Leaves used in native medicine, as a cure for headache, by being placed on the head and wrapped round with a cloth. Bark and young shoots used to dye cotton black by boiling with alum.

Phyllanthus gracilipes (*Miq.*) *Muell. Arg.* in Linnaea 32: 47 (1863) & in DC.: 423 (1866); J. J. Sm.: 98 (1910); Merr.: 327 (1921); Ridley: 204 (1924); Backer & Bakh. f.: 466 (1963); Airy Shaw in Kew Bull. 23: 35 (1969) & 26: 319 (1972); Whitmore: 122 (1973).

Eriococcus gracilis Hassk. in Tijdschr. Nat. Geschied. & Physiol. 10: 143 (1843) & Cat. Pl. Hort. Bot. Bogor. Alter: 243 (1844); *non Phyllanthus gracilis* Roxb. (1832), *nec* (Hassk. sub *Ceramantho*) Baill. ex Muell. Arg. (1863).
Reidia gracilis (Hassk. 1843) Miq., Fl. Ind. Bat. 1(2): 373 (1859).

R. gracilipes Miq., *l.c.*: 374 (1859).
Phyllanthus concinnus Ridley in Journ. Roy. As. Soc. Str. Br. 59: 171 (1911).
P. discofractus Croiz. in Journ. Arn. Arb. 23: 31 (1942).

K3.—Siam, Indochina, Sumatra, Java.
No field data available for Borneo; only one collection seen.

Differs from *P. pulcher* and *P. hullettii* chiefly in the larger leaves, 3–12 cm. long.

Phyllanthus hullettii *Ridley* in Bull. Misc. Inf. Kew 1923: 363 (1923) & Fl. Malay Penins. 3: 204 (1924); Airy Shaw in Kew Bull. 26: 320 (1972).

P. gracilipes sec. Whitmore: 122 (1973), *p.p.*, *vix* (Miq.) Muell. Arg.

SR; SB.—Siam (NW.), Malaya (Malacca).
Shrub to 1 m. high, semi-gregarious on silty sandy soil near sandstone rocks at or below upper flood level, or in forest along river bank, in the neighbourhood of rapids, at about 50 m. alt.

Close to *P. gracilipes*, but leaves only 1·5–3·5 cm. long; evidently semi-rheophytic.

Phyllanthus kinabaluicus *Airy Shaw* in Kew Bull. 29: 294 (1974).

SB.—Endemic.
Shrub of 1 m., by river at 1350–1500 m. alt.

This has an inflated capsule like *P. elegans* Wall. ex Muell. Arg., of SE. Asia, but has considerably smaller leaves, lacking the reddish margin of that species; the stems are shortly rufo-puberulous as in *P. gracilipes*, *P. pulcher* and *P. hullettii*. The seeds are distinctive in that they entirely lack the adpressed transverse fibrils that characterize many species of § *Eriococcus*.

Phyllanthus pachyphyllus *Muell. Arg.*: 353 (1866); Airy Shaw in Kew Bull. 23: 29 (1969) & 26: 321 (1972); Whitmore: 123 (1973).

P. coriaceus Wall. ex Hook. f.: 292 (1887); Ridley: 203 (1924).
? *P. klossii* Ridley in Journ. Fed. Malay States Mus. 10: 114 (1920).
P. campanulatus Ridley in Bull. Misc. Inf. Kew 1923: 362 (1923) & Fl. Malay Penins. 3: 203 (1924); Airy Shaw in Kew Bull. 23: 30 (1969).
P. frondosus var. *rigidus* Ridley: 203 (1924).
? *P. annamensis* Beille: 585 (1927); Merr. & Chun in Sunyatsenia 5: 93 (1940).
? *P. sciadiostylus* Airy Shaw in Kew Bull. 23: 30 (1969).

K1.—SE. Asia, Hainan (?), Malaya.
Shrub at 90 m. alt.; no further field data.

Branchlets sharply quadrangular; leaves rigidly coriaceous, glabrous, rather shining, drying chestnut-brown; flowers mostly in axillary fascicles; capsules very shortly pedicelled.
Only one collection seen from Borneo.

Phyllanthus pulcher *Wall. ex Muell. Arg.* in Linnaea 32: 49 (1863) & in DC.: 421 (1866); Hook. f.: 301 (1887) (excl. synon. *P. pallidifolius*, *P.*

zollingeri, *P. pallidus* & *Reidia glaucescens*); J. J. Sm.: 94 (1910); Ridley: 201 (1924); Beille: 598 (1927); Hend. in Journ. Malayan Br. Roy. As. Soc. 17: 71 (1939); Backer & Bakh. f.: 466 (1963); Airy Shaw in Kew Bull. 26: 322 (1972); Whitmore: 121 (1973).

SR.—SE. Asia, W. Malesia (exc. Philippines).

Small shrub in *kerangas* sand at 15 m. alt. (On limestone in Lower Siam and N. Malaya, *teste* Henderson, *l.c.*).

Differs from *P. hullettii* chiefly in its glabrous ovary and non-rheophytic habitat. Only one specimen seen.

Phyllanthus reticulatus *Poir.* in Lam., Encycl. Méth., Bot. 5: 298 (1804); Muell. Arg.: 344 (1866); Hook. f.: 288 (1887); J. J. Sm.: 63 (1910): Merr. in Philipp. Journ. Sci. 11, C. Bot.: 74 (1916) & Enum.: 327 (1921) & 394 (1923); Ridley: 202 (1924); Beille: 575 (1927); Pax & Hoffm. in Engl. & Harms, Pflanzenf. ed. 2, 19c: 62 (1931); Webster in Journ. Arn. Arb. 38: 57 (1957); Backer & Bakh. f.: 467 (1963); Airy Shaw in Kew Bull. 26: 322 (1972); Whitmore: 123 (1973); *sensu lato.*

Cicca microcarpa Benth., Fl. Hongk.: 312 (1861).

Phyllanthus microcarpus (Benth.) Muell. Arg. in Linnaea 32: 51 (1863) & in DC.: 343 (1866); Merr.: 393 (1923).

P. dalbergioïdes Wall. ex J. J. Sm.: 67 (1910); Merr.: 327 (1921); Ridley: 201 (1924).

Glochidion microphyllum Ridley in Journ. Str. Br. Roy. As. Soc. 59: 173 (1911).

Phyllanthus erythrocarpus Ridley in Bull. Misc. Inf. Kew 1923: 362 (1923) & Fl. Malay Penins. 3: 202 (1924); cf. Hend. in Journ. Malayan Br. Roy. As. Soc. 17: 71 (1939).

Kirganelia lineata sec. Alston in Trimen, Handb. Fl. Ceylon 6 (Suppl.): 259 (1931), *non Zizyphus lineatus* (L.) Willd. *nec Rhamnus lineatus* L. = *Berchemia lineata* (L.) DC.

SR; SB; K1, 1a.—Trop. Africa, India, Ceylon, SE. Asia, S. China, and throughout Malesia.

Scandent shrub or treelet to 3 m. high, once noted from primary riparian forest on silty clay river bank beyond tidal influence, up to 420 m. alt.

Exceedingly variable; distinguished from all other Malesian species by the small, rounded or elliptic, thinly membranous leaves and small globose black or red berries.

Phyllanthus urinaria *L.*, Sp. Pl.: 982 (1753); Muell. Arg.: 364 (1866); Hook. f.: 293 (1887); Merr.: 327 (1921) & 396 (1923); Ridley: 200 (1924); Beille: 586 (1927); Croiz. in Journ. Jap. Bot. 16: 657 (1940); Webster in Journ. Arn. Arb. 38: 194 (1957); Backer & Bakh. f.: 469 (1963); Webster & Airy Shaw in Kew Bull. 26: 91 (1971); Airy Shaw in Kew Bull. 26: 325 (1972); Whitmore: 122 (1973). [*Non* Hutch. in Journ. Linn. Soc., Bot. 42: 133 (1914); *vide P. chamaepeuce* Ridley.]

SR; SB; K1.—Pantropical.

Herb to 60 cm. high, in open and waste places, in forest undergrowth, etc., up to 150 m. alt.; once noted on limestone-derived soil rich in bat guano.

A decoction of the leaves is used in childbirth (Sabah).

Phyllanthus virgatus *Forst. f.*, Fl. Ins. Austr. Prodr.: 65 (1786); Backer & Bakh. f.: 469 (1963); Whitmore: 121 (1973).

P. simplex Retz., Obs. Bot. 5: 29 (1789); Muell. Arg.: 391 (1866) (var. *genuinus*); Hook. f.: 295 (1887); Merr.: 395 (1923); Ridley: 200 (1924); Beille: 578 (1927) (excl. var. *tonkinensis*); Pax & Hoffm. in Engl. & Harms, Pflanzenf. ed. 2, 19c: 64 (1931).
P. simplex var. *virgatus* (Forst. f.) Muell. Arg. in Linnaea 32: 32 (1863) & in DC.: 392 (1866).

K1a.—India, SE. Asia, S. China, and throughout Malesia to Polynesia.
No field data available for Borneo. Only known from a single collection (*Motley* 351) from Bandjermasin.

Pimelodendron *Hassk.*

1 Leaves thin, shining, with prominent nerves; ♂ inflorescence elongate (up to 25 cm. long) and branched; fruit up to 10 cm. in diameter **P. macrocarpum**
1 Leaves thick, dull, with immersed nerves; ♂ inflorescence short (less than 5 cm. long) and simple; fruit up to 3 cm. in diameter:
2 Style-branches 6–8, almost free, thick, obovoid, rounded at the apex, not reflexed . . . . **P. griffithianum**
2 Style-branches 10–12, connate, narrow, ± linear, acute at the apex, reflexed . . . . **P. zoanthogyne**

Pimelodendron griffithianum (*Muell. Arg.*) *Benth.* in Benth. & Hook. f., Gen. Pl. 3: 332 (1880); Hook. f.: 468 (1888); Pax & Hoffm. v: 56 (1912); Ridley: 313 (1924), *pro majore parte*; Whitmore: 125, fig. 12, left and centre (1973).

Stomatocalyx griffithianus Muell. Arg.: 1142 (1866).
Pimeleodendron borneënse Warb. in Engl., Bot. Jahrb. 18: 199 (1893); Pax & Hoffm. v: 56 (1912); Merr.: 347 (1921); **synon. nov.**
? *P. amboinicum* sec. Boerl.: 295 (1900), *non* Hassk.; cf. Merr.: 347 (1921), *in obs.*
P. acuminatum Merr. in Philipp. Journ. Sci. 11, C. Bot.: 74 (1916) & Enum.: 347 (1921), **synon. nov.**
P. papaveroïdes J. J. Sm. in Bull. Jard. Bot. Buitenz. III, 6: 104 (1924), **synon. nov.**

SR; B; SB; K1.—Malaya.
Tree to 30 m. high, in primary Lowland Mixed Dipterocarp forest, pole forest, tropical Heath forest, *alan* terrace or *ramin* peat swamp, on Tertiary sandstone, yellow sandy clay, sandy loam, clay-rich soil, clay loam, deep yellow clay, dacite ridge, basalt ridge, black or brownish soil, or alluvium, up to 1800 m. alt.

The bark is said to have a slight fruity smell; the flowers are slightly scented or odourless.

There is no doubt that *P. borneënse* and *P. acuminatum* are merely forms of this common and not excessively variable species, and, so far as I can judge from specimens distributed from Bogor as *P. papaveroïdes*, the same applies to this taxon also.

For a note on the spelling of the generic name, see *Kew Bull.* 25: 552 (1971).

Pimelodendron macrocarpum *J. J. Sm.* in Bull. Jard. Bot. Buitenz. III, 6: 103 (1924); Airy Shaw in Kew Bull. 25: 551 (1971); Whitmore: 125, fig. 12, right (1973).

P. griffithianum sec. Hook. f.: 468 (1888), *pro minore parte, non* (Muell. Arg.) Benth.

K1.—Sumatra, Malaya.

Tree to 27 m., in forest in low undulating country up to 20 m. alt.

The thin, shining leaves, with clearly raised nerves, the elongate, branched male inflorescences, and the very large fruits, distinguish *P. macrocarpum* sharply from the other species.

Pimelodendron zoanthogyne *J. J. Sm.* in Bull. Jard. Bot. Buitenz. III, 6: 105 (1924).

K3.—Endemic.

Small tree; no further field data.

The connate, reflexed, narrow, acute style-branches distinguish this taxon from *P. griffithianum*, but in the male flower and in vegetative characters the two seem indistinguishable. Further material is very desirable. No other *Pimelodendron* has apparently been collected in W. Indonesian Borneo (**K3**).

Pterococcus *Hassk.*

Pterococcus corniculatus (*Sm.*) *Pax & Hoffm.* ix–xi: 22 (1919); Backer & Bakh. f.: 490 (1963); Airy Shaw in Kew Bull. 26: 327 (1972); Whitmore: 84, 126 (1973).

Plukenetia corniculata Sm. in Nova Acta Upsal. 6: 4 (1799); Muell. Arg.: 722 (1886); Hook. f.: 464 (1888); J. J. Sm.: 526 (1910); Merr. in Philipp. Journ. Sci. 16: 564 (1920) & Enum.: 344 (1921) & 447 (1923).
Pterococcus glaberrimus Hassk. in Flora 25(2), Beibl.: 41 (1842); Ridley: 309 (1924).
Hedraiostylus glaberrimus (Hassk.) Hassk., Cat. Pl. Hort. Bog. Alter: 234 (1844).
Sajorium corniculatum (Sm.) Dietr., Synops. Pl. 5: 331 (1852); Baill., Ét. Gén. Euphorb.: 484 (1858).

SR.—E. Himalaya, Assam, Burma, Siam (SE.), and scattered through W. Malesia to the Moluccas.

Scandent plant in thickets; no further information for Borneo.

Closely related to *Plukenetia* L., differing principally in the very short thick stylar column and in the conspicuously 4-winged or 4-horned capsule. The leaves are membranous, dentate, and abruptly caudate, with a wide basal sinus on the margin of which there are sometimes one or two macular glands.

Ptychopyxis *Miq.*

1 Petiole less than 1 cm. long; tertiary nervation distinctly impressed above; fruits covered with longitudinal ridges and transverse folds **Pt. costata** var. **oblanceolata**

1 Petiole exceeding 1 cm. in length; tertiary nervation not impressed above:

 2 Leaves ample, 5–19 cm. wide, broadly obovate, less frequently oblanceolate:

  3 Leaves very large, 40–55 cm. long; petiole massive, 4–6 mm. thick, with loose papery epidermis, glabrous . . . . **Pt. grandis**

  3 Leaves smaller, up to 36 cm. long; petiole relatively slender, up to 3 mm. thick, epidermis not loose or papery, but finely tomentellous, at least when young:

   4 Leaves drying reddish-brown on both surfaces; ♂ inflorescences shorter and stouter, up to 8 cm. long; flowers apparently larger, strongly fulvous-tomentose; fruits 3–7 cm. long . **Pt. kingii**

   4 Leaves drying reddish-brown above, but a paler and often greenish-brown below; ♂ inflorescences longer and slenderer, up to 12 cm. long; flowers apparently smaller, more shortly tomentellous; fruit up to 3·5 cm. in diam.. . **Pt. arborea** var. **arborea**

 2 Leaves smaller and narrower, 2–7 cm. wide:

  5 Leaves broadest at or below middle, rounded to slightly cordate at base, 5–10 cm. long, 2–4·5 cm. wide; lateral nerves about 6 pairs; ovary bilocular, styles 2 . . . . **Pt. glochidiifolia**

  5 Leaves mostly broadest above middle, narrowed towards base but sometimes rounded and abruptly subauriculate, mostly exceeding 10 cm. in length; lateral nerves about 8–10 pairs; ovary trilocular, styles 3:

   6 Leaves drying brown, sometimes 25 cm. long or more, chartaceo-coriaceous; tertiary nervation mostly rather smooth and relatively inconspicuous below; ♂ inflorescence fairly robust **Pt. javanica**

   6′ Leaves drying brownish, not exceeding 18 cm. in length, ± coriaceous; tertiary nervation sharply elevated and conspicuous below; petioles not more than 1·5 cm. long **Pt. arborea** var. **cacuminum**

   6″ Leaves drying greenish, not exceeding 18 cm. in length, thinly chartaceous; tertiary nervation sharply elevated and conspicuous below; ♂ inflorescence very slender . . **Pt. bacciformis**

Ptychopyxis arborea (*Merr.*) *Airy Shaw* in Kew Bull. 14: 369 (1960) & 16: 347 (1963); Meijer, 7: 53 (1967).

Mallotus arboreus Merr., Pl. Elmer.: 159 (1929).
Baccaurea sp. (No. 2, quoad *Elmer* 21245 & 21672), Merr., *l.c.*: 153.

Var. **arborea**:

SR; SB; K1.—Endemic.
Tree to 25 m. high, in primary Mixed Dipterocarp forest, sometimes marshy,

on clay, or loam with coral limestone, or sand and limestone, or pure limestone, or basalt ridges, up to 600 m. alt.

Differs from *Pt. kingii*, from which it is perhaps scarcely specifically distinct, in its somewhat less robust habit, in its longer and slenderer male inflorescences, up to 12 cm. long, with smaller flowers, and in its smaller fruits, up to 3·5 cm. long.

Fruit said to be edible, with a sweet jelly-like aril.

Var. **cacuminum** *Airy Shaw* in Kew Bull. 20: 28 (1966).

K1.—Endemic.

Tree of 15 m., in forest on narrow stony mountain ridge at 1600 m. alt.

Petioles only 1–2 cm. long; leaves narrowly elliptic, with strong venation; male inflorescence up to 4 cm. long. Known from a single gathering which could possibly represent an untypical high-mountain form of *Pt. javanica*.

Ptychopyxis bacciformis *Croiz.* in Journ. Arn. Arb. 23: 49 (1942); Airy Shaw in Kew Bull. 14: 371 (1960) & 16: 348 (1963); Meijer, 7: 53 (1967).

Coelodepas sp. ? Merr., Pl. Elmer.: 156 (1929).

Pt. poilanei Croiz. in Journ. Arn. Arb. 23: 50 (1942); Airy Shaw in Kew Bull. 14: 372 (1960).

SR; SB; K1.—Indochina.

Tree to 35 m. high, in primary Mixed Dipterocarp forest on blackish or stony clayey soil up to 300 m. alt.

Distinguished from *Pt. javanica* by the slightly thinner leaves, not more than 18 cm. long, usually drying green, with more sharply elevated nervation beneath, and by the very slender male inflorescence.

The white aril on the seeds is said to be sweet, resembling [the fruit of ?] *Passiflora foetida* L.

Ptychopyxis costata *Miq.*, Fl. Ind. Bat., Suppl.: 402 (1861); Hook. f.: 455 (1887) & in Hook. Ic. Pl. 18: t. 1703 (1887); Boerl.: 254, 293 (1900); Ridley: 295 (1924); Croiz. in Journ. Arn. Arb. 23: 48 (1942); Airy Shaw in Kew Bull. 14: 366 (1960).

Var. **oblanceolata** *Airy Shaw* in Kew Bull. *l.c.* (1960).

SR (C.).—Malaya.

Tree of 15 m., on slope in Mixed Dipterocarp forest at unknown alt.

The very short petioles, impressed tertiary nervation and large ovoid fruits with longitudinal ridges and transverse or irregular folds distinguish *Pt. costata* sharply from the remainder of the genus. Var. *oblanceolata* differs from var. *costata* in having oblanceolate leaves only 3·5–8 cm. broad. The above record is based upon a specimen recently collected in the Buan Forest Reserve, Bintulu, *Paul Chai* S. 32110, agreeing perfectly with the Malayan plant.

Ptychopyxis glochidiifolia *Airy Shaw* in Kew Bull. 14: 373 (1960); Meijer, 7: 53 (1967).

SR; B; K1.—Endemic.

Tree to 26 m. high, in Mixed Dipterocarp forest on sandy loam on low hills and ridges up to 50 m. alt.

Very distinct in its small, few-nerved, narrowly ovate leaves, shortly racemose female inflorescence, and bilocular ovary and fruit.

Ptychopyxis grandis *Airy Shaw* in Kew Bull. 14: 367 (1960).

SR.—Endemic.

Tree to 15 m. high, in primary Mixed Lowland Dipterocarp forest or very old secondary forest on sandstone or sandy clay loam up to 1000 m. alt.

Leaves very large, 40–55 cm. long, 8–22 cm. broad, with a massive petiole 5–15 cm. long and 4–6 mm. thick, covered with a loose glabrous papery epidermis; fruit large, rostrate-fusiform, 6–9 cm. long.

Ptychopyxis javanica (*J. J. Sm.*) *Croiz.* in Journ. Arn. Arb. 23: 49 (1942); Airy Shaw in Kew Bull. 14: 371 (1960) & 16: 348 (1963); Backer & Bakh. f.: 479 (1963); Airy Shaw in Kew Bull. 26: 327 (1972); Whitmore: 127 (1973).

Podadenia javanica J. J. Sm.: 387 (1910); Pax & Hoffm. vii: 21 (1914).

Ptychopyxis angustifolia Gage in Rec. Bot. Surv. Ind. 9: 248 (1922); Ridley: 296 (1924); Croiz., *l.c.* 48 (1942).

SB; ? KI.—Malaya, ? Sumatra, Java.

No ecological data for Borneo (but cf. *Pt. arborea* var. *cacuminum*, p. 189 *supra*).

Leaves relatively narrow, 10–25 cm. long, 2–7 cm. wide, thinly coriaceous, drying brownish.

Ptychopyxis kingii *Ridley*: 296 (1924); Airy Shaw in Kew Bull. 14: 368 (1960); Meijer, 7: 53 (1967); Airy Shaw in Kew Bull. 27: 86 (1972); Whitmore: 127 (1973).

Mallotus arboreus var. *platyphyllus* Merr. in Papers Mich. Acad. Sci. 19: 161 (1933); cf. Kew Bull. 27: 86 (1972).

SR; SB.—Malaya, Sumatra.

Tree to 24 m. high, in primary Lowland Dipterocarp forest on sandstone or ultrabasic soil or well-drained clay alluvium up to 540 m. alt.

Differs from *Pt. arborea* in the usually more robust habit, in the shorter and stouter male inflorescences, up to 8 cm. long, with larger and more strongly tomentose flowers, and in the larger fruits, up to 7 cm. long.

Fruit said to be edible.

Richeriella *Pax & Hoffm.*

Richeriella malayana *Hend.* in Gard. Bull. Str. Settlem. 7: 122, t. 32 (1933) & in Journ. Malayan Br. Roy. As. Soc. 17: 71 (1939); Airy Shaw in Kew Bull. 25: 489, 492 (1971) & in Hook. Ic. Pl. 38: t. 3703 (1974); Whitmore: 128 (1973).

Var. **malayana**: capsule 0·5–1 cm. in diameter.

SR; KI.—Malaya.

Small tree to 10 m. high, in forest on limestone rocks or large boulders with intervening igneous-derived soil, or on loam and limestone, or on reddish-yellow slightly sandy clay, up to 200 m. alt.

Var. **macrocarpa** *Airy Shaw* in Kew Bull. 25: 490 (1971): capsule 1·5–2 cm. in diameter.

SR.—Endemic.
Small tree to 12 m. high, on limestone at 60 m. alt.

Richeriella (which is related to *Securinega* and not to *Richeria*) is recognizable by its rather large, smooth, laurel-like leaves and delicate, branched or subsimple, axillary or ramiflorous inflorescences, bearing very minute, sessile or shortly pedicelled flowers.

Ricinus *L.*

Ricinus communis *L.*, Sp. Pl.: 1007 (1753); Muell. Arg.: 1017 (1866); Hook. f.: 457 (1887); J. J. Sm.: 537 (1910); Pax & Hoffm. ix–xi: 119 (1919); Merr.: 344 (1921) & 447 (1923); Ridley: 309 (1924); Gagnep.: 327 (1925); Merr. in Philipp. Journ. Sci. 29: 384 (1926); Corner: 274 (1940); Backer & Bakh. f.: 492 (1963); Airy Shaw in Kew Bull. 26: 328 (1971).

SB; K3.—Widely cultivated in all tropical countries; perhaps native in north-east tropical Africa (Somaliland, N. Kenya, etc.), preferring stream-beds and soils with high nitrogen or saline content.
Shrub to 2·5 m. high, in secondary forest on sandy soil or rocky ground up to 180 m. alt.

The castor-oil plant. Leaves and stems said to be used by the Kinabatangan Dusun (Sabah) as a remedy for stings and bites of insects or animals; also for stomach-ache.

Sapium *P. Br.*

1 No glands at apex of petiole; inflorescence unisexual, somewhat branched; ripe fruit baccate, bilocular . . . . **S. baccatum**
1 Usually 1–2 small but conspicuous glands at apex of petiole; inflorescence usually bisexual, almost unbranched; fruit a trilocular dehiscent capsule . . . . **S. discolor**

Sapium baccatum *Roxb.*, Fl. Ind. 3: 694 (1832); Wight, Ic. Pl. Ind. Or. 6: 6 (1853); Hook. f.: 470 (1888); Pax & Hoffm. v: 240 (1912); Ridley: 315 (1924); Gagnep.: 400 (1926); Corner: 276 (1940); Airy Shaw in Kew Bull. 26: 329 (1971); Medway in Biol. Journ. Linn. Soc. 4: 131 (1972); Whitmore: 128, 129 (1973).

Excoecaria affinis Griff., Notulae 4: 486 (1851); Muell. Arg.: 1223 (1866).
Sapium populifolium Wall. ex Wight, Ic. Pl. Ind. Or. 6: t. 1940 (1853).
Stillingia baccata (Roxb.) Baill., Ét. Gén. Euphorb.: 513 (1858).
St. paniculata Miq., Fl. Ind. Bat., Suppl.: 461 (1861).
Excoecaria baccata (Roxb.) Muell. Arg.: 1211 (1866).
Carumbium baccatum (Roxb.) Kurz, For. Fl. Brit. Burma 2: 412 (1877).

K3.—E. Himalaya, Assam, S. China, SE. Asia, Andamans, Malaya, Sumatra.

Tree at 410 m. alt.; no further field data for Borneo.

The absence of glands at the apex of the petiole, the branched unisexual inflorescence, and the bilocular ovary and baccate fruit, distinguish this species from *S. discolor*. Very scarce in Borneo, only one specimen seen: *Neth. Ind. For. Serv.* bb. 26437.

Sapium discolor (*Champ. ex Benth.*) *Muell. Arg.* in Linnaea 32: 121 (1863); Hook. f.: 469 (1888); Pax & Hoffm. v: 239 (1912); Ridley: 316 (1924); Gagnep.: 399 (1926); Corner: 276 (1940); Airy Shaw in Kew Bull. 26: 329 (1972); Whitmore: 129 (1973).

Stillingia discolor Champ. ex Benth. in Hook. Journ. Bot. & Kew Garden Misc. 6: 1 (1854).

Excoecaria discolor (Champ. ex Benth.) Muell. Arg.: 1210 (1866).

SR; SB; K3.—Siam, Indochina, S. China, Malaya, Sumatra, Banka.

Tree to 22 m. high, in primary or secondary forest on dark rocky soil from very low altitudes up to 600 m.

Differs from *S. baccatum* in the usual presence of 1–2 small but conspicuous glands at the apex of the petiole, in the usually simple bisexual inflorescence, and in the trilocular ovary and dehiscent capsular fruit.

Sapium indicum Willd.: see **Excoecaria indica** (*Willd.*) *Muell. Arg.*

[SAPIUM SEBIFERUM (*L.*) *Roxb.* Recorded for Sabah by Hutchinson (apud Gibbs) in *Journ. Linn. Soc., Bot.* 42: 136 (1941), on the basis of *Gibbs* 2659. The specimen, however, is *Homalanthus populneus* (Geisel.) Pax; *vide* p. 136, *supra*. A similar confusion of these two rather similar plants is referred to by Merrill, *Enum. Philipp. Fl. Pl.* 2: 461 (1923). Very strangely, however, the Gibbs specimen and the reference to Hutchinson's record were cited by Merrill, *Bibliogr. Enum. Born. Pl.*: 348 (1921), under *Sapium indicum* Willd.! *S. sebiferum* is a native of China and Formosa, but is often cultivated; I have seen no cultivated material from Borneo.]

Sauropus *Bl.*

1 ♀ flowers and fruit borne on exceptionally long pedicels (3–11 cm.); stipules white-margined; capsules up to 2 cm. in diameter; ♂ flowers very small, star-shaped; stems sometimes angled . **S. macranthus**

1 ♀ flowering and fruiting pedicels shorter; stipules not conspicuously pale-margined:

2 Inflorescences racemose, mostly borne on lower part of stems near ground; tertiary nervation ± transverse, close and parallel, recalling that of *Rinorea* . . . . **S. micrasterias**

2 Inflorescences fasciculate, axillary; tertiary nervation inconspicuous:

3 Stems ± compressed and angled or keeled:

4 Leaves exceeding 3 cm. in length; ♂ flowers with hooded or concave tepals, keeled within; anthers large, erect; capsule thick-walled **S. pierrei**

4 Leaves small, not exceeding 3 cm. in length; ♂ flowers minute; capsule small, thin-walled . . **Synostemon bacciformis**

3 Stems not compressed or keeled; ♂ flowers discoid, flat; anthers small, horizontal; capsule rather thin-walled:

5 Leaves, ♀ flowers and fruiting calyx larger, the latter up to 3 cm. in diameter . . . . **S. rhamnoïdes**

5 Leaves, ♀ flowers and fruiting calyx smaller, the latter up to 1·5 cm. in diameter . . . . **S. androgynus**

Sauropus androgynus (*L.*) *Merr.* in Philipp. Bur. For. Bull. 1: 30 (1903) & in Journ. Str. Br. Roy. As. Soc. 76: 92 (1917); Pax & Hoffm. xv: 217 (1922); Merr.: 329 (1921) & 405 (1923); Beille: 645 (1927); Backer & Bakh. f.: 471 (1963); Airy Shaw in Kew Bull. 26: 333 (1972); Whitmore: 131 (1973).

Clutia androgyna L., Mant. 1: 128 (1767).
Sauropus albicans Bl., Bijdr.: 596 (1825); Muell. Arg.: 240 (1866); Hook. f.: 332 (1887); Ridley in Journ. Str. Br. Roy. As. Soc. 59: 176 (1911) & Fl. Malay Penins. 3: 220 (1924).
S. sumatranus Miq., Fl. Ind. Bat., Suppl.: 446 (1861); Ridley: 220 (1924).
S. parviflorus Pax & Hoffm. xv: 218 (1922).

SR; SB; K3.—W. Peninsular India and Ceylon to S. China, SE. Asia, and throughout W. Malesia to Celebes and Moluccas; ? New Guinea.
Shrub or small tree to 2 m. high, in thickets, forest clearings, disturbed ground, etc., sometimes on limestone, up to 540 m. alt.

Cultivated as a vegetable.

Sauropus macranthus *Hassk.*, Retzia 1: 166 (1855) & Hort. Bog. s. Retziae ed. nova: 52 (1858); Muell. Arg.: 240 (1866); Backer & Bakh. f.: 471 (1963); Airy Shaw in Dansk Bot. Arkiv 25(2): 34, fig. 13 (1967) & in Kew Bull. 26: 336 (1972); Whitmore: 131 (1973).

S. spectabilis Miq., Fl. Ind. Bat., Suppl.: 446 (1861); Muell. Arg.: 240 (1866); Pax & Hoffm. xv: 219 (1922); Ridley: 220 (1924); Beille: 647 (1927); Hend. in Journ. Malayan Br. Roy. As. Soc. 17: 72 (1939); Adelbert & Meeuse apud Backer in Blumea 5: 508 (1945).
S. macrophyllus Hook. f.: 333 (1887), *pro parte*; Pax & Hoffm. xv: 226 (1922), *pro parte*.
? *S. forcipatus* Hook. f.: 334 (1887); Pax & Hoffm. xv: 218 (1922); Ridley: 220 (1924).
Glochidion umbratile Maiden & Betche in Proc. Linn. Soc. N.S.W. 30: 370 (1905).
Sauropus robinsonii Merr. in Philipp. Journ. Sci. 7: 407 (1912) & Enum.: 405 (1923); Pax & Hoffm. xv: 220 (1922).
S. wichurae Muell. Arg. ex Pax & Hoffm. xv: 222 (1922).
S. grandifolius Pax & Hoffm. xv: 220 (1922) [*non* sec. Beille: 648 (1927)].
S. longipedicellatus Merr. in Sunyatsenia 2: 34 (1934).

SB.—NE. India, S. China, SE. Asia, and scattered through Malesia to N. Australia.
Sprawling shrub to 4 m. high, in riverside forest at 1000–1500 m. alt.

The large capsules borne on long stout fruiting pedicels (up to 11 cm. long) are usually sufficient for the recognition of this species. It is remarkable that in Borneo it appears to be confined to Mt. Kinabalu.

Sauropus micrasterias *Airy Shaw* in Kew Bull. 14: 354 (1960) & 16: 344 (1963).

SR.—Endemic.

Shrub to 1 m. high, in rocky primary forest on flat wet alluvium (yellow clay) in dense shade, among scattered limestone boulders at foot of limestone hill, up to 360 m. alt.

Throwing up erect stems from a long creeping base; leaves produced in the upper portion of the stems, inflorescences mostly from the lower portion near the ground. Inflorescences racemose, bearing male and female flowers, which are bright pink, deep red or dull raspberry-red in colour. The species is related to *S. bonii* Beille and *S. suberosus* Airy Shaw, of SE. Asia and Malaya.

Sauropus pierrei (*Beille*) *Croiz.* in Journ. Arn. Arb. 21: 494 (1940); Airy Shaw in Kew Bull. 23: 55 (1969), *in clavi.*

Breyniopsis pierrei Beille: 630, fig. 75/1–9 (1927); Pax & Hoffm. in Engl. & Harms, Pflanzenf. ed. 2, 19c: 59 (1931).

SB.—Indochina (Cochinchina).

Shrub or small tree to 2·5 m. high, in primary hill forest up to 150 m. alt.

Two collections from Sabah represent the first record of this scarce species from outside Indochina: Tawau Distr., Ulu Balong Forest, *Aban Gibot* SAN 29625; Lahad Datu Distr., Silabukan Forest Reserve, *Sitiol* SAN 33435.

Distinguished from the superficially similar *S. rhamnoïdes* by the flattened, keeled stem, shorter fruiting pedicels and thick-walled fruit, and especially by the male flowers, which have hooded tepals, keeled on the inside, and large erect anthers (§ *Hemisauropus*).

Sauropus rhamnoïdes *Bl.*, Bijdr.: 596 (1825); Muell. Arg.: 240 (1866); J. J. Sm.: 191 (1910); Merr.: 329 (1921) & Pl. Elmer.: 139 (1929); Pax & Hoffm. xv: 219 (1922); Backer & Bakh. f.: 471 (1963).

SR; SB; K1.—Philippines, Java.

Shrub (sometimes scandent) or small tree to 9 m. high, in damp primary Mixed Dipterocarp forest on black or clay-rich soil, or on basalt, or among massive limestone boulders (with some residual soils) in ravines and gorges, occasionally in disturbed forest near villages, etc., from low levels up to 300 m., ascending to 1500 m. on Kinabalu.

Distinguished from *S. androgynus* by the mostly larger leaves and very much larger female flowers and fruiting calyx.

Sebastiania *Spreng.*

1 ♂ inflorescences axillary; ♀ flower and capsule on greatly elongate pedicels; forest tree or shrub with large glossy leaves, usually glaucous beneath (§ *Sarothrostachys*) . . . . **S. borneënsis**

1 ♂ inflorescences terminal or leaf-opposed; ♀ flower and capsule on very short pedicels; herbaceous plant of seashores, etc., with small narrowly oblong leaves which are not or scarcely glaucous beneath (§ *Elachocroton*) . . . . **S. chamaelea**

Sebastiania borneënsis *Pax* in Pax & Hoffm. v: 122 (1912); Merr.: 347 (1921); Airy Shaw in Kew Bull. 14: 396 (1960) & in Hook. Ic. Pl. 38: t. 3723 (1974); Whitmore: 131 (1973).

SR; B; SB.—Malaya (*teste* Whitmore).

Shrub or small tree to 14 m. high, in primary Lowland Dipterocarp forest on clay-rich soil, or in submontane Heath forest on sandy soil overlying Tertiary sandstone, or on brown soil or yellow sandy clay, up to 1200 m. alt.

Leaves elliptic, entire or vaguely sinuate, chartaceous, glabrous, glossy above with finely raised nerves, usually conspicuously glaucous below, up to 20 × 9 cm.; male inflorescences spicate, in axillary fascicles, 0·5–2·5 cm. long, with numerous minute yellow flowers subtended by small red bracts; female inflorescences 1–3-flowered, the flowers and fruits borne on greatly elongate, finally more or less rigid pedicels (to 17 cm. long).

Two closely related species have been described by van Steenis, one of which, *S. lanceifolia* v. Steen., from the Lingga archipelago, is perhaps not specifically distinct from *S. borneënsis*.

Sebastiania chamaelea (*L.*) *Muell. Arg.*: 1175 (1866); Hook. f.: 475 (1888); Boerl.: 268 (1900); Pax & Hoffm. v: 116 (1912) (var. *asperococca* (F. Muell.) Pax, = var. *chamaelea*); Merr. in Philipp. Journ. Sci. 11, C. Bot.: 76 (1916) & Enum.: 347 (1921) (sphalm. '*chaemela*') (var. *asperococca*); Ridley: 317 (1924); Gagnep.: 454 (1926); Backer & Bakh. f.: 498 (1963); Airy Shaw in Kew Bull. 26: 339 (1972); Whitmore: 131 (1973).

Tragia chamaelea L., Sp. Pl.: 981 (1753).
Microstachys chamaelea (L.) Juss., Euph. Gen. Tent.: 49 (1824).
Cnemidostachys chamaelea (L.) Spreng., Syst. 3: 835 (1826).
Elachocroton asperococcus F. Muell. in Hook. Journ. Bot. & Kew Garden Misc. 9: 17 (1857).
Stillingia chamaelea (L.) Baill., Ét. Gén. Euphorb.: 516 (1858).
S. asperococca (F. Muell.) Baill., *l.c.*: 517 (1858).
Cnemidostachys linearifolia Miq., Fl. Ind. Bat., Suppl.: 460 (1861).
Excoecaria chamaelea (L.) Baill. in Adansonia 6: 323 (1867).

SR; SB.—Widespread from India and Ceylon to S. China, and throughout Malesia (except the Philippines) to Australia and the Pacific.

Spreading herb up to 30 cm. high, with long woody taproot, on seashores, beaches, waste sandy ground, etc., at very low altitudes.

Totally different from *S. borneënsis* in general appearance. The small capsule with two dorsal rows of orange teeth on each loculus is characteristic.

The chequered taxonomic history of this little plant—it has been referred to seven different genera—indicates something of the difficulty of finding satisfactory generic limits in the tribe *Hippomaneae*.

Securinega *Juss.*

Securinega virosa (*Roxb. ex Willd.*) *Baill.* in Adansonia 6: 334 (1866) (excl. descr.); Pax & Hoffm. in Engl. & Harms, Pflanzenf. ed. 2, 19c; 60 (1931); Backer & Bakh. f.: 466 (1963); Airy Shaw in Kew Bull. 25: 493 (1971), *in clavi*, & 26: 340 (1972); Whitmore: 131 (1973).

Phyllanthus virosus Roxb. ex Willd., Sp. Pl. 4: 578 (1805).
Xylophylla obovata Willd., Enum. Hort. Berol.: 329 (1809).
Flueggea microcarpa Bl., Bijdr.: 580 (1825); Hook. f.: 328 (1887); Boerl.: 216 (1900); Ridley: 216 (1924); Beille: 528 (1927).
F. virosa (Roxb. ex Willd.) Baill., Ét. Gén. Euphorb.: 593 (1858); J. J. Sm.: 51 (1910); Merr.: 390 (1923); Gage in Journ. As. Soc. Beng. 75(5); 525 (1936); Corner: 255 (1940).
Securinega microcarpa (Bl.) Muell. Arg.: 434–436, *passim* (1866).
S. obovata (Willd.) Muell. Arg.: 449 (1866).
Flueggea obovata (Willd.) Wall. ex F.-Vill., Novis. App.: 189 (1880).

K1a.—Widespread in Trop. Africa and Asia, eastward to China and Japan; frequent in SE. Asia, Malaya, the Philippines and Java, but very scarce in Sumatra and Borneo; in E. Malesia apparently scarce in Celebes, the Moluccas and Lesser Sunda Is., but frequent in New Guinea, and again in N. Australia and Queensland.

No field data available for the one Bornean specimen seen (Hayoep, 1908, *Hubert Winkler* 2126). According to Merrill *S. virosa* is a plant of dry thickets in the Philippines, but in Java, according to J. J. Smith, it occurs very generally in primary forest, especially on stream-banks, and indifferently on clay or limestone. Its extreme scarcity in Borneo (and Sumatra) is thus difficult to explain. Winkler's locality is in the Meratus mountains, NE. of Bandjermasin.

Spathiostemon *Bl.*

Spathiostemon javensis *Bl.*, Bijdr.: 622 (1825); Airy Shaw in Kew Bull. 16: 357 (1963), *in obs.*, & 20: 45 (1966); Meijer, 7: 29 (1967); Whitmore: 132 (1973).

Adelia javanica Miq., Fl. Ind. Bat. 1(2): 388 (1859).
Homonoia javensis (Bl.) Muell. Arg. in Linnaea 34: 200 (1865) & in DC.: 1022 (1866); J. J. Sm.: 240 (1910) & in Nova Guinea 8: 240 (1910); Merr. in Philipp. Journ. Sci. 11, C. Bot.: 283 (1916); Pax & Hoffm. ix–xi: 117 (1919); Merr.: 344 (1921) & 447 (1923) & Pl. Elmer.: 161 (1929); Backer & Bakh. f.: 492 (1963).
Polydragma mallotiforme Hook. f.: 457 (1887) & in Hook. Ic. Pl. 18: t. 1701 (1888); Boerl.: 256 (1900); Ridley: 310 (1924).
Mallotus eglandulosus Elm. in Leafl. Philipp. Bot. 1: 313 (1908).
M. calvus Pax & Hoffm. vii: 195 (1914); cf. Airy Shaw in Kew Bull. 20: 45 (1966).

SR; SB; K1.—Malaya (v. rare), Philippines, Java, and scattered through E. Malesia to E. New Guinea.

Tree to 17 m. high, or sometimes a scrambling shrub, in primary forest, frequently on limestone, but also on loam, sandstone, blackish rocky soil, brown or yellowish soil, sometimes on periodically inundated swampy land; ascending to 600 m. alt. on Kinabalu.

Very Mallotoid in appearance, and in fact scarcely differing from *Mallotus* (especially § *Axenfeldia*) except in the fascicled or phalanged arrangement of the stamens. The species appears to be particularly abundant in Sabah.

Sumbaviopsis *J. J. Sm.*

Sumbaviopsis albicans (*Bl.*) *J. J. Sm.*: 357 (1910); Elm., Leafl. Philipp. Bot. 4: 1304 (1911); Pax & Hoffm. vi: 14 (1912) & vii: 424 (1914); Merr.: 428 (1923) & 4: 93 (1926); Gagnep.: 418 (1926) (incl. var. *disperma* Gagnep.); Merr., Pl. Elmer.: 156 (1929): Hend. in Gard. Bull. Str. Settlem. 7: 125 (1933); Airy Shaw in Kew Bull. 14: 357 (1960); Backer & Bakh. f.: 477 (1963); Meijer, 7: 26 (1967); Airy Shaw in Kew Bull. 26: 341 (1972); Whitmore: 132 (1973).

Adisca? albicans Bl., Bijdr.: 611 (1825).
Rottlera? albicans (Bl.) Hassk., Hort. Bogor.: 238 (1844), *quoad synon. tantum, excl. descr.*
Croton albicans (Bl.) Reichb. f. & Zoll. in Verhand. Natuurk. Vereen. Ned. Ind. 1: 21 (1856) & in Linnaea 28: 322 (1856); Boerl.: 283 (1900).
Sumbavia macrophylla Muell. Arg. in Flora 47: 482 (1864) & in DC.: 727 (1866); Hook. f.: 408 (1887); Pax & Hoffm. vi: 12 (1912).
Coelodiscus speciosus Muell. Arg. in Linnaea 34: 154 (1865) & in DC: 758 (1866); Hook. f.: 426 (1887).
Cephalocroton? albicans (Bl.) Muell. Arg.: 760 (1866); Scheff. in Ann. Mus. Bot. Lugd.-Bat. 4: 120 (1868).
Mallotus speciosus (Muell. Arg.) Pax & Hoffm. vii: 205 (1914).
Doryxylon albicans (Bl.) Balakr. in Bull. Bot. Surv. Ind. 9: 58 (1967).

SR; SB; Ki.—Assam, SE. Asia, and scattered through W. Malesia north to Palawan.

Tree to 21 m. high, in primary forest on sandstone or limestone soils up to 255 m. alt.

A very characteristic plant from its large, thin, long-petioled, distantly repand-dentate, shortly peltate leaves, minutely and densely white-hoary or ochraceous beneath and glabrous above when mature, its stellate-tomentose, elongate, racemose bisexual inflorescences (female below, male above), its petaliferous male flowers, and its rather large, subglobose, ochraceous-tomentellous capsules.

The fact that *S. albicans* has received nine different generic placements—three of them from Mueller himself—is an indication of the taxonomic perplexity to which the plant has given rise. Superficially the toothing of the leaves and their white-hoary undersurface recall *Cladogynos*, whilst the bisexual racemose inflorescences and petaliferous flowers suggest those of *Croton*. The genus, however, is most closely related to *Doryxylon* Zoll., of the Philippines and Lesser Sunda Islands, *Thyrsanthera* Gagnep., of Siam and Indochina, and *Chrozophora* Neck. ex Juss., widespread in the hotter and drier parts of the Old World from the Mediterranean and tropical Africa to South-East Asia.

Suregada *Roxb. ex Rottl.*

1 Leaves 4–9(–12) cm. long, elliptic or oblong, coriaceous, shining, with conspicuously elevate-reticulate nervation; always associated with limestone rock . . . . **S. calcicola**
1 Leaves not as above; not growing on limestone . . . **S. glomerulata**

Suregada calcicola *Airy Shaw* in Kew Bull. 23: 129 (1969).

SR.—Endemic.
Small tree to 5 m. high, on limestone slopes with deep litter layer, or on vertical limestone cliffs, up to 240 m. alt.

The rather small, shining, coriaceous leaves, with strongly reticulate venation, coupled with the (for *Suregada*) unusual limestone habitat, distinguish this species rather clearly.

Suregada glomerulata (*Bl.*) *Baill.*, Ét. Gén. Euphorb.: 396 (1858).

Erythrocarpus glomerulatus Bl., Bijdr.: 605 (1825).
Gelonium glomerulatum (Bl.) Hassk., Cat. Hort. Bogor.: 237 (1844); Muell. Arg.: 1128 (1866); Hook. f.: 460 (1887); Boerl.: 294 (1900); J. J. Sm.: 594 (1910); Pax & Hoffm. iv: 18 (1912); Merr.: 346 ('*glomeratum*') (1921) & 455 (1923) & Pl. Elmer.: 164 (1929); Ridley: 311 (1924); Corner in Gard. Bull. Str. Settlem. 10: 299 (1939) & Ways. Trees: 255 (1940); Backer & Bakh. f.: 497 (1963).
G. papuanum Pax in Engl. iv: 20 (1912); Airy Shaw in Kew Bull. 16: 367 (1963); **synon. nov.**
Doryalis macrodendron Gilg in Engl., Bot. Jahrb. 55: 283, fig. 5 (1918), **synon. nov.**; cf. Airy Shaw. *l.c.*
? *G. borneënse* Pax & Hoffm. in Mitt. Inst. Allg. Bot. Hamburg 7: 230 (1931), *e descr.*
Suregada multiflora sec. Whitmore: 133 (1973), *p.p.*, *an* (Juss.) Baill.?

SR; SB; K1, ?3.—Throughout Malesia to the Northern Territory of Australia.
Shrub or tree to 14 m. high, in primary mixed Lowland Dipterocarp forest, sometimes swampy, on yellow clayey soil or brown or black sandy soil, or on coral rock near shore, ascending to 1500 m. on Kinabalu.

Protean and bewildering in its variation. It is difficult to believe that the extremes can be conspecific, but they appear to be linked by innumerable transitions. One specimen from Sabah (*J. S. & Aban* SAN 30066) has large oblong-elliptic leaves distantly crenate distally. Certain specimens from Kinabalu tend to approach *S. calcicola*. The rather distinct coastal coral rock form (*Jaswir Singh* SAN 26299) has small, broadly obovate leaves, rounded at the apex; it occurs also in Malaya and elsewhere.

Syndyophyllum *Lauterb. & K. Schum.*

Syndyophyllum excelsum *Lauterb. & K. Schum.*, Fl. Deutsch. Schutzgeb. Südsee: 403, t. 12 (1901); Pax & Hoffm. iii: 105, fig. 33 (1911) & in Engl. & Harms, Pflanzenf., ed. 2, 19c: 172 (1931); Airy Shaw in Kew Bull. 14: 393 (1960); Meijer, 7: 25 (1967).

Subsp. **occidentale** *Airy Shaw* in Hook. Ic. Pl. 38: t. 3722 (1974).

'*Excoecaria macrophylla* J. J. Sm.?' sec. Merr. in Philipp. Journ. Sci. 29: 387 (1926), *non* J. J. Sm.

SR; SB (incl. Banggi I.); **K1.**—Sumatra. (Subsp. *excelsum* in New Guinea.)

Tree to 30 m. high, in primary forest on stony black soil, often by streams, or on limestone with loam, up to 75 m. alt.

Syndyophyllum is distantly related to *Moultonianthus* and *Erismanthus*, differing from both in the fine, parallel, transverse tertiary nervation of the leaves and in the flexuous, pendulous, usually bisexual inflorescences. The flowers are borne in dense fascicles consisting either of numerous males and 1(–2) females, or entirely of males. The stamens are considerably exserted at anthesis, with slender filaments and narrowly cordate, apiculate anthers. The general aspect of the female flowers and fruit somewhat recalls *Alchornea*, *Wetria*, etc.

Subsp. *occidentale* differs from the type subspecies in New Guinea in the puberulous rather than tomentellous indumentum, in the much longer inflorescences, and in the ten rather than five stamens, with longer filaments and glabrous rather than puberulous anthers.

Synostemon *F. Muell.*

Synostemon bacciformis (*L.*) *G. L. Webster* in Taxon 9: 26 (1960), *in adnot.*; Backer & Bakh. f.: 471 (1963); Airy Shaw in Kew Bull. 26: 343 (1972); Whitmore: 133 (1973).

Phyllanthus bacciformis L., Mant. Alt.: 294 (1771).
Agyneia bacciformis (L.) Juss., Euphorb. Gen. Tent.: 24, t. 6 (1824); Muell. Arg.: 238 (1866); Hook. f.: 285 (1887); Boerl.: 211 (1900) (sphalm. '*bucciformia*'); Pax & Hoffm. xv: 213 (1922); Ridley: 198 (1924); Beille: 642 (1927); Gage in Journ. As. Soc. Beng. 75(5): 524 (1936); Merr. & Chun in Sunyatsenia 5: 91 (1940); van Steenis in Bull. Bot. Gard. Buitenz. III, 17: 410 (1948).
Phyllanthus goniocladus Merr. & Chun in Sunyatsenia 2: 260, t. 51 (1935).

SB.—Mauritius, India, Ceylon, SE. Asia, S. China, Malaya, Banka, Java, Celebes.

Spreading herb of 30–50 cm., on sandy beaches at sea-level.

Only one specimen seen from Borneo: Sabah, Tanjong Aru, *Shah & Kadim* MS. 975. The genus is scarcely distinct from *Sauropus*.

Tapoïdes *Airy Shaw*

Tapoïdes villamilii (*Merr.*) *Airy Shaw* in Kew Bull. 14: 474 (1960) & 20: 412 (1966) & in Hook. Ic. Pl. 37: t. 3632 (1967); Meijer, 7: 30 (1967).

Ostodes villamilii Merr. in Journn. Str. Br. Roy. As. Soc. 76: 92 (1917) & Enum.: 345 (1921).

SR (NE.); **SB.**—Endemic.

Tree to 9 m. high, in primary Lowland Dipterocarp forest, from very low alt. near mangroves up to 60 m. alt. on ridges.

The genus *Tapoïdes* is related to *Elateriospermum* Bl., *Vernicia* Lour. and *Loerzingia* Airy Shaw, differing especially in its dioecious flowers, calyx closed in bud and splitting at anthesis into three irregular segments each bearing one to two membranous lobules at the apex, five free veiny petals, conspicuous irregularly lobed annular disk and six to eight short pubescent stamens. *T. villamilii* is evidently quite a scarce element in the forests of NE. Sarawak and Sabah.

Trewia *L.*

Trewia nudiflora *L.*, Sp. Pl.: 1193 (1753); Muell. Arg.: 953 (1866); Hook. f.: 423 (1887); Boerl.: 248 (1900); J. J. Sm.: 390 (1910); Pax & Hoffm. vii: 140 (1914); Merr.: 431 (1923); Gagnep.: 343 (1925); Backer & Bakh. f.: 481 (1963) (*Trevia*); Airy Shaw in Kew Bull. 20: 405 (1966) & 23: 79 (1969) & 26: 343 (1972); Whitmore: 134 (1973).

Tetragastris ossea Gaertn., Fruct. 2: 130 (1788).
Rottlera indica Willd. in Götting. Journ. Naturwiss. 1: 8, t. 3 (1797).
Trewia macrophylla Roth, Nov. Pl. Sp.: 373 (1821) [*non* Bl. (1825)].
T. macrostachya Klotzsch in Ergebn. Bot. Reise Pr. Waldem.: 117, t. 23 (1862).
Mallotus cardiophyllus Merr. in Philipp. Journ. Sci. 7: 398 (1912).

KI.—India, Ceylon, SE. Asia and S. China; scattered through W. Malesia, but everywhere rare or extremely local.

Deciduous dioecious tree to 45 m. high, in forest on low ridges, on loam soil with coral limestone, or loam on sandstone, up to 100 m. alt.

Trewia is closely related to *Mallotus*, especially § *Rottleropsis*, differing in its large male flowers, few-flowered female inflorescences, greatly elongate styles and large globose drupaceous fruit. The flowers appear precociously or with the young leaves.

For a note on the spelling of the generic name, see Kew Bull. 20: 406 (1966).

The related *Neotrewia cumingii* (Muell. Arg.) Pax & Hoffm., widespread and common in the Philippines, south to Mindanao, could well be expected to occur in Sabah.

Trigonopleura *Hook. f.*

Trigonopleura malayana *Hook. f.*: 399 (1887) & in Hook. Ic. Pl. 18: t. 1753 (1888); Pax & Hoffm. iii: 95, fig. 30 (1911); Ridley: 263 (1924); Symington in Bull. Misc. Inf., Kew, 1937: 319 (1937); Stern in Am. Journ. Bot. 54: 671 (1967); Whitmore: 134 (1973).

Alsodeia dubia Elm., Leafl. Philipp. Bot. 8: 2875 (March 1915).
Trigonopleura philippinensis Merr. in Philipp. Journ. Sci. 10, C. Bot.: 275 (Aug. 1915).
T. dubia (Elm.) Merr. in *op. cit.* 11, C. Bot.: 77 (1916); Pax & Hoffm. xiv (Euph.-Addit. vi): 42 (1919); Merr., Enum.: 453 (1923).
? *T. borneënsis* Merr., *l.c.*: 76 (1916) & Enum.: 345 (1921), *e descr.*
Peniculifera penangensis Ridley in Journ. Roy. As. Soc. Str. Br. 82: 173 (1920) & Fl. Malay Penins. 1: 290 (1922); cf. Symington, *l.c. supra.*

SR; SB; K1, ?3.—N. Malaya, Sumatra, Banka, Philippines.
Tree to 39 m. high, in primary Mixed Dipterocarp forest on sandstone, sandy loam, sandy clay, clay-rich shale-derived soil, black soil, or yellow podsolic soil, up to 1380 m.

Flowers heavily scented. Fresh leaves or leaves dried over fire used in Sabah as *gambier* for chewing with *sireh.*
The following specimen has appreciably larger capsules than usual, 13–14 mm. long, with unusually thick endocarp, and may eventually merit taxonomic recognition.
SR.—First Division: Kuching Distr.; Arboretum, Semengoh Forest Reserve, in primary Lowland Dipterocarp forest, alt. 90 m., 30 Nov. 1961, *Rosli & Galau* S. 15747:—Tree No. 642, 35 cm. diameter.
Trigonopleura has no connexion with *Trigonostemon*, with which it has often been associated, but is quite closely related to *Chaetocarpus*, from which it differs chiefly in the possession of petals and in the smooth, non-bristly or warty but wrinkled pericarp of the fruit, in which the three broad hyaline septa often persist on the columella after dehiscence. Both genera make an evident approach to *Casearia* (*Flacourtiaceae*).

Trigonostemon *Bl.*

1 Petals orange or yellow (cf. also *T. laevigatus*):
  2 Leaves coriaceous, dark brown when dry, not or scarcely trinerved at base; inflorescence dense-flowered, rather rigid; flowers small, petals 2–3 mm. long . . . . **T. sumatranus**
  2 Leaves membranaceous, green or ochraceous when dry, usually clearly trinerved at base; inflorescence very lax-flowered, slender and usually pendulous (but cf. var. *elegantissimus*); flowers rather large, petals 4–5 mm. long . . . . **T. viridissimus**
    3 Inflorescence delicate, fewer-flowered, pendulous var. **viridissimus**
    3 Inflorescence more robust, many-flowered, ± erect var. **elegantissimus**

1 Petals black, deep purple or reddish:
  4 Inflorescence cauliflorous, very slender, lax, much-branched **T. capillipes**
  4 Inflorescence axillary or leaf-opposed, more robust, pseudo-racemose or shortly cymose:
    5 Leaves usually clearly trinerved at base, sometimes subopposite or subverticillate; inflorescence shortly cymose, floral bracts inconspicuous; petals occasionally yellow . . . **T. laevigatus**
    5 Leaves usually not clearly trinerved, not subopposite; inflorescence (at least the ♀) pseudo-racemose, with conspicuous floral bracts:
      6 Capsule sparsely or densely verrucose; leaves usually ochraceous when dry:
        7 Capsule sparsely verrucose above; pedicels longer and slenderer, up to 2 cm. long . . . . **T. elmeri**
        7 Capsule densely verrucose all over; pedicels shorter and stouter, up to 1 cm. long . . . . **T. ionthocarpus**
      6 Capsule smooth, not verrucose:
        8 Leaves spathulate-oblanceolate, up to 55 cm. long, very shortly and stoutly petioled; petals narrowly spathulate-oblong, densely papillose on the upper surface . . . . **T. sandakanensis**
        8 Leaves elliptic or ovate or oblong; petioles longer and thinner; petals obovate, not papillose:
          9 Leaves narrowly oblong; inflorescence short, slender, few-flowered, the terminal flower ♀, the remainder ♂; floral bracts inconspicuous . . . . . **T. salicifolius**
          9 Leaves elliptic or ovate; inflorescence longer and more robust, either unisexual, or bisexual with ♂ and ♀ together along the rhachis; floral bracts often very conspicuous:
            10 Petioles very elongate (up to 25 cm. long); plant glabrous or sparsely pilose; inflorescences unisexual, the males with a long bare basal portion . . . **T. merrillianus**
            10 Petioles much shorter (up to 5 cm. long); plant usually softly tomentellous or villous, rarely almost glabrous; inflorescences bisexual, with conspicuous bracts and fruiting sepals . . . . **T. villosus**
              11 Bracts lanceolate, acute . . . . var. **villosus**
              11 Bracts ovate, obtuse . . . . var. **borneënsis**

Trigonostemon capillipes (*Hook. f.*) *Airy Shaw* in Kew Bull. 20: 413 (1966) & 26: 345 (1972); Whitmore: 135 (1973).

Dimorphocalyx capillipes Hook. f.: 404 (1888); Pax & Hoffm. iii: 33 (1911); Ridley: 266 (1924).

Trigonostemon diffusus Merr. in Sarawak Mus. Journ. 3: 525 (1928); Jabl. in Brittonia 15: 164 (1963).

SR.—S. Penins. Siam, N. Malaya (Perlis).

Small tree to 6 m. high, in Mixed Dipterocarp forest on granodiorite up to 750 m. alt.

The slender, much-branched, cauliflorous inflorescence is very characteristic. Flowers rich deep purplish-crimson.

Trigonostemon elmeri *Merr.*, Pl. Elmer.: 162 (1929); Jabl. in Brittonia 15: 162 (1963).

? *T. longifolius* sec. Whitmore: 136 (1973), quoad loc. 'Borneo', *non* Baill.

SB; K1, 3?—Endemic.
Shrub or tree to 13 m. high, locally common in undergrowth of primary or secondary forest (sometimes swampy), on stony or rocky hillsides, on black, brown or yellow sandy soil, up to 150 m. alt.

Close to *T. longifolius* Baill., of Malaya and Sumatra, differing in the usually longer, slenderer, less reflexed fruiting pedicels, and in the presence of scattered warts on the apex of the capsule. Petals deep purplish-black.

Trigonostemon ionthocarpus *Airy Shaw* in Kew Bull. 21: 407 (1968); Meijer, 10: 231 (1968).

SR (?); **SB.**—Endemic.
Shrub or small tree to 4 m. high, in primary forest on blackish soil up to 120 m. alt.

Differs from *T. elmeri* in its densely verrucose capsule and shorter, stouter fruiting pedicels. The Sarawak material has thicker leaves with almost immersed venation, longer inflorescences and long, slender fruiting pedicels; it may be specifically distinct, though the densely warty capsules seem identical.

Trigonostemon laevigatus *Muell. Arg.* in Flora 47: 538 (1864) & in DC.: 1111 (1866); Boerl.: 232, 284 (1900); Pax & Hoffm. iii: 94 (1911); Merr.: 345 (1921) (*T. laevig. alter*); Ridley: 652 (1924); Jabl. in Brittonia 15: 167–8 (1963); Airy Shaw in Kew Bull. 25: 346 (1972); Whitmore: 135 (1973).

T. anomalus Merr. in Philipp. Journ. Sci. 16: 569 (1920) & Enum.: 451 (1923).
? *T. petelotii* Merr. in Univ. Calif. Publ. Bot. 10: 425 (1924), *e descr.*

SR; SB; K1.—Andamans, Lower Siam, Malaya, ? Indochina, Philippines.
Shrub or tree to 11 m. high, in primary or secondary Mixed Dipterocarp forest on yellow sandy clay or loam, brown soils or basalt-derived soils, up to 280 m. alt.

A distinct but very variable species, the leaves often subopposite, usually more or less trinerved at the base (the basal nerves sometimes submarginal), petioles mostly short (rarely up to 4 cm. long), the inflorescences shortly cymose, terminal or leaf-opposed; petals shortly bilobed, red or carmine, rarely yellow.

Trigonostemon merrillianus *Airy Shaw* in Kew Bull. 25: 549 (1971).

Dimorphocalyx (?) *borneënsis* Merr. in Philipp. Journ. Sci. 11, C. Bot.: 73 (1916) & Enum.: 345 (1921), *non Trigonostemon borneënsis* Merr. (1929).

SR (C.).—Endemic.
Small tree to 5 m. high, in primary Lowland Mixed Dipterocarp forest on clay loam or clay-rich soil up to 15 m. alt.

Only known from three localities in central Sarawak (Third Division). Petioles often very elongate (up to 25 cm. long), lamina elliptic (up to 26 × 9 cm.), remotely and obscurely denticulate or subcrenate; inflorescences elongate, the males up to 20 cm. long, with the flowers borne in a few subterminal dense coralliform cymose clusters, the females up to 50 cm. long, with the flowers borne distantly in the upper portion subtended by conspicuous ovate bracts. Sepals deep pink or purplish, petals black.

Trigonostemon salicifolius *Ridley* in Bull. Misc. Inf., Kew, 1923: 360 (1923) & Fl. Malay Penins. 3: 264 (1924); Jabl. in Brittonia 15: 152, fig. 1 (1963).

T. verticillatus var. *salicifolius* (Ridl.) Whitmore: 136 (1973).

B.—Malaya (Selangor).

Shrub or small tree of unknown stature (to 5 m. high in Malaya), on low terrace in *kerangas* at 15 m. alt. (up to 270 m. in Malaya).

The record is based upon a single collection only: Temburong-pandaruan, 3 Sept. 1958, *Brunig* S. 1137. The species is closely related to *T. verticillatus* (Jack) Pax, but seems distinct in its elongate, narrowly oblong, obscurely and remotely denticulate leaves (up to 28 × 3 cm.). The racemes are short, slender and few-flowered; the flowers have deep purple petals and five stamens.

Trigonostemon sandakanensis *Jabl.* in Brittonia 15: 159–161, figs. 5 & 6 (1963).

SB.—Endemic.

Shrub to 1·5 cm. high, in wet parts of primary forest by streams in low undulating country up to 21 m. alt.

A remarkable species. Leaves crowded towards tips of branches, very large, spathulate-oblanceolate (up to 55 × 13 cm.), subentire, finely pilosulous on both surfaces, very shortly and stoutly petioled. Inflorescences pseudo-racemose, up to 26 cm. long, with a stout puberulous rhachis; flowers in fasciculiform cymules subtended by a conspicuous spreading narrowly subulate bract. Sepals five, red or reddish-purple; petals five, paler, violet-pink, elongate, narrowly spathulate, densely papillose on the upper surface; stamens five, rarely three, anther-thecae shortly cornute.

Trigonostemon sumatranus *Pax & Hoffm.* iii: 90 (1911); Merr. in Papers Mich. Acad. Sci. 20: 101 (1935); Jabl. in Brittonia 15: 163 (1963); Stern in Am. Journ. Bot. 54: 671 (1967); Whitmore: 135 (1973).

SR; SB; K1, 3.—Sumatra, Malaya.

Tree to 12 m. high, in primary Lowland Dipterocarp forest on sandstone or on loam with lime up to 400 m. alt.

Usually recognizable by the large, much-branched, pyramidal, bisexual, frequently terminal inflorescences, with numerous small male and scattered larger female flowers. The petals are orange or ochraceous. Leaves coriaceous, glabrous, distantly and shallowly denticulate; petioles variable in length.

Trigonostemon villosus *Hook. f.*: 397 (1887); Pax & Hoffm. iii: 88 (1912); Ridley: 265 (1924); Jabl. in Brittonia 15: 158 (1963); Whitmore: 135 (1973).

T. tomentellus Pax & Hoffm. iii: 89 (1912).

Var. **villosus**: bracts lanceolate, very acute.

SR; SB.—Malaya.

Shrub or small tree to 2 m. high, in primary hill forest on black soil or on limestone scree up to 210 m. alt.

Branchlets, petioles, midrib and inflorescences shortly and softly fulvous-tomentellous (rarely almost glabrous—Sabah, *Muin Chai* SAN 31638); leaves membranous-chartaceous; inflorescences axillary, pseudo-racemose, with conspicuous lanceolate bracts. Male flowers very small; female flowers with large lanceolate sepals often almost as large as the bracts. Petals black or deep purple. Capsule tomentellous.

Var. **borneënsis** (*Merr.*) *Airy Shaw*, comb. & stat. nov.

Trigonostemon borneënsis Merr., Pl. Elmer.: 162 (1929); Jabl. in Brittonia 15: 162 (1963); Whitmore: 136 (1973).

SB; K1.—Endemic.

Erect shrub or shrub-like tree to 2 m. high, in dense primary forest near streams up to 300 m. alt.

Differs from var. *villosus* in the ovate, obtuse bracts.

Trigonostemon viridissimus (*Kurz*) *Airy Shaw* in Kew Bull. 25: 545 (1971); Whitmore: 135 (1973).

Sabia viridissima Kurz in Journ. As. Soc. Bengal 41(2): 304 (1872); Hook. f., Fl. Brit. Ind. 2: 3 (1876); Kurz, For. Fl. Brit. Burma 1: 301 (1877).

Blachia viridissima (Kurz) King in Journ. As. Soc. Bengal 65(2): 455 (Mat. Fl. Mal. Penins. 8: 741) (1896), *in obs.*

Trigonostemon ovatifolius J. J. Sm.: 583 (1910); Backer & Bakh. f.: 495 (1963).

T. membranaceus Pax & Hoffm. iii: 91 (1911).

T. macgregorii Merr. in Philipp. Journ. Sci. 16: 566 (1920) & Enum.: 452 (1923).

Neotrigonostemon diversifolius Pax & Hoffm. in Notizbl. Bot. Gart. Berlin 10: 385 (1928) & in Engl. & Harms, Pflanzenf. ed. 2, 19c: 169 (1931).

Kurziodendron viridissimum (Kurz) Balakr. in Bull. Bot. Surv. Ind. 8: 68, figs. 1–7 (1966).

Var. **viridissimus**: inflorescence very slender, pendulous, lax, few-flowered.

SR; SB.—Burma, Andamans, W. Malesia, Lesser Sunda Is.

Shrub or tree to 9 m. high, in primary Mixed Lowland Dipterocarp forest, once noted on basalt, up to 270 m. alt.

Leaves membranaceous, trinerved at base, green or ochraceous when dry. Inflorescences very laxly pyramidal, very slender, pendulous; petals yellow or ochraceous.

Var. **elegantissimus** (*Airy Shaw*) *Airy Shaw*, comb. & stat. nov.

Trigonostemon elegantissimus Airy Shaw in Kew Bull. 20: 48 (1966); Whitmore: 135 (1973).

K1.—S. Malaya (*teste* Whitmore).
Tree to 5 m. high, in forest on sandstone at 100 m. alt.

Only differs from var. *viridissimus* in the much more robust, much-branched, many-flowered inflorescence, which is apparently more or less erect rather than pendulous. The supposed differences in the size of the leaves and petals do not hold.

Wetria *Baill.*

Wetria insignis (*Steud.*) *Airy Shaw* in Kew Bull. 26: 350 (1972) & 27: 87 (1972); Whitmore: 136 (1973).

Trewia macrophylla Bl., Bijdr.: 612 (1825), *non* Roth (1821) [quae = *T. nudiflora* L.].
T. insignis Steud., Nomencl. ed. 2, 2: 698 (1841), *pro nom. nov.*; Pax & Hoffm. vii: 142 (1914).
Wetria trewioïdes Baill., Ét. Gén. Euphorb.: 409 (1858), *pro nom. nov.*; Boerl.: 249 (1900); Ridley: 282 (1924).
Pseudotrewia macrophylla [(Bl.)] Miq., Fl. Ind. Bat. 1(2): 414 (1859).
Alchornea blumeana Muell. Arg. in Linnaea 34: 167 (1865) & in DC.: 900 (1866), *pro nom. nov.* (*non A. macrophylla* Mart. *nec A. trewioïdes* (Benth.) Muell. Arg.).
Agrostistachys pubescens Merr. in Philipp. Journ. Sci. 4, Bot.: 274 (1909); Pax & Hoffm. vi: 99 (1912).
Wetria macrophylla [(Bl.)] J. J. Sm.: 471 (1910); Pax & Hoffm. vii: 219 (1914); Merr.: 341 (1921) & 437 (1923) & in Philipp. Journ. Sci. 24: 115 (1924) & 29: 383 (1926); Airy Shaw in Kew Bull. 14: 473 (1960) & 16: 353 (1963); Backer & Bakh. f.: 485 (1963); Meijer, 7: 28 (1967).
Trigonostemon forbesii Pax in Pax & Hoffm. iii: 88 (1911); Jabl. in Brittonia 15: 165 (1963).

SR; SB; K1.—Burma, Penins. Siam. W. Malesia, New Guinea (Papua).
Tree to 23 m. high, in primary forest, sometimes common in undergrowth, frequently on coral limestone, sometimes on loam or sandstone, or on brown or black stony soil, once noted on Tertiary granodiorite, up to 165 m. alt.

The elongate, oblanceolate, sharply dentate, very short-petioled leaves, with very numerous parallel lateral nerves, and the very elongate inflorescences, make this an unmistakable species. It is closely related to *Alchornea*.

Appendix I
Stilaginaceae

Antidesma *L.*

1 Drupes very small, 2–2·5 mm. diam., glabrous, with lateral styles; inflorescences much-branched, slender, with minute flowers . . . **A. venenosum**

1 Drupes larger:
 2 Leaves distinctly cordate at the base (cf. also *A. leucopodum* var. *platyphyllum*); styles terminal:
  3 Leaves 13–21 cm. long, coriaceous, shining, acute or shortly acuminate; petiole 1·5–3 cm. long . . . . **A. cordatum**
  3 Leaves 3–10 cm. long, membranous or chartaceous, often rounded at the apex; petiole 0·5–2 cm. long . . . . **A. ghaesembilla**
 2 Leaves not distinctly cordate at the base:
  4 Inflorescences often at least partly cauliflorous; styles terminal:
   5 ♀ inflorescences often very elongate (10–16 cm. or more); tertiary nerves usually closely parallel, at right angles to the midrib:
    6 Drupes small, lenticular, 4 mm. long, with 4–6 styles; pubescence less strong . . . . **A. leucopodum** (with vars.; see below)
    6 Drupes large, oblong, 10 mm. long, with 8–10 styles; pubescence dense . . . . **A. polystylum**
   5 ♀ inflorescences short, 3–6 cm. long; leaves stiffly coriaceous:
    7 Venation of leaves, on the undersurface, finely tessellated; inflorescences fascicled, simple; pubescence of inflorescence whitish, not rusty . . . . **A. thwaitesianum**
    7 Venation of leaves not tessellated; inflorescences often branched; pubescence ferrugineous . . . . **A. coriaceum**
  4 Inflorescences usually axillary or terminal:
   8 Rheophytic or at least markedly stenophyllous plants (leaves approx. 5–8 times as long as broad):
    9 Stipules ± ovate, often persistent; leaves less than 17 mm. wide . . . . **A. linearifolium**
    9 Stipules subulate or very narrowly elliptic:
     10 Leaves 1–2 cm. wide . . **A. hosei** var. **angustatum**
     10 Leaves up to 5 cm. wide:
      11 Infructescence up to 50 cm. long; leaves coriaceous, up to 30 cm. long . . . . **A. riparium**

11 Infructescence up to 10 cm. long; leaves thinner, up to 25 cm long:
  12 Drupe ± rhomboid or fusiform, not flattened; leaves mostly ± membranous (very near *A. montanum*) **A. salicinum**
  12 Drupe narrowly ovoid-oblong, flattened; leaves chartaceous to coriaceous:
    13 Glabrous, or young parts minutely grey-puberulous; midrib raised on the upper surface; ovary and drupe minutely puberulous **A. stenophyllum**
    13 Young parts thinly rufous-pubescent; midrib impressed on the upper surface; ovary rufous-tomentellous . . . . **A. stenocarpum**

8 Plants neither rheophytic nor markedly stenophyllous:
  14 Midrib distinctly raised on the upper surface; leaves 20–40 cm. long; stipules narrowly subulate; infructescence 30–40 cm. long; ovary glabrous; drupe 10 mm. long **A. pendulum**
  14 Midrib flat or channelled above:
    15 Stipules broad, conspicuous, ± persistent:
      16 Leaves large, 20–35 cm. long, glossy, usually olivaceous when dry; ♀ inflorescence up to 30 cm. long **A. stipulare**
      16 Leaves medium-sized, up to 15 cm. long, less glossy, usually reddish-brown when dry; ♀ inflorescence up to 10 cm. long . . . . **A. neurocarpum**
    15 Stipules narrow, subulate or lanceolate, often very caducous:
      17 Drupe small, 5–8 mm. long:
        18 Petiole 3–7(–12) mm. long; drupe often white-pustulate when dry; leaves mostly membranous **A. montanum**
        18 Petiole 8–15 mm. long, slender; drupe not white-pustulate; leaves firmly chartaceous **A. cuspidatum** *var.*
      17 Drupe larger, 8–15 mm. long, usually ± compressed, rarely white-pustulate; leaves usually chartaceous to coriaceous:
        19 Leaves 5–15(–18) cm. long:
          20 Plants ± ferrugineous-pubescent **A. tomentosum** var. **bangueyense**
          20 Plants lacking any rufous indumentum, ± intermediate between *A. neurocarpum* and *A. coriaceum* . . . . **A. hosei**
            21 Drupe 10–15 mm. long, compressed parallel to the suture . . . . . var. **hosei**
            21 Drupe 5–11 mm. long, rounded or compressed at right angles to the suture var. **microcarpum**
        19 Leaves up to 30 cm. long:

22 Primary nerves 10–20 pairs; plant often with strong rufous indumentum
A. tomentosum var. **tomentosum**

22 Primary nerves 5–10 pairs; leaves almost glabrous:

23 ♀ inflorescence up to 24 cm. long; leaves up to 28 cm. long, olivaceous-green when dry
A. montis-silam

23 ♀ inflorescence up to 5 cm. long; leaves up to 23 cm. long, reddish-brown when dry
A. brachybotrys

N.B. The above key does not take account of the unconfirmed record for *A. bunius*; see below.

Antidesma brachybotrys *Airy Shaw* in Kew Bull. 26: 457 (1972); Whitmore: 57 (1973).

SR; K3.—Malaya (Johore).
Treelet to 2 m. high, in Mixed Dipterocarp forest (sometimes swampy), on clayey loam, below 100 m. alt.

A robust species, almost glabrous, near *A. helferi* Hook. f. (Lower Burma, Lower Siam, N. Malaya, Philippines), but leaves up to 23 × 9 cm. drying brown; female inflorescences less than 5 cm. long; female calyx urceolate; fruits broadly ovoid, compressed, almost symmetrical, up to 14 × 9 mm.

Antidesma bunius (*L.*) *Spreng.*, Syst. Veg. 1: 826 (1825); Muell. Arg.: 262 (1866); Hook. f.: 358 (1887); J. J. Sm.: 270 (1910); Pax & Hoffm. xv: 160 (1922); Merr.: 412 (1923) & in Philipp. Journ. Sci. 29: 380 (1926); Gagnep.: 524 (1926); Corner: 233 (1940); Backer & Bakh. f.: 458, 460 (1963); Airy Shaw in Kew Bull. 26: 353 (1971); Whitmore: 55 (1973), *in obs.*

Stilago bunius L., Mant.: 122 (1767).
Antidesma ciliatum Presl, Epim. Bot.: 234 (1851).
A. cordifolium Presl, *l.c.* (1851).
A. bunius var. *cordifolium* (Presl) Muell. Arg.: 262 (1866).
Sapium crassifolium Elm., Leafl. Philipp. Bot. 2: 485 (1908).
Antidesma crassifolium (Elm.) Merr. in Philipp. Journ. Sci. 7, Bot.: 383 (1912); Pax & Hoffm. xv: 139 (1922).

SB (Banggi I.).—S. India, Ceylon, E. Himalaya, SE. Asia, S. China, and throughout Malesia (exc. Malaya and mainland Borneo) to New Guinea and Queensland.
Shrub or tree, in open places; no further information for Borneo.

I have not seen the solitary gathering (*Castro & Melegrito* 1519) upon which Merrill's record (*l.c.*, 1926) of *A. bunius* for Banggi Island was based. The species could well be expected to occur there, since it is widespread in the Philippines, down to Mindanao, though it is apparently not yet known from Palawan. There is, however, just a possibility that a specimen of *A. thwaitesianum* Muell. Arg. might be represented. This species has been much confused with *A. bunius* in the past, and has in fact been collected on Banggi, near the coast, in recent years (*Fabia* SAN 25755, 12 June 1961; cf. *Kew Bull.* 26: 464

(1972)); but I have not yet seen a genuine specimen of *A. bunius* from that island. The opportunity of examining Castro & Melegrito's gathering will be awaited with interest.

Antidesma cordatum *Airy Shaw* in Kew Bull. 26: 465 (1972); Whitmore: 56 (1973).

SR; SB.—Malaya (Selangor).

Tree to 18 m. high, scarce in primary hill Dipterocarp forest up to 210 m. alt.

Closely related to *A. coriaceum* Tul., from which it is distinguished by the rounded or cordate base of the leaves and 9–16 pairs of nerves.

Antidesma coriaceum *Tul.* in Ann. Sci. Nat. III, 15: 204 (1851); Muell. Arg.: 252 (1866); Pax & Hoffm. xv: 164 (1922); Ridley: 233 (1924); Pax & Hoffm. in Mitt. Inst. Bot. Hamburg 7: 224 (1931); Corner: 233 (1940); Meijer, 7: 32 (1967); Whitmore: 57 (1973).

A. puncticulatum Miq., Fl. Ind. Bat., Suppl.: 468 (1861); Muell. Arg.: 266 (1866); **synon. nov.**

A. fallax Muell. Arg. in Linnaea 34: 68 (1865) & in DC.: 253 (1866); Hook. f.: 359 (1887).

Aporosa griffithii Hook. f.: 353 (1887); Pax & Hoffm. xv: 91 (1922); cf. Ridley, *l.c. supra*.

Antidesma pachyphyllum Merr. in Philipp. Journ. Sci. 11, C. Bot.: 58 (1916) & Enum.: 332 (1921).

A. nitens Pax & Hoffm. xv: 136 (1922), **synon. nov.**

SR; B; SB; K1, 3.—Malaya, Sumatra, Banka.

A small tree to 12 m. (rarely 24 m.) high, frequent in understorey of mixed primary lowland peat-swamp forest (*kerangas* or *padang*), on black or sandy soil or giant podsol, at very low altitudes (rarely to 255 m.).

Branches often whitish; leaves coriaceous, very 'laurel'-like (i.e. *Prunus laurocerasus* L.), smooth and shining, usually drying plumbeous above and brown below; young parts ferrugineous, soon glabrescent; inflorescences 1–5 cm. long, sometimes branched at the base, persistently shortly ferrugineous-tomentellous; flowers almost sessile; drupe compressed, symmetrical, 4–6 mm. long, styles terminal.

Antidesma cuspidatum *Muell. Arg.* in Linnaea 34: 67 (1865) & in DC.: 252 (1866); Hook. f.: 360 (1887); Merr. in Philipp. Journ. Sci. 11, C. Bot.: 54 (1916) & Enum.: 332 (1921); Pax & Hoffm. xv: 129 (1922); Ridley: 232 (1924); Corner: 233 (1940) (fig. 64, showing leaf, not good; petiole much too short); Whitmore: 58 (1973).

Var. **borneënse** *Airy Shaw*, var. nov., fructu majore 6–7 mm. diametro minus compresso.

SR.—First Division: *Sine loc.*, 1865–68, *Beccari* 1256. 'Sarawak river', 1890, *Sitam* for *Haviland* 'b.x.s.b.'. On the hill near Bau, 16 Dec. 1892, *Haviland* '= 1762':—Small tree. Without locality or date, *Native Collector* 507. Kuching District, 12th mile, Penrissen Road, hill slope, near tree no.

4547, alt. . . ., 1 Sept. 1966, *Banyeng & Sibat* S. 24915 (type, K):—Small tree 7·5 m. high, 25 cm. girth; purplish-red fruits. Sampadi Forest Reserve, 26th mile, Bau/Lundu Road, hillside, on yellow loam, alt. 195 m., 19 June 1968, *Ilias Paie* S. 24949:—Tree 7·5 m. tall, 25 cm. girth; fruits green, reddish when ripe.

The fruits of typical *A. cuspidatum*, of Malaya, are only 4–5 mm. in diameter, and much compressed, almost lenticular. In the above small population in south-western Sarawak they are conspicuously larger and relatively much less compressed. *A. cuspidatum* is distinguished from most forms of the closely related *A. montanum* by its longer and slenderer petioles.

It should be noted that the above-cited *Haviland* '= 1762' is in fact different from *Haviland* 1762, which is a form of *A. montanum* Bl. (*A. phanerophlebium* Merr.).

Antidesma ghaesembilla *Gaertn.*, Fruct. 1: 189 (1788); Muell. Arg.: 251 (1866); Hook. f.: 357 (1887); J. J. Sm.: 287 (1910); Hutch. in Journ. Linn. Soc., Bot. 42: 134 (1914); Merr.: 332 (1921) & 414 (1923); Pax & Hoffm. xv: 155 (1922); Ridley: 230 (1924); Gagnep.: 505 (1926); Backer & Bakh. f.: 458 (1963); Meijer, 7: 32 (1967); Airy Shaw in Kew Bull. 26: 353 (1972); Whitmore: 56 (1973).

A. pubescens Roxb., Pl. Corom. 2: 35, t. 167 (1798).
A. frutescens Jack in Mal. Misc. 2: 91 (1822); Pax & Hoffm. xv: 157 (1922); cf. Merr. in Journ. Arn. Arb. 33: 216 (1952).
A. paniculatum Bl., Bijdr.: 1128 (1825).

SB; K1, 1a.—W. Himalaya and Ceylon to S. China and throughout Malesia to Bismarck Archip. and N. Australia.

Tree usually of about 6 m., rarely to 28 m. in height, locally common in secondary vegetation in open grassland (*lalang*, etc.), by stream-sides, etc., on brown or sandy soil, more rarely in primary forest; mostly at low altitudes, occasionally up to 600 m.—The almost complete absence of this widespread species from Sarawak and Indonesian Borneo is remarkable.

Antidesma hosei *Pax & Hoffm.* xv: 138 (1922); Airy Shaw in Kew Bull. 28: 271 (1973); Whitmore: 57 (1973).

? *A. clementis* Merr. in Journ. Str. Br. Roy. As. Soc. 76: 90 (1917), *e descr.*; *non* Merr. in Philipp. Journ. Sci. 9, C. Bot.: 465 (1914).
A. plumbeum Pax & Hoffm. xv: 133 (1922), **synon. nov.**
A. digitaliforme sec. Meijer, 7: 32 (1967), *non* Tul.

Nearly 50 gatherings of this polymorphic complex are represented at Kew. The range of variation is so great that it is difficult to believe that they are all conspecific, but at present it seems impossible to find any natural dividing lines between them. *A. plumbeum* is a sample from the stenophyllous end of the range. As will be seen below, the range of soil and habitat is equally great; the plant is, however, evidently a calcifuge. The complex may be said to lie more or less midway between *A. neurocarpum* Miq. and *A. coriaceum* Tul. It is especially close to the former, but usually differs consistently in its much narrower stipules and sparser rufous indumentum, which is often indeed completely lacking. *A. coriaceum* differs in its almost sessile flowers (both male and female) and usually smaller fruits; the petioles are often also appreciably longer.

Var. **hosei**: drupe 10–15 mm. long, compressed parallel to the suture.—Airy Shaw, *l.c.*

SR; B; SB; K1 ?—Malaya; Celebes.

Shrub or tree to 18 m. high, in *kerangas* or primary Lowland or Mixed Dipterocarp or Dipterocarp–*Nothofagus* or *Quercus* forest, or in submontane peat-swamp forest dominated by *Dacrydium beccarii*, or in Riparian forest, in seasonal swamps, by stream-sides, on hillsides and ridges, on sandstone, or shale-derived soil, or riverine alluvium, or shallow peat or white sand overlying alluvium, or yellowish sandy soil, or blackish or brownish soil, or ultrabasic soil, from low altitudes up to 1680 m.

Var. **microcarpum** *Airy Shaw* in Kew Bull. 28: 271 (1973).

SR; B; SB.—Malaya.

Tree to 10·5 m. high, in primary forest, mossy forest, or submontane peat-swamp forest dominated by *Dacrydium beccarii*, on shallow peat overlying alluvium, up to 4500 m. alt.

Distinguished by the much smaller fruits, 5–11 mm. long, which are rounded or compressed at right angles to the suture, and often have a finely granular surface.

Var. **angustatum** (*Airy Shaw*) *Airy Shaw*, comb. nov.

A. neurocarpum var. *angustatum* Airy Shaw in Kew Bull. 28: 270 (1973).

SR.—Endemic.

Treelet, stature unknown, in Mixed Dipterocarp forest on shale ridge at 400 m. alt.

Leaves only 1–2 cm. broad, long-acuminate. The narrow stipules indicate that this stenophyllous form should be attached to *A. hosei* rather than *A. neurocarpum*, but it must be admitted that these two species run each other very close.

Antidesma leucopodum *Miq.*, Fl. Ind. Bat., Suppl.: 465 (1861); Pax & Hoffm. xv: 136 (1922); Meijer, 7: 32 (1967); Airy Shaw in Kew Bull. 23: 280 (1969) & 26: 356 (1972) & 28: 273 (1973); Whitmore: 56 (1973).

Var. **leucopodum**: leaves rarely exceeding 22 × 7 cm., not puncticulate; ovary and disk pilose; drupe 4 mm. long.

A. cauliflorum W. W. Sm. in Notes Roy. Bot. Gard. Edinb. 8: 313 (1915); Merr.: 331 (1921); Pax & Hoffm. xv: 122 (1922); *pro parte*.
A. cauliflorum Merr. (*pro sp. nov.*) in Journ. As. Soc. Str. Br. 76: 89 (1917).
A. trunciflorum Merr.: 333 (1921) (*nom. nov.* pro *A. cauliflorum* Merr.).
A. hirtellum Ridley in Bull. Misc. Inf. Kew 1923: 366 (1923) & Fl. Malay Penins. 3: 229 (1924); Airy Shaw in Kew Bull. 23: 278 (1969).

SR; B; SB; K1.—Malaya, Sumatra, Anamba Is.

Shrub or small tree to 18 m. high, in primary Lowland or Mixed Dipterocarp forest (also in secondary forest), on yellow sandy clay or clay-rich alluvium or black or brown soil, or sandstone, or loam on sandstone, or limestone rocks on low sandy ridges, up to 1500 m. alt.

Extensively cauliflorous, even to the base of the trunk, but also producing inflorescences from the branchlets; bark whitish; leaves drying olivaceous-green, often with numerous primary nerves and closely parallel transverse secondary nerves; inflorescences (especially the female) often greatly elongated; young parts, nerves and inflorescences thinly rusty-pubescent; drupes small, 4 mm. long, flattened, symmetrical, with terminal styles.

Var. **platyphyllum** *Airy Shaw* in Kew Bull. 28: 273 (1973).

A. cauliflorum W. W. Sm., *l.c. supra, pro parte.*

SR (NE.); **B.**—Endemic.

Shrub or tree to 4·5 m., in understorey of disturbed primary forest, on yellow sandy loam, at 45 m. alt.

Leaves 30–40 × 10–12 cm., base sometimes rounded or subcordate, very densely and minutely white-puncticulate above; drupes 7–8 mm. long.

Var. **kinabaluense** *Airy Shaw* in Kew Bull. 28: 273 (1973).

SB.—Endemic.

Tree to 15 m., in forest at 750–1950 m. alt.

Ovary and disk glabrous; disk very tumid; drupe 7–8 mm. long, glabrous.

Antidesma linearifolium *Pax & Hoffm.* xv: 130 (1922); Meijer, 7: 30 (1967); Airy Shaw in Kew Bull. 23: 282 (1969), *in obs.*, & 28: 275 (1973), *in obs.*

SR; SB.—Endemic.

Shrub of 1–3 m., locally common along upper part of floodable area of river bank, on shale, or between boulders along stream in relict forest, altitudes not indicated.

Leaves narrow, willow-like, 7–16 mm. broad, long-acuminate, almost glabrous; stipules conspicuous, narrowly ovate, subpersistent; inflorescences short; ovary glabrous; drupe subobliquely ovoid, slightly flattened, 4–8 mm. long, with terminal or subterminal styles.

Related to *A. stenophyllum* Merr., which differs in its much broader leaves (up to 3 cm.) and very narrow stipules. Both species are related to *A. neurocarpum* Miq. and *A. hosei* Pax & Hoffm.

Antidesma montanum *Bl.*, Bijdr.: 1124 (1825); Muell. Arg.: 264 (1866); J. J. Sm.: 276 (1910); Hallier in Meded. Rijks Herb. 1910: 7 (1911); Pax & Hoffm. xv: 158 (1922) (*q.v.* for detailed synonymy); Ridley: 231 (1924); Merr. in Philipp. Journ. Sci. 29: 381 (1926); Gagnep.: 515 (1926); Corner: 234 (1940); Adelbert & Meeuse *apud* Backer in Blumea 5: 507 (1945); Backer & Bakh. f.: 458 (1963); Meijer, 7: 31 (1967); Airy Shaw in Kew Bull. 26: 358 (1972); Whitmore: 58 (1973).

A. pubescens Bl., Bijdr.: 1123 (1825), *non* Roxb. (1798).

A. pubescens var. *moritzii* Tul. in Ann. Sci. Nat. III, 15: 215 (1851).
A. moritzii (Tul.) Muell. Arg. in Linnaea 34: 67 (1865) & in DC.: 252 (1866); Hook. f.: 362 (1887); Stapf in Trans. Linn. Soc., Bot. 4: 225 (1894); Merr.: 332 (1921).
A. palawanense Merr. in Philipp. Journ. Sci. 9, C. Bot.: 467 (1914); Pax & Hoffm. xv: 131 (1922); **synon. nov.**
A. phanerophlebium Merr. in Philipp. Journ. Sci. 11, C. Bot.: 59 (1916); Pax & Hoffm. xv: 160 (1922); Meijer, 7: 31 (1967); **synon. nov.**
A. pseudo-montanum Pax & Hoffm. xv: 163 (1922), **synon. nov.**

SR; B; SB; K1–3.—SE. Asia throughout W. Malesia to Celebes and the Lesser Sunda Is.
Shrub or small tree to 14 m., recorded from a great variety of soils (clay, loam, sandstone, 'mor' soil, porphyry, frequent on limestone) and formations (mixed peat-swamp forest, Heath woodland, *kerangas*, *alan bunga* forest, Mixed Dipterocarp forest, river banks, hilltops, seashore, etc.) up to 1050 m. alt.

Very variable. Stems usually finely puberulous; leaves usually membranous, elliptic-oblong, with conspicuous lateral nerves, stipules subulate; male inflorescences branched, slender, puberulous; female inflorescences often simple; fruits small, rhombic-ovoid or obovoid, rugose.
The rather well-marked form described as *A. phanerophlebium* has more coriaceous leaves, with strongly impressed nerves, and rather larger fruits. It occurs commonly in Sarawak, but passes insensibly into the common form. The short petioles are usually sufficient to distinguish *A. montanum* from the closely related *A. cuspidatum*.

Antidesma montis-silam *Airy Shaw* in Kew Bull. 28: 269 (1973).

SB.—Endemic.
Tree to 15 m. high, in primary forest (sometimes swampy) on blackish ultrabasic soil, up to 600 m. alt.

A large-leaved (up to 28 × 9 cm.) species, almost glabrous except for the inflorescences, closely related to *A. pendulum* Hook. f., but distinguished by the midrib being impressed, not raised, above, and by the thinly puberulous fruits.

Antidesma neurocarpum *Miq.*, Fl. Ind. Bat., Suppl.: 466 (1861); Muell. Arg.: 253 (1866); Hutch. in Journ. Linn. Soc., Bot. 42: 135 (1914); Merr.: 332 (1921); Pax & Hoffm. xv: 136 (1922); Meijer, 7: 30, tab. opp. p. 130 (1967); Airy Shaw in Kew Bull. 26: 358 (1972); Whitmore: 57 (1973).

A. alatum Hook. f.: 358 (1887); Pax & Hoffm. xv: 119 (1922); Ridley: 227 (1924); Corner: 232 (1940).
A. rubiginosum Merr. in Philipp. Journ. Sci. 11, C. Bot.: 61 (1916); Pax & Hoffm. xv: 163 (1922).
? *A. hallieri* Merr., *l.c.*: 57 (1916); Pax & Hoffm. xv: 148 (1922); *e descr.*
A. inflatum Merr. in Journ. As. Soc. Str. Br. 76: 91 (1917) & Enum.: 332 (1921).

SR; B; SB; K1–3.—Peninsular Siam, Malay Penins., Sumatra, Banka.
Shrub or tree to 21 m., in primary Lowland or Mixed or Upper Dipterocarp forest, pole forest, Heath forest, submontane forest, '*blukar* and *ladang*

terrein' along river banks, on basalt (frequent), yellow sandy podsol, clay-rich soil, shale, rocky blackish soil, brown soil, yellow-brown or red-brown clay loam or sandy loam, or sandstone, or 'sand and limestone' (once only), up to 1650 m. alt.

Usually recognizable by the reddish foliage (when dry), thin rufous indumentum of the young parts and inflorescences, persistent broadly or narrowly ovate stipules, and largish flattened or inflated fruits. Specimens with narrower stipules and very reduced indumentum are sometimes difficult to distinguish from the closely related *A. hosei* Pax & Hoffm. Like the latter species, *A. neurocarpum* is apparently a calcifuge.

Antidesma pendulum *Hook. f.*: 356 (1887); Pax & Hoffm. xv: 164 (1922); Ridley: 227 (1924); Whitmore: 56 (1973).

SR.—Malaya, Sumatra.
Small tree 3–6 m. high, in primary Lowland or Mixed Dipterocarp forest on leached yellow sandy clay or clay loam, at very low altitudes.

Leaves very large, 20–40 cm. long, chartaceous, glabrous, with lax venation, the midrib raised above, and narrowly subulate stipules; female inflorescence and infructescence very elongate, 30–40 cm. long, almost glabrous, flowers sessile; ovary glabrous; fruits somewhat obliquely or almost symmetrically ovate, flattened, 10 mm. long, 8 mm. wide, glabrous. The raised midrib and glabrous fruits distinguish this from the related *A. montis-silam* Airy Shaw. There is sometimes an approach to *A. riparium* Airy Shaw, which differs in its very much smaller fruits (5–6 mm. long).

Antidesma polystylum *Airy Shaw* in Kew Bull. 26: 460 (1972).

SB; ?K3.—Endemic.
Tree to 8 m. high, in primary forest on brown or blackish soil up to 210 m. alt.

Closely related to *A. leucopodum* Miq., but differing in the strong ochraceous tomentellum of the lower leaf-surface and inflorescences, and especially in the large oblong fruits (10 × 4 mm.), crowned with 8–10 styles.

Antidesma riparium *Airy Shaw* in Kew Bull. 23: 282 (1969).

SR.—Endemic.
Treelet to 4·5 m. high, in clayey soil on river bank, probably low alt.

More or less intermediate between *A. pendulum* Hook. f. and *A. stenophyllum* Merr., differing from the former in its much smaller fruits, and from the latter in the much larger leaves (up to 28 × 5 cm.), the lateral nerves of which show no conspicuous anastomosis, in the deciduous stipules, and in the elongate infructescences (up to 50 cm. long).

Antidesma salicinum *Ridley*: 228 (1924); Airy Shaw in Kew Bull. 28: 276 (1973), *in obs.*; Whitmore: 56 (1973).
A. salicifolium sec. Hook. f.: 366 (1887), *an* Miq.? (*non* Presl 1849).

K2, 3; SB?.—Malaya, Sumatra.
Shrub in swampy forest by river at 5 m. alt.

Scarcely differing from the variable *A. montanum* Bl. except in the narrow leaves, and perhaps better treated as a rheophytic form of that species.

The specimen forming the basis of the doubtful record for Sabah (above) (*Ampuria* SAN 40815) has more firmly chartaceous leaves than the remainder, but seems otherwise almost identical.

Antidesma stenocarpum *Airy Shaw* in Kew Bull. 23: 281 (1969) & 28: 275 (1973), *in obs.*

SR.—Endemic.

Tree to 4·5 m. tall, on river banks, probably low alt.

A rheophytic plant with elongate (15–25 cm.), narrow (to 3·5 cm.), long-acuminate, thinly chartaceous leaves with numerous (15–19) lateral nerves and subulate stipules; the young parts, nerves and inflorescences thinly rufous-pubescent; male flowers sessile, female flowers and fruits shortly pedicelled, the latter flattened, narrowly ovoid-oblong, 8–9 mm. long and 4 mm. wide, conspicuously elongate-areolate and minutely puberulous, obtuse at the apex with subterminal styles.

Related to the *A. tomentosum* complex, with which it seems to be connected through *A. tomentosum* var. *rivulare* (Merr.) Pax & Hoffm.

Antidesma stenophyllum *Merr.* in Philipp. Journ. Sci. 11, C. Bot.: 62 (1916) & Enum.: 333 (1921); Pax & Hoffm. xv: 138 (1922); Airy Shaw in Kew Bull. 23: 282 (1969) & 28: 275 (1973), *in obs.*

SR.—Endemic.

No field data available.

A rheophyte, completely glabrous, or the young parts and stipules minutely grey-puberulous; leaves up to 20 × 3 cm., long-acuminate, the midrib raised on the upper as well as the lower surface, chartaceous or thinly coriaceous, nerves 8–10 pairs; male flowers sessile; fruits very similar to those of *A. stenocarpum.*

Antidesma stipulare *Bl.*, Bijdr.: 1125 (1825); Muell. Arg.: 249 (1866); J. J. Sm.: 261 (1910); Merr.: 333 (1921); Pax & Hoffm. xv: 142 (1922); Ridley: 226 (1924); Backer & Bakh. f.: 458, 460 (1963); Meijer, 7: 30 (1967); Whitmore: 56 (1973).

? *A. diepenhorstii* Miq., Fl. Ind. Bat., Suppl.: 466 (1861); Muell. Arg.: 251 (1866); J. J. Sm.: 262 (1910), *in obs.*; Pax & Hoffm., *l.c.*
A. amboinense Miq. in Ann. Mus. Bot. Lugd.-Bat. 1: 218 (1864).
A. auritum sec. Stapf in Trans. Linn. Soc. II, Bot. 4: 225 (1894), Merr.: 331 (1921), *non* Tul.
A. cordato-stipulaceum Merr. in Philipp. Journ. Sci. 4, C. Bot.: 275 (1909).
A. grandistipulum Merr. in Philipp. Journ. Sci. 11, C. Bot.: 56 (1916), & Enum.: 332 (1921), **synon. nov.**
A. sarawakense Merr., *l.c.*: 57 (1916), & Enum.: 333 (1921), **synon. nov.**
A. kunstleri Gage in Rec. Bot. Surv. India 9: 225 (1922).
A. stenophyllum (sphalm. '*tsenoph.*') Gage, *l.c.*, *non* Merr. (1916).
A. urophyllum Pax & Hoffm. in Mitt. Inst. Bot. Hamburg 7: 224 (1931), *e descr.*, **synon. nov.**

SR; B; SB; K1, 3.—Malaya; Sumatra (?); Philippines; Java; Moluccas (Amboina, Buru).

Shrub to 3·5 m. high, in primary Mixed Dipterocarp forest or submontane forest or *Agathis* forest or *kerangas*, on shallow peat overlying podsolized soils, on friable granodiorite-derived soil, waterlogged, sandy, acid soil, sandstone, dacite, basalt, black rocky soil, or limestone (once), up to 1650 m. alt.

A smooth, glossy, almost completely glabrous plant. Leaves oblong or elongate-elliptic, up to 35 × 8 cm., chartaceous to coriaceous, olivaceous or fuscous when dry; stipules usually large, ovate, often cordate at base, usually persistent; male inflorescence rather short and inconspicuous, with sessile flowers; female inflorescence and infructescence usually elongate, up to 30 cm. or more; drupes large, 10–15 mm. long, obliquely ovate, compressed, thickened at the base, long-pedicelled.

Typical *A. stipulare* can usually be distinguished from glabrescent forms of *A. tomentosum* Bl. by the mostly narrower, more coriaceous, more glossy leaves, with more widely spaced nerves, and by the broadly or more narrowly ovate, but not subulate, stipules.

f. **amboinense** (*Miq.*) *J. J. Sm.*: 262 (1910), *in adnot.*

A. amboinense Miq. in Ann. Mus. Bot. Lugd.-Bat. 1: 218 (1864).

SR; SB.—Moluccas.

Shrub or small tree to 5 m., in primary Mixed Dipterocarp forest on friable granodiorite-derived soil or limestone up to 600 m. alt. The solitary record from limestone was from the Bidi Cave in Sarawak (*Clemens* 20595).

Differs from the typical plant in the narrowly lanceolate, strongly acuminate stipules. *A. batuense* J. J. Sm. in Ic. Bogor. 4: 251, t. 380 (1914), Pax & Hoffm. xv: 138 (1922), from the Batu Islands, west of Sumatra, may belong here, but it appears to have completely sessile fruits. I have, however, seen no material.

Antidesma thwaitesianum *Muell. Arg.*: 263 (1866); Alston in Trimen, Handb. Fl. Ceylon 6 (Suppl.): 262 (1931); Airy Shaw in Kew Bull. 26: 360 & 462 (1972); Whitmore: 57 (1973).

A. bunius var. *thwaitesianum* (Muell. Arg.) Trimen, Syst. Cat. Fl. Pl. Ceyl.: 81 (1885) & Handb. Fl. Ceyl. 4: 43 (1898).

A. bunius sec. Hook. f.: 358–9 (1887); Pax & Hoffm. xv: 160 (1922); ? Merr. in Philipp. Journ. Sci. 29: 380 (1926); *non* (L.) Spreng.

A. coriaceum sec. Gagnep.: 513 (1926), *non* Tul.

A. castroi Merr. ex Meijer, 7: 32 (1967), *anglice, in clavi.*

SB; K1.—Ceylon; SE. Asia; N. Malaya; SW. Philippines (Palawan).

Tree to 25 m. high, in primary forest on level land, or on swampy flat behind seashore, near mangrove, on black or brown sandy or peaty soil, at very low altitudes up to 40 m.

Closely related to *A. coriaceum* Tul., from which it differs in the finely tessellated venation of the leaf-undersurface and in the whitish, not rusty, pubescence of the inflorescences, which are apparently always fascicled, never panicled. The absence of the plant from the western and southern parts of Borneo is noteworthy.

See also note under *A. bunius*.

Antidesma tomentosum *Bl.*, Bijdr.: 1126 (1825); Muell. Arg.: 248 (1866); J. J. Sm.: 264 (1910); Merr. in Philipp. Journ. Sci. 11, C. Bot.: 62 (1916) & Enum.: 333 (1921); Pax & Hoffm. xv: 116 (1922); Ridley: 226 (1924); Merr. in Sarawak Mus. Journ. 3: 525 (1928); Corner: 234 (1940); Backer & Bakh. f: 459 (1963); Meijer, 7: 31 (1967); Whitmore: 55 (1973).

Var. **tomentosum**:

A. cumingii Muell. Arg.: 249 (1866); Pax & Hoffm. xv: 120 (1922); Merr., Enum.: 413 (1923); **synon. nov.**
A. persimile Kurz in Journ. Bot. Brit. & For. 13: 330 (1875); Hook. f. (sphalm. '*perserrula*'): 365 (1887); Pax & Hoffm. xv: 119 (1922); Ridley: 229 (1924); **synon. nov.**
A. kingii Hook. f.: 356 (1887).
A. longipes Hook. f.: 355 (1887); Pax & Hoffm. xv: 121 (1922); Ridley: 229 (1924); **synon. nov.**
A. membranifolium Elm., Leafl. Philipp. Bot. 1: 313 (1908).
A. gibbsiae Hutch. in Journ. Linn. Soc., Bot. 42: 134 (1914); Pax & Hoffm. xv: 114 (1922); **synon. nov.**
? *A. foxworthyii* [sic] Merr. in Philipp. Journ. Sci. 11, C. Bot.: 55 (1916) & Enum.: 332 (1921), *e descr.*
A. rivulare Merr. in Philipp. Journ. Sci. 11, C. Bot.: 60 (1916) & Enum.: 333 (1921), **synon. nov.**
A. tomentosum var. *rivulare* (Merr.) Pax & Hoffm. xv: 117 (1922).
A. sumatranum Pax & Hoffm. xv: 120 (1922), *e descr.*, **synon. nov.**
A. caudatum Pax & Hoffm. in Mitt. Inst. Bot. Hamb. 7: 223 (1931), *e descr.*, **synon. nov.**
A. tomentosum var. *giganteum* Pax & Hoffm., *l.c.*: 224 (1931), *e descr.*

SR; B; SB; K1.—Sumatra, Malaya, Java, Celebes.
Shrub or small tree to 5 m. high, in primary hill forest or ridge Dipterocarp forest, on limestone, igneous soil (dyke) on limestone, lime and sandstone, Tertiary sandstone, yellow sandy clay, black, greyish or brown sandy soil, up to 1500 m. alt.

Usually a large-leaved plant, but very variable, generally distinguishable from *A. stipulare* Bl. by its broader, thinner leaves, strong rufous indumentum and narrowly subulate stipules, but forms occur with narrow leaves or with practically no indumentum, and then the subulate stipules are almost the only reliable point of distinction. The species seems ecologically indifferent as to soil.

Var. **bangueyense** (*Merr.*) *Airy Shaw*, comb. & stat. nov.

A. subolivaceum Elm., Leafl. Philipp. Bot. 4: 1272 (1911); Pax & Hoffm. xv: 113 (1922); Merr., Enum.: 418 (1923); **synon. nov.**
A. bangueyense Merr. in Philipp. Journ. Sci. 24: 114 (1924) & 29: 380 (1926); Meijer, 7: 31 (1967).

SB.—Philippines (Palawan, Sulu).
Shrub or tree to 9 m. high, in primary forest on black or brown soil or sandstone, up to 210 m. alt.

A rather distinct form with smaller leaves and only 5–10 instead of 10–20 pairs of lateral nerves; particularly frequent on the island of Banguey (Banggi) and in the neighbouring district of Kudat.

Antidesma venenosum *J. J. Sm.* in Ic. Bogor. 4: 41, t. 313 (1910); Merr. in Philipp. Journ. Sci., 11, C. Bot.: 55 (1916) & Enum.: 333 (1921); Pax & Hoffm. xv: 150 (1922); Meijer, 7: 32 (1967).

SR; SB; K1.—Endemic.

Shrub or small tree to 9 m. high, in primary Mixed Dipterocarp forest or secondary forest, on yellow, brown or black soils, on clay-rich or sandy soils, or on loam containing lime, up to 900 m. alt.

A slender species, with densely minutely puberulous branchlets and inflorescences; leaves membranous or chartaceous, oblong or oblong-obovate, rounded or truncate at the base, caudate at the apex, 7–12(–17) cm. long; petiole slender, 5–13 mm. long; stipules narrowly subulate; inflorescences slender, much-branched, with minute flowers; drupes very small, 2–2·5 mm. diam., compressed, glabrous, with lateral styles.

Antidesma venenosum cannot be confused with any other Bornean species, and in the whole Malesian region only *A. microcarpum* Elm. (*A. maesoïdes* Pax & Hoffm.) has equally small fruits, but these are rhombic-ovoid and not compressed.

Appendix II
Pandaceae

1 Inflorescence racemose, pendulous or erect, sometimes greatly elongate; drupes often flattened and transverse (broader than long), or subglobose (and then 1·5–3 cm. diam.), with a hard thick endocarp **Galearia** *Zoll. & Mor.*
1 Inflorescence fasciculate, axillary; drupes subglobose, less than 1 cm. diam., with a relatively thin endocarp . **Microdesmis** *Planch.*

Galearia *Zoll. & Mor.*

(Key adapted from Forman in *Kew Bull.* 26: 156–7 (1971))

1 Petals ± straight and flat, tomentellous; fruits subglobose [or strongly dorsiventrally compressed]*, orbicular [or 3–4-angled] in outline, thick-walled with a thick, much perforated endocarp, usually with 3–4 loculi; minor veins of leaf mostly at right angles to midrib (Subgen. *Orthopetalum*) . . . . **G. maingayi**
1 Petals in male flowers deeply concave or cucullate, glabrous except for apex; fruits laterally compressed, endocarp thin, unperforated, 1(–2)-locular; minor veins irregularly reticulate (Subgen. *Galearia*):
  2 Leaves narrowly oblong, 2–3·5 cm. broad, more than 5 times as long as broad . . . . **G. stenophylla**
  2 Leaves ± elliptic, usually much broader than above:
  3 Floral bracts insignificant, less than 2 mm. long . . **G. fulva**
  3 Floral bracts elongate, linear, 10–25 mm. long . **G. aristifera**

Galearia aristifera *Miq.*, Fl. Ind. Bat., Suppl.: 471 (1861); Pax & Hoffm. iii: 102 (1911); Merr.: 344 (1921); Pax & Hoffm. in Mitt. Inst. Bot. Hamburg 7: 230 (1931); Forman in Kew Bull. 26: 159 (1971); Whitmore: 98 (1873).

Bennettia aristifera (Miq.) Muell. Arg.: 1039 (1866).
Galearia leptostachya Pax in Pax & Hoffm. iii: 102 (1911); Merr.: 345 (1921) ('*-stachys*').

SR; K3.—Sumatra, Malaya.

Small tree to 11 m., in primary Mixed or Lowland Dipterocarp forest on yellow sandy clay up to 90 m. alt.

* Characters in square brackets refer to the non-Bornean, E. Malesian species, *G. celebica* Koord.

Clearly distinguished from the other species by the conspicuous elongate linear floral bracts, 10–25 mm. long. The fruit has not yet been collected.

Galearia fulva (*Tul.*) *Miq.*, Fl. Ind. Bat. (1(2): 430 (1859); Hook. f.: 378 (1887); Pax & Hoffm. iii: 101 (1911); Ridley: 257 (1924); Forman in Kew Bull. 26: 160 (1971), *q.v.*; Airy Shaw in Kew Bull. 26: 362 (1971); Whitmore: 98 (1873).

Cremostachys fulva Tul. in Ann. Sci. Nat. III, 15: 261 (1851).
Bennettia fulva (Tul.) Muell. Arg. in Linnaea 34: 205 (1865) & in DC.: 1037 (1866).
Galearia phlebocarpa (R. Br.) Miq., *l.c.* (1859); Pax & Hoffm. iii: 100 (1911); Merr.: 345 (1921); Pax & Hoffm. in Mitt. Inst. Bot. Hamburg 7: 229 (1931); Meijer, 7: 45 (1967).
G. sessiliflora Merr. in Journ. Str. Br. Roy. As. Soc. 86: 320 (1922) & Pl. Elmer.: 161 (1929).
G. dolichobotrys Merr. in Philipp. Journ. Sci. 29: 384 (1926); Meijer, *l.c.*

For extensive further synonymy *vide* Forman, *l.c.* (1971); also in *op. cit.* 14: 311 (1960) & 20: 309 (1966).

SR; SB; K1, 1a.—SE. Asia, Malaya, Sumatra, Philippines.
Shrub or tree to 12 m. high, in primary or secondary Mixed Dipterocarp forest or *kerangas*, on black, brown or white sand, dark brown or brownish-white soil or sandy loam, basalt, granodiorite-derived soil, tufa plateau, or rocks in moist ravine, up to 1200 m. alt.

An exceedingly variable species, in density and colour of indumentum, in size, shape and texture of leaves, in length of inflorescence, in length of pedicels and in shape and pubescence of fruit. The broad leaves and small floral bracts, however, distinguish it clearly from *G. stenophylla* and *G. aristifera*, respectively, the only other species of Subgen. *Galearia* occurring in Borneo.
The leaves are said to be used in Sabah as a vegetable.

Galearia maingayi *Hook. f.*: 377 (1887); Pax & Hoffm. iii: 103 (1911); Ridley: 255 (1924); Forman in Kew Bull. 14: 311–13 (1960) & 20: 312–13 (1966) & 26: 158 (1971); Meijer, 10: 233–4 (1968); Whitmore: 98 (1973).

SR; B; K1, 3.—Sumatra, Malaya.
Large tree to 35 m. high, in primary forest on yellow sandy loam or riverine alluvium up to 300 m. alt.

The characters of Subgen. *Orthopetalum*, given in the key, distinguish *G. maingayi* sharply from the other Bornean species. The relatively massive, smooth, subglobose drupe is very distinctive. The rhachis of the inflorescences is shorter, more robust and more rigid than in Subgen. *Galearia*.

Galearia stenophylla *Merr.* in Journ. Str. Br. Roy. As. Soc. 86: 320 (1922); Meijer, 7: 45 (1967); Forman in Kew Bull. 26: 159 (1971); Whitmore: 98 (1973).

G. lancifolia Ridley in Bull. Misc. Inf. Kew 1926: 476 (1926).

SR; SB.—Malaya.
Shrub or small tree to 6 m., in primary hill forest up to 400 m. alt.

The very narrow leaves are sufficient to distinguish *G. stenophylla* from *G. fulva*, its nearest relative.

Microdesmis *Planch.*

Microdesmis caseariifolia *Planch.* in Hook. Ic. Pl. 8: t. 758 (1848), *in adnot.*; Muell. Arg.: 1041 (1866); Hook. f.: 380 (1887); Boerl.: 226 (1900); Pax & Hoffm. iii: 106 (1911); Hallier in Meded. Rijks Herb. 1910: 11 (1911); Hutch. in Journ. Linn. Soc., Bot. 42: 135 (1914); Merr.: 345 (1921) (sphalm. '*Trigonostemon laevigatus*', prior, l. 10) (cf. indicem, p. 618) & 451 (1923); Ridley: 258 (1924); Gagnep.: 460 (1926); Hend. in Journ. Malay Br. Roy. As. Soc. 17: 71 (1939); Stern in Am. Journ. Bot. 54: 670 (1967); Meijer, 10: 234 (1968); Airy Shaw in Kew Bull. 26: 362 (1971); Whitmore: 119 (1973).

Tetragyne acuminata Miq., Fl. Ind. Bat., Suppl.: 463 (1861); Muell. Arg.: 1254 (1866); cf. Hallier, *l.c.*
Microdesmis philippinensis Elm., Leafl. Philipp. Bot. 4: 1300 (1911).

SB; Kia.—S. China, SE. Asia, W. Malesia (exc. Java).
Shrub or tree, sometimes scrambling, to 11 m. high, in primary or secondary forest (sometimes swampy), on brown or yellowish sandy soil or sandstone ridges, or on stony black soil, from low levels up to 450 m. alt.

The Bornean distribution is very curious: no records for Sarawak or Brunei, numerous gatherings from Sabah, and only one old specimen (Bandjermasin, 1857–8, *Motley* 334) from SE. Indonesian Borneo. Apparently almost never noted on limestone; Henderson's records (*l.c. supra*) from the Langkawi Islands (?) and Perlis in N. Malaya are exceptional.
Sometimes used by Muruts for posts in house construction, etc.

Appendix III
Species occurring on limestone[1] in Borneo

* = occurring almost exclusively on limestone.
† = occurring very frequently on limestone.
Unmarked = occurring occasionally on limestone.

(M) = recorded from limestone in Malaya by Henderson, Flora of the Limestone Hills of the Malay Peninsula, in *Journ. Malayan Branch R. As. Soc.* 17: 68–72 (1939).

†Acalypha caturus
Agrostistachys longifolia
*Alchornea rugosa
Aporusa falcifera
A. frutescens
A. grandistipula
A. nitida
A. prainiana
Baccaurea bracteata
B. hookeri
†B. lanceolata (M)
B. racemosa
B. stipulata
B. sumatrana
Blumeodendron kurzii (M)
Botryophora geniculata
Bridelia glauca
Claoxylon indicum
†C. stapfianum
†Cleidion javanicum (M, as *Cephalomappa* sp.)
*Cleistanthus beccarianus
†C. brideliifolius
*C. celebicus
C. hirsutulus
C. macrophyllus
C. megacarpus
C. myrianthus
C. podopyxis
Croton argyratus (M)
†C. griffithii
C. oblongus
*Dimorphocalyx luzoniensis var. trichocarpus
Drypetes fusiformis
D. longifolia
D. microphylla
D. neglecta
D. polyalthioïdes
D. rhakodiskos
Endospermum peltatum
*Euphorbia cf. lacei
*Excoecaria borneënsis
Fahrenheitia pendula
*Glochidion brunneum
G. calospermum
G. lutescens
G. singaporense
Homalanthus populneus (M, as *H. populifolius*)
†Mallotus dispar (M)
M. floribundus (M)
*M. havilandii
M. korthalsii
M. lackeyi
M. miquelianus

[1] Including various soils (sands, loams, clays, etc.) noted as containing an admixture of lime.

M. moritzianus
†M. oblongifolius
*M. resinosus
M. sumatranus
†Margaritaria indica
Neoscortechinia forbesii
N. nicobarica
Phyllanthus urinaria
†Ptychopyxis arborea
†Richeriella malayana (M)
Sauropus androgynus
†S. micrasterias
S. rhamnoïdes
†Spathiostemon javensis
Sumbaviopsis albicans
*Suregada calcicola
Syndyophyllum excelsum
ssp. occidentale
Trewia nudiflora
Trigonostemon sumatranus
T. villosus
†Wetria insignis

Antidesma leucopodum
var. leucopodum
†A. montanum (M)
A. neurocarpum
A. stipulare
†A. tomentosum (M)
A. venenosum

ADDENDUM

Bridelia adusta *Airy Shaw*, sp. nov. (*supra*, p. 63).

Diagnosis: foliis elliptico-oblongis coriaceis 7–9-nervibus supra dense tenuiter reticulato-venulosis nitidulis subtus glaucis siccitate ochraceo- vel intense cinnamomeo-brunneis (quasi deustis) marginibus revolutis distinctissima.—SARAWAK: Baram Distr., Gunong Api, 1 Oct. 1971, *Anderson* S. 30854 (type, K).

Index

Page numbers in bold print refer to accepted taxa from Borneo, those in italics refer to synonyms.

Printed in England for Her Majesty's Stationery Office
By Butler & Tanner Ltd, Frome and London
Dd 506704 K6 10/75